SOLID EARTH GEOMAGNETISM

DEVELOPMENTS IN EARTH AND PLANETARY SCIENCES

SOLID EARTH GEOMAGNETISM

Tsuneji Rikitake and Yoshimori Honkura

Developments in Earth and Planetary Sciences

05

Terra Scientific Publishing Company
Tokyo, Japan

D. Reidel Publishing Company

A MEMBER OF THE
KLUWER ACADEMIC PUBLISHERS GROUP

Dordrecht / Boston / Lancaster

Library of Congress Cataloging in Publication Data

Main entry under title:

Rikitake, Tsuneji, 1921–
Solid earth geomagnetism.

(Developments in earth and planetary sciences ; 05)
1. Magnetism, Terrestrial. I. Honkura, Yoshimori, 1945– . II. Title. III. Series.
QC815.2.R58 1985 538'72 85-18372
ISBN 90-277-2120-3 (Reidel)

Published by Terra Scientific Publishing Company (TERRAPUB),
307 Shibuyadai-haim, 4-17 Sakuragaoka-cho, Shibuya-ku, Tokyo 150, Japan,
in co-publication with D. Reidel Publishing Company, Dordrecht, Holland

Sold and distributed in the U.S.A. and Canada
by Kluwer Boston Inc.,
190 Old Derby Street, Hingham, MA 02043, U.S.A.,
in Japan by Terra Scientific Publishing Company (TERRAPUB),
307 Shibuyadai-haim, 4-17 Sakuragaoka-cho, Shibuya-ku, Tokyo 150, Japan

In all other countries, sold and distributed
by Kluwer Academic Publishers Group,
P. O. Box 322, 3300 AH Dordrecht, Holland

D. Reidel Publishing Company is a member of the Kluwer Academic Publishers Group

Printed in Japan

PREFACE AND INTRODUCTORY REMARKS

More than forty years have passed since Professors Chapman and Bartels published the famous book entitled "Geomagnetism" (1940). The book has long been the only text on the subject in spite of the dramatic developments in modern studies on geomagnetism. The book doubtless played a vitally important role in the advancement of this branch of geophysics but many new findings should nowadays be added to those described in the book.

In recent years, a number of books on rock magnetism, paleomagnetism, electromagnetic induction within the earth and the like have been published. Although much of the development in these particular topics of geomagnetic study can be learned from these books, it is desirable, especially for students undertaking geophysics courses, to have a comprehensive textbook that covers overall developments in modern geomagnetism.

One of the authors (Rikitake), who has had experience of giving lectures on geomagnetism to graduate students at the University of Tokyo as well as at the Tokyo Institute of Technology, has long been aware of the necessity of having an adequate textbook of reasonable size. At the same time, the book should cover all the important disciplines of modern geomagnetism. It was fortunate that another author (Honkura) joined the Department of Applied Physics, Tokyo Institute of Technology in 1980, so the two authors planned to write such a textbook.

The authors felt, however, that the book should be concerned only with geomagnetism related to the solid part of the earth. In the beginning the possibility of inviting someone to cover geomagnetism related to the upper atmosphere, the space surrounding the earth, the magnetosphere and so on, was considered. However, such a plan had to be rejected because of the difficulties in cooperative work to be completed in a limited time. The authors thus confined themselves to "Solid Earth Geomagnetism" most topics of which are within their capacity.

Chapter 1 will deal with the main magnetic field of the earth. Although the subject is one of the classical problems in geomagnetism and has been studied since the times of W. Gilbert and C. F. Gauss, there have been very many new findings in recent years. Starting from the early history of geomagnetism, measuring techniques, classical and modern, will be briefly reviewed in this chapter. Much of the distribution of the geomagnetic field disclosed by these measurements will be presented in the form of a world magnetic chart and as the coefficients of spherical harmonic functions representing the geomagnetic field. Special mention will be made of the newly-adopted International Geomagnetic Reference Field (IGRF) and the satellite magnetic surveys including the MGST(6/80) field model derived using a satellite called MAGSAT.

Various features of geomagnetic secular variation such as the decrease in the earth's dipole moment, westward drift of the non-dipole field, northward shift of the dipole and the like will be presented in Chapter 2. A remarkable observation, that the dipole field would vanish within a period of 1200 years, should the present rate of decrease in the dipole moment continue, will be pointed out, as based on the newly-obtained data.

Chapter 3 will be reserved for a brief summary of paleomagnetism which provides evidence of geomagnetic field reversals, polar wandering and continental drift in the past. In view of the existing reference books on rock magnetism and paleomagnetism (NAGATA, 1961; IRVING, 1964; STACEY and BANERJEE, 1974), no details of the subject will be dealt with in this chapter except for the points that are necessary for understanding the subsequent chapters.

It is planned in Chapter 4 to present an up-to-date review on the dynamo theory of the geomagnetic field in fair detail. At the moment most geomagneticians believe that this is the only theory which can account for the origin of geomagnetism. Starting from the semi-classical models due to Elsasser, Bullard, Parker and others, rather modern ideas put forward by Braginskii, Gubbins, Busse and others will be outlined. Special mention will also be made of the theory of turbulent dynamos as developed mostly by Krause, Rädler, Steenbeck and others. By studying these contributions, it appears that we arrive at the possibility that the geomagnetic field is produced and maintained by some sort of dynamo action prevailing in the earth's core. Theories of non-steady dynamos will also be reviewed with special reference to geomagnetic reversal. It will be emphasized that the Rikitake model for coupled disk dynamos, which was in the beginning thought to be only a remote analog of the actual dynamo process in the core, reflects some essence of dynamo action. Towards the end of the chapter, theories of non-dipole field and electromagnetic coupling between the core and mantle will be presented although they are by no means complete.

Recent magnetic field measurements of the moon and other planets by magnetometers on board space probes were highly important for studying geomagnetism. As the generation of the geomagnetic field is intrinsically correlated to the internal structure of the earth, whether or not other planets, whose internal structures could be guessed to some extent from astronomical and other observations, have a planetary magnetic field, provides a key point for discussing the origin of geomagnetism. Chapter 5 summarizes our knowledge of the magnetic fields of the moon and other planets as observed by the Mariner, Mars, Venera, Pioneer, Voyager and Apollo missions. Magnetic measurements on the lunar surface by the magnetometers set up there by the Apollo missions will be described in some detail, together with the magnetic properties of lunar rocks brought back to the earth by the missions. It will be emphasized that the earth is the only celestial body, among the terrestrial planets, which has a large-scale planetary magnetic field. The fact is of utmost importance for discussing the origin of geomagnetism.

In addition to the planetary magnetic field, which has been the main concern of the preceding chapters, local anomalies of the geomagnetic field are sometimes of geophysical importance. Recent development of air-borne magnetometers enables us to conduct accurate aeromagnetic surveys over geological formations, volcanoes and so on as will be described in Chapter 6. This chapter will also cover the marine magnetic anomalies as disclosed by measurements with ship-borne proton precession magnetometers. The amplitude of magnetic anomaly measured on the surface of the deep ocean is often so large that it cannot readily be accounted for by any conceivable intensity contrast of magnetization in rocks forming the sea bottom. In order to overcome such a difficulty, the idea of sea-floor spreading has been put forward. It is assumed that the mantle material welling up to the oceanic ridge axis, and subject to cooling there, is magnetized in the direction of the geomagnetic field at that time and that the geomagnetic field undergoes frequent reversals. The sea floor, which is spread from the ridge axis because the newly-formed floor cannot accumulate there, is characterized by magnetizations of alternating polarity occurring successively, in a direction approximately perpendicular to the ridge axis. In that case the strong magnetic anomalies observed over the sea surface could be reasonably understood. In Chapter 6, the way in which the study of marine magnetic anomalies contributed to the development of sea-floor spreading and consequently to plate tectonics will also be described.

Geophysicists and seismologists have recently been paying much attention to local anomalous changes in the geomagnetic field associated with crustal movement and seismic and volcanic activities. As will be stressed in Chapter 7, possibilities of earthquake prediction on the basis of a precursory seismomagnetic effect have even been suggested. Geomagnetic studies in this line will be described in this chapter under the title "tectonomagnetism" as proposed by NAGATA (1971).

In contrast to the geomagnetic field arising from internal processes inside the earth, Chapter 8 will aim at describing changes in the geomagnetic field produced by electric currents induced in the earth in association with geomagnetic changes external to the earth. These changes are geomagnetic solar and lunar variations, geomagnetic storm, substorm, bay, micropulsation and the like. Analyses of these changes have an important bearing on inferring the electrical conductivity within the earth.

In order to interpret the relations between the geomagnetic changes of external and internal origins as presented in Chapter 8, we have to rely on theories of electromagnetic induction in uniform and non-uniform spheres as will be demonstrated in Chapter 9. Theories of electromagnetic induction in a thin spherical sheet will be described in the chapter, too.

The analyses in Chapter 8 and the theories in Chapter 9 enable us to study the electrical conductivity in the earth and moon on the basis of electromagnetic induction within these celestial bodies. Recent studies of the overall distribution of electrical conductivity in the earth and moon will be presented in Chapter 10 in a summarized form. In addition to electromagnetic induction by geomagnetic changes of external origin, it is necessary to analyse geomagnetic secular variations in order to infer the electrical conductivity in the lower mantle of the earth. Only a very crude approach will be able to be made to estimating the core's conductivity, as will be seen in the chapter. The temperature distributions and possible phase changes that certainly affect the distribution of electrical conductivity within the earth and moon will also be discussed briefly.

When a portion of the earth whose extent is smaller than the typical wavelength of a geomagnetic change is considered, the earth may be treated as a conductor having a plane boundary. In Chapter 11, electromagnetic induction in such plane conductors will be studied.

One of the most important findings in geomagnetic studies in the past thirty years is certainly the discovery of local conductivity anomalies in the earth's crust and mantle as will be summarized in Chapter 12. After the discovery of the Central Japan and North Germany anomalies with arrays of geomagnetic variometers in the early 1950's, a great many anomalies have been found in various areas of the world. Some of them are certainly caused by electric currents induced in a highly-conducting sea. However, many of them turned out to be correlated to conductors embedded in the earth's crust and mantle. It will be useful for those who want to know the present-day state of this topic to refer to this chapter.

This book is planned as one of the volumes of the series "Developments in Earth and Planetary Sciences" (DEPS) published jointly by the Terra Scientific Publishing Company and the D. Reidel Publishing Company, the Netherlands. The authors are thankful to Mr. K. Oshida of the Terra Scientific Publishing Company for his help in the course of drafting the manuscripts. A part of the publishing costs is covered by the funds given to the authors by the Japanese Ministry of Education. The authors are grateful to the ministry for this financial help.

T. Rikitake
Y. Honkura

Acknowledgments: The authors are grateful to the following publishers for permission to reproduce the figures listed:

American Association for the Advancement of Science: Fig. 3.5.
American Geophysical Union: Figs. 3.6, 3.7, 3.8, 7.2, 12.28.
Elsevier Scientific Publishing Co.: Figs. 6.3, 12.25.
Geological Society of America: Fig. 3.7.
Macmillan Journals Ltd.: Figs. 2.2, 2.10.
Pergamon Press Ltd.: Fig. 8.5.
Royal Astronomical Society: Figs. 4.11, 4.12, 4.13, 12.4, 12.13, 12.19, 12.20, 12.22.
Springer-Verlag Inc.: Fig. 12.12.
The Royal Society: Figs. 4.7, 4.9, 4.18.

CONTENTS

CHAPTER 1

CHAPTER 1

MAIN MAGNETIC FIELD OF THE EARTH

1.1 *Early History of Geomagnetism*

1.1.1 *South-pointing cart and magnetic compass*

The fact that a magnetic compass points to the north has long been known in China. There is a famous legend that an emperor used a "south-pointing cart" in a battlefield as early as in B.C. 2634. The cart was so constructed that the arm of a doll mounted on it always pointed to the south, so that the emperor could accurately surmise the direction to which his enemy had retreated. Although it is not known whether such a legend is true or not, a model of a south-pointing cart is actually exhibited in the History Museum in Peking, as can be seen in Fig. 1.1 which was taken by one of the authors (RIKITAKE) in 1978 with special permission of the museum.

It is said that a south-pointing cart was used in later years as a ceremonial pilot car when an emperor went out from his palace. Christian missions, that visited China around the 17th century, seem likely to have imagined that the cart was operated by a magnet, so that it was reported to Europe that a magnet had been extensively used in China. To tell the truth, however, a south-pointing cart seems likely to be a device consisting of pulleys and gears. The doll's arm is to be set initially so as to point to the south. When the cart changes its direction by a certain angle on its way, the doll is rotated by the same angle in the opposite direction by the action of the pulley and gear mechanism, so that the doll's arm always points to the south.

Although it is disappointing that the south-pointing cart is not operated by a magnet, something about a "south-pointing ladle" was written in a book published in the 1st century in A.D. It seems likely, however, that the north-pointing power of a magnet was firmly established later by Shen Kue (1030–94) (see Fig. 1.2) in his famous book in which very clear descriptions about magnetic needles or south-pointing needles can be seen. It is written in the book that such a needle was hung by a thin string or supported on a finger nail or an edge of a bowl.

FIG. 1.1. South-pointing cart at the History Museum in Peking, China. The photo was taken by one of the authors (Rikitake) in 1978 with special permission from the museum.

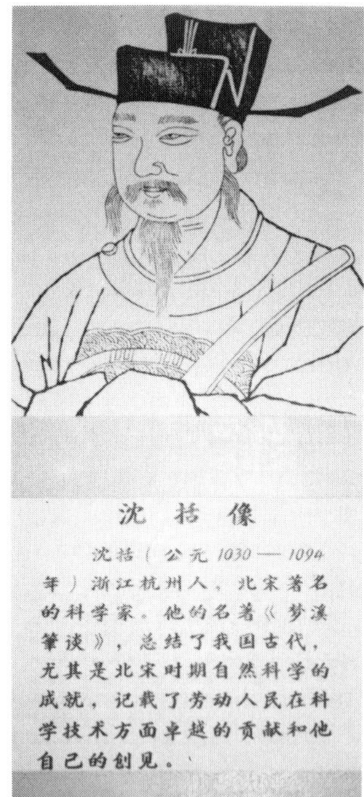

FIG. 1.2. Portrait of Shen Kue at the History Museum in Peking, China. The photo was taken by one of the authors (Rikitake) in 1978 with special permission from the museum.

In Europe, the earliest and most reliable literature about the north-pointing power of a magnet is said to have been presented by a priest called A. Neckam (1157–1217).

From classical literature, it is clear that a magnetic compass was in practical use in China as early as in the beginning of the 12th century. That is also the case for Europe in the 13th century.

It is surprising that magnetic compasses are still extensively used for navigation even today in spite of the highly developed gyrocompass, radio wave and satellite navigation techniques. Probably, this is due to the simple construction of a magnetic compass, so that it very seldom fails.

A magnet does not generally point to the true north. Such a fact was pointed out by Shen Kue in China in the 11th century. However, C. Columbus (c.a. 1466–1506) was the first who clearly reported that the direction in which a magnet points is different from place to place. It was reported by him that the magnet, which had been pointing to a direction slightly deviating from the north toward the east, tended to point to a direction deviating toward the west at a location of 28°N and 28°W on his voyage to America across the Atlantic Ocean in 1492. The distribution of the direction pointed by a magnet became fairly well known in Europe by the end of the 16th century. The deviation angle from the true north is called the magnetic declination.

1.1.2 Inclination of a magnet

The fact that a magnet, which can rotate about a horizontal axis, tends to incline from the horizontal plane was discovered by G. Hartmann in Germany in 1544 and R. Norman in England in 1576. This was named the magnetic inclination or dip and it became known that inclination increases with increasing latitude, by the end of the 16th century.

1.1.3 Gilbert and his experiment

William Gilbert (1544–1603) is well known as a court physician to Queen Elizabeth the 1st in England. He was also very much interested in physical phenomena associated with magnets. He made a spherical magnet from a natural magnet or lodestone which he called the "terrella." As can be seen in Fig. 1.3, he then studied the behavior of small magnets placed around the terrella. It was concluded from his experiments that the small magnet is almost parallel to the surface of terrella at its equator. As the small magnet approaches the poles, the inclination angle of the magnet increases. The behavior of the small magnet thus disclosed could explain the changes in the magnetic inclination with latitude as known in Europe at the time. Gilbert at last reached a conclusion that the earth is a huge magnet. It seems likely that the then-known phenomena related to magnets could be accounted for by

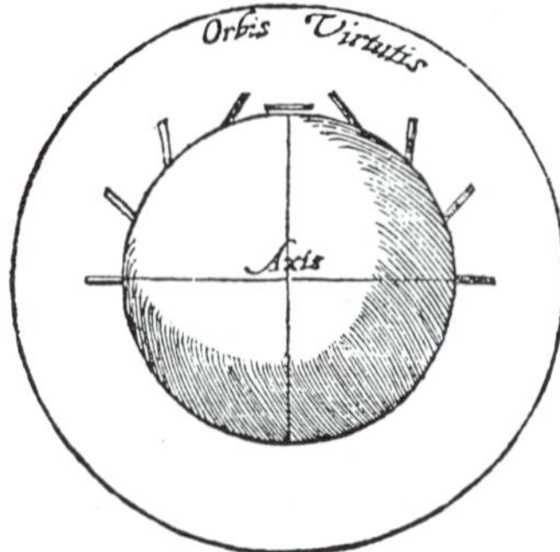

FIG. 1.3. Gilbert's "terrella" model (GILBERT, 1600, 1893, 1958).

such an idea. It was even made clear that the north pole of a magnet is attracted by the earth's south pole, and the south pole of a magnet is attracted by the earth's north pole. It is important that Gilbert proved that the attractive and repulsive forces between magnetic poles having different and the same signs are the origin of the north-pointing power of a magnet. Gilbert's work was published originally in Latin (GILBERT, 1600, 1893, 1958). Fortunately, an English translation of the book is now available, so that we can see how Gilbert proceeded in his research.

1.1.4 Gauss and his general theory of geomagnetism

In contrast to the discovery of magnetic declination and inclination, it took some time to discover the distribution of intensity of the geomagnetic field because of difficulties in measurement. A method of relative measurement based on the period of oscillation of a magnetic needle became fashionable toward the end of the 18th century, so that it was discovered that the higher the latitude, the larger is the intensity of the geomagnetic field. A method of absolute measurement, which is much more reliable than the relative one, was introduced by Carl F. Gauss (1777–1855) in the 1830's. He also published a famous paper (GAUSS, 1839) in which he separated the field arising from origins inside the earth from that arising from origins outside the earth. Geomagnetic data, then available at 91 stations, were used for the analysis. It was concluded by the study that the geomagnetic field of internal origin is far larger than that of external origin. It was also found that the geomagnetic field can be approximately represented by that of a small bar magnet placed at the center of the earth, or a uniformly magnetized sphere, although there are some fields of more complex distribution.

As this book is not particularly intended to be a history of geomagnetic study, no further account of early work will be mentioned. Those who are particularly interested in such a history may refer to CHAPMAN and BARTELS (1940).

1.2 Distribution of the Geomagnetic Field

The distribution of the geomagnetic field over the earth has been gradually unveiled through the efforts of geomagneticians over the past few centuries. Although it has been difficult to conduct accurate measurements of the geomagnetic field over the sea, the development of air-borne and sea-borne magnetometers has provided a powerful means of observing the geomagnetic field over the sea in recent years. It has also become fashionable in the latter half of this century to measure the magnetic field in the space surrounding the earth by magnetometers mounted on man-made satellites. Even the fields in interplanetary space and on the moon and other planets have become the target of measurement.

1.2.1 Elements of the geomagnetic field

As magnetic field is a vector quantity, it is necessary to measure the direction and intensity in order to specify the geomagnetic field at a location. It has been customary to measure the horizontal intensity (H), magnetic declination (D) and inclination (I) in a classical observation of the geomag-

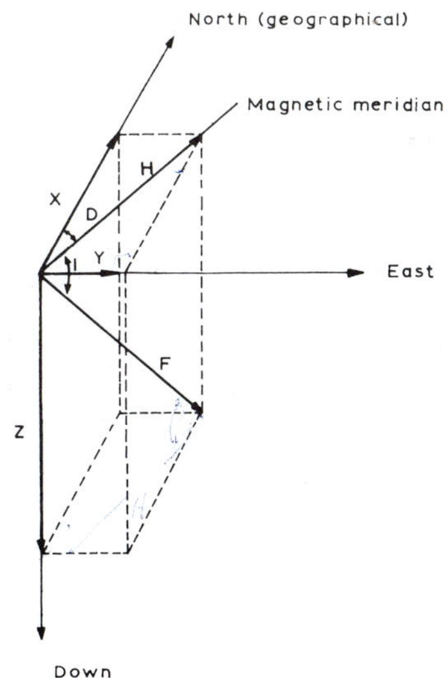

FIG. 1.4. Elements of the geomagnetic field.

netic field. The three quantities are called the three elements of the geomagnetic field.

As can be seen in Fig. 1.4, we may use other elements, which are independent from one another, for specifying the geomagnetic field. It has now become fashionable to make use of the combination of the total intensity (F) and H and D because an accurate measurement of F can easily be made by a proton precession magnetometer, developed in recent years. In polar regions, the combination of the northward (X), eastward (Y) and downward (Z) components is often used.

It is apparent that the following relations hold good between the geomagnetic elements shown in Fig. 1.4:

$$
\begin{gathered}
H = F\cos I,\ X = H\cos D,\ Y = H\sin D,\ Z = F\sin I \\
X^2 + Y^2 + Z^2 = F^2,\ X^2 + Y^2 = H^2 \\
\cos D = X/\sqrt{X^2 + Y^2},\ \sin I = Z/\sqrt{X^2 + Y^2 + Z^2},\ \tan I = Z/H.
\end{gathered}
\tag{1.1}
$$

1.2.2 Measurement of the geomagnetic field

In order to measure the geomagnetic field, a variety of magnetometers have been used since the early centuries. Only a very brief account of instruments for measuring the geomagnetic field will be given below because this book is not particularly aimed at the instrumental side of geomagnetic study.

(1) Classical measurement of magnetic declination and inclination

One of the simplest ways for measuring the geomagnetic declination is to use a magnet hung by a thin string. It is customary to mount such a magnet system on a horizontal circle on which a telescope is also provided. Such a declinometer has long been used in classical study of geomagnetism.

A dip circle is one of the simplest means for measuring the magnetic inclination or dip. A magnetic needle which can rotate freely about a horizontal axis, is first brought into a plane perpendicular to the magnetic meridian. After that, the system is rotated by 90 degrees in the horizontal plane. In this case the inclination of the magnet from the horizontal plane is identified as the magnetic inclination. It is known, however, that the accuracy of measurement by a dip circle cannot be smaller than 1 min of arc because of friction at the axis.

The accuracy of inclination measurements was dramatically improved by the introduction of an earth-inductor by W. Weber (1804–91) in 1837. It then became possible to achieve an accuracy of $0.1'$.

(2) Classical measurement of geomagnetic field intensity

When a small magnet having a magnetic moment M is hung by a thin string, period of free oscillation T_A of the magnet at a point A is proportional to $1/\sqrt{MH_A}$, where H_A is the horizontal intensity of the geomagnetic field there. Similarly, period T_B at another point B is proportional to $1/\sqrt{MH_B}$ with the same proportionality constant. We therefore have

$$H_A/H_B = T_B^2/T_A^2$$

so that H_B is obtained relative to H_A which is regarded as a reference field.

The relative measurement of geomagnetic field intensity was popular around the 18th century. However, the magnetic moment of a magnet changes when it is subjected to a mechanical shock, and it is well known that the moment decreases with increase in temperature. In order to avoid such defects, Gauss proposed a new method of absolute measurement as already stated in the last section.

The product MH of magnetic moment M of a small magnet and horizontal intensity H is first obtained by means of an oscillation experiment. Next, the magnet is placed near another magnet which is deflected by the magnetic field of the former magnet. Such a deflection experiment leads to measurement of M/H. It is therefore possible to solve M and H independently by combining the results of both experiments. It appears to the authors that a number of magnetic observatories, are still conducting absolute measurements on the basis of the modified Gauss method. However, it should be borne in mind that measurements of this kind take a long time to carry out, so that measurements become difficult on magnetically disturbed days.

(3) Modern methods of geomagnetic measurement

An earth-inductor has now been improved in such a way that a small coil of diameter 2 cm or so is rotated with a speed of 12 turns per second or so with a suitable gearing-up mechanism. The electromotive force induced in the coil is amplified with a high-gain amplifier. A non-magnetic ball-bearing made of Cu-Be alloy is often used. When the axis of the rotating coil is so adjusted that no electromotive force is induced in the coil, the axis is then aligned in the direction of geomagnetic force. In such a way, the declination and inclination can be measured at the same time.

A number of measuring techniques which rely on the magnetic field which is produced by electric currents controlled by a potentiometer referring to a standard cell have been introduced since 1930's for measurements of geomagnetic field intensity.

A pair of Helmholtz coils, which produce a fairly uniform magnetic field in the center of the coils, is provided for the Geographical Survey Institute (GSI) type magnetometer (Fig. 1.5) which is widely used as an observatory as well as field-survey instrument in Japan. The declination and inclination are first measured by the sensor, which is a small rotating coil. After that, an

FIG. 1.5. Geographical Survey Institute magnetometer. The photo was taken by one of the authors (Honkura).

electric current controlled by a standard-cell is applied to the coils, so that an artificial magnetic field is produced around the sensor in addition to the geomagnetic field. We then measure the new inclination. We can then estimate the intensity of any components of geomagnetic field through some trigonometric calculations because the newly-applied artificial field is well known. Since the introduction of the proton precession magnetometer, it has become usual even for the GSI magnetometer to measure D and I by the coil sensor and F by a proton precession magnetometer.

(4) Proton precession magnetometer

A new technique for measuring the geomagnetic field intensity, which is based on a principle completely different from those in classical measurements, was introduced in 1950's. When an electric current of several amperes is made to flow in a coil wound around a small bottle containing water, the magnetic moment of the hydrogen ions or protons in the water tend to be directed toward the direction of the magnetic field produced by the electric current. We next cut off the current very suddenly and all the small proton magnets tend to perform precessional motion around the geomagnetic field. When the geomagnetic field is uniform enough, the synthesized result of such movements is equivalent to that of a magnet in the coil. If the magnetizing coil is then used as a pick-up coil, some voltage of the order of microvolts is excited in it. Such a voltage can readily be amplified and the frequency f of precessional motion is counted by a frequency counter.

FIG. 1.6. Proton precession magnetometer at the Peking Standard Seismological Observatory. The photo was taken by one of the authors (Rikitake) in 1978.

It is known that f is strictly proportional to geomagnetic total intensity F and that the proportionality constant is determined only by the magnetic moment of a proton which is an atomic constant. When F and f are measured in units of gammas (see (7) in this section) and Hz, we have

$$F = 23.4874\,f. \tag{1.2}$$

Since (1.2) is based on an atomic constant only, no temperature and humidity corrections are necessary for a measurement, so that it can be said that the proton precession magnetometer provides a really absolute measurement of the geomagnetic field intensity. The measurement is of course dependent on the accuracy of frequency measurement. But it is known that an accuracy amounting to 10^{-6} or so is quite easily achieved by the present-day quartz oscillator in that kind of measurement. Figure 1.6 shows the sensor of a proton precession magnetometer at the Peking Standard Seismological Observatory, China.

One advantage of a proton precession magnetometer is that it is insensitive to not-too-fast shaking, so that it can be mounted on ships, aeroplanes, rockets, satellites and so on. It is necessary, however, to avoid noises arising from carriers themselves. It is customary, therefore, to tow the sensor of a proton precession magnetometer at a distance of more than 200 m behind a ship for a sea-borne magnetometer. The invention of the proton precession magnetometer opened a new phase of magnetic surveys over the sea which had previously been extremely difficult to conduct.

(5) Optical pumping and super-conducting magnetometers

In recent years, magnetometers based on quite a new principle have been developed. One of them is the optical pumping magnetometer based on the Zeeman effect and quantum electronics. Another is called the Superconducting Quantum Interference Device (SQUID). Although it is said that a magnetic field as small as 0.01 gammas (see (7) in this section) can be continuously observed by these magnetometers, they are not as popular as those stated in the above paragraphs because of various difficulties. No detailed account of these magnetometers will be given in this book.

(6) Geomagnetic variometer

It is sometimes convenient to have magnetometers which measure only the deviation from the average values of geomagnetic components. Magnetometers that measure such deviations are called variometers.

For the purpose of recording temporal changes in the geomagnetic field, an optical variometer has been widely used since the time of Gauss. Deflection of a bar magnet from its equilibrium state is detected as that of a light beam reflected by a mirror attached to the magnet. It is necessary, for measuring changes in the declination, to use a torsionless string to hang the magnet. When an optical length of 3 m is used, it is possible to have a 2 mm deflection on the recording paper for a change of 1 min of arc.

In the case of a variometer for the horizontal component, the string is twisted in such a way that the magnet is placed approximately perpendicularly to the magnetic meridian. In order to measure changes in the vertical component, a magnet is placed almost horizontally with the help of supporting strings and counter balances. The deflection of the magnet about the horizontal axis is converted into horizontal deflection of the light beam by means of a prism.

It has been customary to record deflections in the three components simultaneously, on a sheet of photographic paper wound on a drum. A darkened room is therefore required for the observations. In recent years, however, deflections of the light beam have been detected by a suitable means which makes an electric current flow through Helmholtz coils attached to the variometer in such a way that the magnetic field produced by the current compensates the deflection of the magnet. The changes in electric current, which are proportional to the geomagnetic changes, can be recorded elsewhere. Figure 1.7 shows a set of three component variometers at the Yatsugatake Magnetic Observatory, Earthquake Research Institute, University of Tokyo, Japan.

Since 1960, flux-gate magnetometers have become extensively used as geomagnetic variometers. When an a.c. voltage of several hundred Hz is applied to the primary coil wound around a core made of magnetically highly-

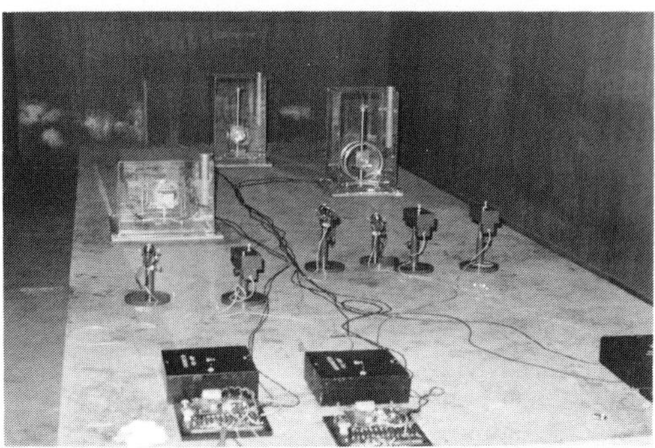

FIG. 1.7. Optical geomagnetic variometers at the Yatsugatake Magnetic Observatory, University of Tokyo, Japan. The photo was taken by one of the authors (Honkura).

permeable material such as permalloy, the voltage induced in the secondary coil is distorted because of the saturation characteristics of the core. The amplitude of the biharmonic voltage involved is proportional to the magnetic bias field in which the core is placed. As no dark room is required for the flux-gate system, this type of magnetometer is currently becoming very fashionable for geomagnetic variometers.

Flux-gate magnetometers are also widely used as instruments mounted on rockets, satellites and space probes. It has now become apparent that the sensitivity increases greatly when a ring core is used.

(7) Unit of magnetic field intensity

When a force of 1 dyne acts on a magnetic pole of 1 electromagnetic unit (e.m.u.) which is placed in a magnetic field, the intensity of the field is defined as 1 oersted (Oe). It is customary in geomagnetism, however, to use a unit called the gauss (Γ) instead of Oe. Although Γ is originally the unit for magnetic induction, there is practically no difference between magnetic field and magnetic induction because the magnetic permeability of the air is almost exactly equal to unity in e.m.u. It is sometimes convenient to use the 'gamma' γ which is defined by

$$1\gamma = 10^{-5}\,\Gamma.$$

In the M.K.S.A. unit system, which is nowadays fairly popular, we use A/m, Wb and N for the units of magnetic field, intensity of magnetic pole and force. In that case, we have the following relations:

$$1\ \text{Wb} = 10^8/4\pi\ \text{e.m.u.}$$

$$1\ \text{A/m} = 4\pi/10^3\ \text{Oe}.$$

The SI unit has gradually become popular in recent years. Wb/m^2, the unit of magnetic induction, is called the Tesla (T) which is related to the electromagnetic unit by

$$1\Gamma = 10^{-4}\text{T},\ 1\,\gamma = 10^{-9}\text{T} = 1\ \text{nT (nanotesla)}.$$

In SI units, magnetization $\boldsymbol{M}$ is defined as

$$\boldsymbol{B} = \mu_0(\boldsymbol{H} + \boldsymbol{M})$$

and hence the unit of magnetization is the same as that of $\boldsymbol{H}$; that is, A/m. On the other hand, magnetic polarization $\boldsymbol{J}$ is defined as

$$\boldsymbol{B} = \mu_0\boldsymbol{H} + \boldsymbol{J}$$

and its unit is T. In electromagnetic units, magnetization $\boldsymbol{M}$ is defined as

$$\boldsymbol{B} = \boldsymbol{H} + 4\pi\boldsymbol{M}$$

and its unit is emu/cm^3. Hence conversion from $\boldsymbol{M}$ in e.m.u. to $\boldsymbol{J}$ in SI and further to $\boldsymbol{M}$ in SI becomes

$$1\ \text{emu/cm}^3 = 4\pi \times 10^{-4}\,\text{T},\ \ 1\ \text{emu/cm}^3 = 10^3\ \text{A/m}.$$

A more detailed list of comparisons and conversion factors is given in the

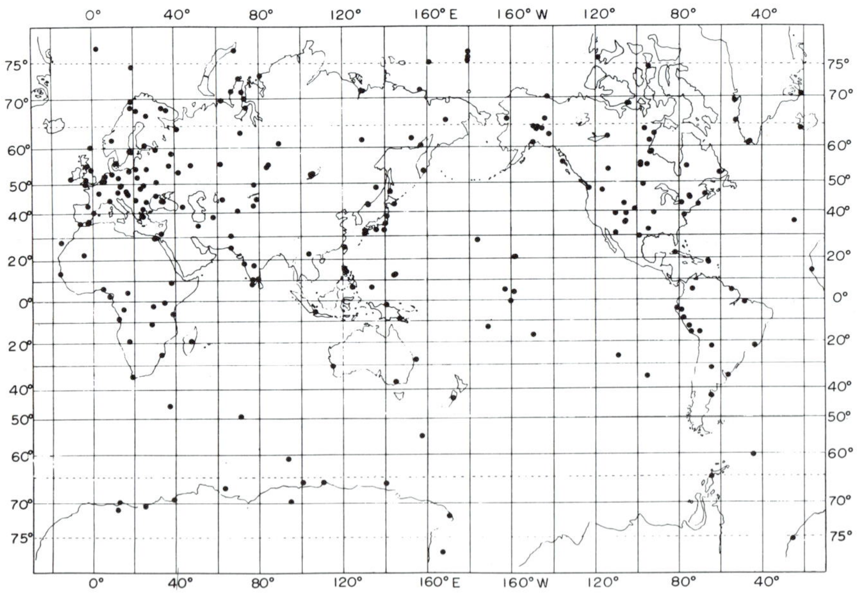

FIG. 1.8. The world distribution of magnetic observatories.

article by PAYNE (1981).

1.2.3 Magnetic survey

An operation to measure the geomagnetic field over an area using a magnetometer is called a magnetic survey. Even today, there are remote areas where no detailed land magnetic surveys have been carried out. However, it has become possible to cover many such areas by aeromagnetic surveys.

Magnetic surveys over the sea have so far been very difficult to perform because of the continuous movement of the ships that carry the magnetometers. In addition, the influence of magnetism arising from the large amount of iron of which a ship is composed was highly detrimental. Since the introduction of the ship-towed proton precession magnetometer, however, it has become easy to measure the geomagnetic total intensity over the ocean, although there are still difficulties in measuring the components of the geomagnetic field. As will be shown in a later chapter, outstanding geomagnetic anomalies, which have something to do with the sea-floor spreading hypothesis, were disclosed by ocean magnetic surveys.

Magnetic surveys by artificial satellites are currently becoming popular. An account on satellite surveys will be given in Section 1.6.

1.2.4 Magnetic observatory

It is customary for the results of magnetic surveys to be reduced to those which would have been observed at a specified epoch. This is quite necessary for constructing a magnetic chart that represents the distribution of an element of the geomagnetic field at a certain epoch. In order to perform such a reduction, we have to rely on continuous observations of the geomagnetic field which are carried out at permanent magnetic observatories.

Since the 1957–58 International Geophysical Year, an international project to examine the physical behavior of the earth and its surroundings, some 200 magnetic observatories have been in operation all over the world. A list of magnetic observatories (International Geophysical Data Library, Science Council of Japan, 1980) is presented in Table 1.1 although it is not known whether or not all the observatories listed are working fully. The world distribution of magnetic observatories is illustrated in Fig. 1.8, on the basis of Table 1.1.

It is usual for a magnetic observatory to have at least two huts for absolute magnetometers as well as for variometers. These huts or buildings should be non-magnetic and changes in room temperature should be very small. Figure 1.9 shows the absolute measurement hut of the Yatsugatake Magnetic Observatory. The hut is constructed from non-magnetic material which is also good for thermal insulation. Figure 1.10 is a view of the absolute

TABLE 1.1. Geomagnetic observatories.

CSAGI No.	Station	Geographic Lat.	Geographic Long.	Geomagnetic Lat.	Geomagnetic Long.
A001	Arctica I	N 77°48′–86°18′	E 03°57′–160°23′	76°	
A003	Arctica II	N 83°50′–86°21′	E 193°08′–323°55′	76°–81°	
	Arctica III	N 75°32′–79°06′	E 179°13′–179°23′	68°–71°	219°–213°
	Arctica IV	N 75°50′–81°19′	E 179°59′–194°26′	69°–75°	219°–215°
	Arctica V	N 75°11′–78°02′	E 160°52′–171°54′	66°–70°	209°–211°
	Arctica VI	N 76°50′–81°19′	E 179°59′–194°26′	69.7°–74.9°	217.8°–215.0°
A009	Tikhaya Bay	N 80°20′	E 52°48′	71.3°	153.5°
A009	(A 188 Heiss Is.)	N 80°30′	E 58°00′	71.0°	156.1°
A009	(A 188 Heiss Is.)	N 80°37′	E 58°03′	71.1°	156.3°
A005	Alert	N 83°30′	W 62°30′	85.7°	168.7°
A010	Murchison Bay	N 80°03′	E 18°15′	75.1°	138.0°
A017	Barentsburg (Pyramida)	N 78°38′	E 16°23′	74.4°	133.0°
A019	Arctic Ice Floe	N 81°38′–85°23′		75°	
A020	C. Chelyuskin	N 77°43′	E 104°17′	66.1°	176.5°
A021	Thule	N 77°29′	E 290°50′	89.2°	357.5°
A028	Mould Bay	N 76°12′	W 119°24′	79.0°	256.3°
A030	Resolute Bay	N 74°42′	W 94°54′	83.0°	287.0°
A031	Bear Is.	N 74°31′	E 19°01′		
A033	Dixon Is.	N 73°33′	E 80°34′	62.8°	161.7°
A037	Tixie Bay	N 71°38′	E 128°53′	60.3°	191.3°
A037	Tixie Bay	N 71°35′	E 129°00′	60.2°	191.3°
A039	Point Barrow	N 71°18′	E 203°15′	68.4°	240.7°
A044	Kap Tobin	N 70°25′	W 21°58′	74.5°	88.6°
A045	Barter Is.	N 70°08′	W 143°40′	69.9°	252.6°
A047	Tormosö	N 69°40′	E 18°57′	67.02°	117.17°
A049	Godhavn	N 69°14′	W 53°31′	80.0°	33.1°
A050	Murmansk	N 68°57′	E 33°03′	64.0°	126.8°
A054	Abisco	N 68°28′	E 18°54′	65.99°	115.63°
A057	Lovozero	N 67°59′	E 35°05′	62.8°	127.3°
A060	Kiruna	N 67°50′	E 20°25′	65.2°	116.0°
A065	Sodankylä	N 67°22′	E 26°39′	63.7°	120.4°
A070	Kotzebue	N 66°40′	W 162°30′	63.4°	242.0°
	Sukkertoppen	N 65°25′	W 52°54′	76.3°	29.1°
A073	Fort Yukon	N 66°34′	W 145°18′	66.6°	256.4°
A077	Cape Wellen	N 66°10′	E 169°09′	61.7°	236.8°

TABLE 1.1 (continued)

CSAGI No.	Station	Geographic		Geomagnetic	
		Lat.	Long.	Lat.	Long.
A092	College	N 64°52′	E 212°10′	64.6°	256.1°
A099	Baker Lake	N 64°20′	W 96°02′	73.9°	314.7°
A101	Reykjavik (Leirvogur)	N 64°11′	W 21°42′	70.3°	71.6°
A102	Big Delta	N 64°00′	W 145°44′	64.2°	258.9°
A107	Healy	N 63°51′	W 148°58′	63.5°	256.2°
A118	Northway	N 62°58′	W 141°57′	64.1°	263.4°
A121	Srednikan	N 62°26′	E 152°19′	52.9°	210.4°
A122	Yellowknife	N 62°26′	W 114°24′	69.0°	292.8°
A123	Dombas	N 62°04′	E 09°07′	62.1°	100.7°
A124	Yakutsk	N 62°01′	E 129°40′	50.8°	193.7°
A130	Anchorage	N 61°14′	E 210°08′	60.9°	257.8°
A132	Julianehaab	N 60°43′	E 313°58′	71.0°	35.9°
A134	Nurmijärvi	N 60°31′	E 24°39′	57.8°	112.9°
A140	Lerwick	N 60°08′	W 01°11′	62.5°	89.0°
A145	Churchill	N 58°30′	W 94°12′	68.6°	322.5°
A145	Churchill	N 58°48′	W 94°06′	68.9°	322.5°
A148	Great Whale	N 55°16′	W 77°48′	66.8°	347.2°
A149	Sitka	N 57°04′	E 224°41′	60.0°	275.0°
A154	Meanook	N 54°37′	W 113°20′	61.9°	300.7°
	St. John's	N 59°04′	E 18°56′	58.5°	21.2°
	Ottawa	N 45°24′	W 75°33′	57.0°	351.5°
	Minsk	N 54°06′	E 26°31′	51.5°	110.5°
	Narssarssuaq	N 61°12′	E 314°36′	71.2°	36.8°
	Melville	N 53°17′	E 299°26′	64.6°	11.9°
	New Alesund	N 78°59′	E 11°59′	75.3°	131.5°
	Cambridge Bay	N 69°01′	E 255°00′	77.0°	295.3°
	Eskimo Point	N 61°06′	E 265°56′	71.0°	322.4°
	Gillam	N 56°21′	E 265°19′	66.3°	324.5°
	Norway House	N 53°59′	E 262°10′	64.0°	321.5°
	Rankin Inlet	N 62°48′	E 267°40′	72.8°	323.5°
	Thompson	N 55°46′	E 262°10′	65.3°	326.6°
	Winnipeg	N 49°38′	E 262°52′	59.4°	324.2°
B009	Lovö	N 59°21′	E 17°50′	58.0°	106.1°
B012	Beloit	N 39°29′	E 261°52′	49.4°	324.7°
B014	Borok	N 58°02′	E 38°58′	52.9°	123.4°
B019	Sverdlovsk	N 56°44′	E 61°04′	48.3°	140.8°
B022	Tomsk	N 56°28′	E 84°56′	45.7°	159.7°
B028	Kazan	N 55°50′	E 48°51′	49.2°	130.5°
B035	Moscow	N 55°29′	E 37°19′	50.8°	120.7°
B038	Eskdalemuir	N 55°19′	W 03°12′	58.5°	83.2°
B044	Hel	N 54°37′	E 18°49′	53.4°	103.9°
B054	Shatsk	N 53°59′	E 41°51′	48.6°	123.9°

TABLE 1.1 (continued)

CSAGI No.	Station	Geographic		Geomagnetic	
		Lat.	Long.	Lat.	Long.
B058	Wingst	N 53°45′	E 09°44′	54.3°	95.3°
B071	Witteveen	N 52°49′	E 06°40′	54.1°	91.5°
B089	Swider	N 52°07′	E 21°15′	50.6°	104.8°
B095	Niemegk	N 52°04′	E 12°41′	52.2°	96.8°
B098	Valentia	N 51°56′	W 10°15′	56.7°	73.8°
B114	Rude Skov	N 55°51′	E 12°27′	55.8°	98.8°
B119	Hartland	N 51°00′	W 04°29′	54.7°	79.3°
B125	Kiev	N 50°43′	E 30°18′	47.5°	112.4°
B132	Manhay	N 50°18′	E 05°41′	52.0°	89.1°
B136	Dourbes	N 50°06′	E 04°36′	52.0°	88.0°
B143	Pruhonice	N 49°59′	E 14°33′	49.9°	97.5°
B145	Lvov	N 49°54′	E 23°45′	48.0°	106.1°
B145	Lvov (Ujgorod)	N 48°35′	E 22°22′	47.1°	104.3°
B159	Victoria	N 48°31′	W 123°25′	54.3°	292.7°
B162	Wien-Kobenzl	N 48°16′	E 16°19′	47.9°	98.4°
B163	Furstenfeldbruck	N 48°10′	E 11°17′	48.8°	93.6°
B168	Chambon la Forêt	N 48°01′	E 02°16′	50.5°	84.6°
B191a	Tihany	N 46°54′	E 17°54′	46.3°	99.3°
B191b	Nagycenk	N 47°36′	E 16°42′	47.1°	99.0°
B222	Ottawa	N 45°12′	W 75°30′	56.8°	351.6°
B239	Castellaccio	N 44°26′	E 08°56′	45.7°	89.7°
B249	Agincourt	N 43°47′	W 79°16′	55.2°	347.0°
B259	Leningrad	N 59°57′	E 30°42′	56.1°	117.6°
B261	Casper	N 42°51′	W 106°18′	51.6°	314.3°
B267	Logrono	N 36°51′	W 02°28′	40.7°	75.5°
B267	Logrono	N 42°27′	W 02°30′	46.7°	77.4°
B269	Carrollton	N 39°22′	E 266°28′	49.8°	330.2°
B295	Boulder	N 40°08′	W 105°14′	49.1°	316.3°
B305	Leadville	N 39°17′	E 253°43′	48.1°	315.4°
B318	Fredericksburg	N 38°12′	E 282°30′	49.7°	349.6°
B321	Ponta Delgada	N 37°46′	W 25°39′	45.7°	51.0°
B328	Budkov	N 49°05′	E 14°01′	49.1°	96.6°
B349	Stonyhurst	N 53°48′	W 00°12′	56.5°	85.3°
B440	Price	N 39°37′	W 110°47′	47.8°	310.2°
	Volozhin	N 54°06′	E 26°31′	51.4°	110.6°
	Newport	N 48°16′	W 117°07′	55.2°	299.8°
	Weston	N 42°23′	E 288°41′	53.9°	357.1°
	Eielson	N 64°40′	E 213°00′	64.6°	257.5°
	Loring AFD	N 46°55′	W 68°07′	58.5°	01.5°
	Arkhangelsk	N 64°35′	E 40°30′	58.9°	129.0°
	Kaliningrad	N 54°42′	E 20°32′	53.1°	105.9°
	Magadan	N 60°07′	E 157°01′	51.5°	215.6°
	Marion	S 46°53′	E 37°51′	−49.2°	95.1°

TABLE 1.1 (continued)

CSAGI No.	Station	Geographic		Geomagnetic	
		Lat.	Long.	Lat.	Long.
	Podkamennaya Tunguska	N 61°36′	E 90°00′	50.7°	165.0°
C001	Petropavlovsk	N 53°06′	E 158°38′	44.5°	218.5°
C016	Yuzhno Sakhalinsk	N 46°57′	E 142°43′	36.7°	206.6°
C018	Odessa	N 46°47′	E 30°53′	43.6°	111.3°
C026	Simferopol	N 44°50′	E 34°04′	41.1°	113.5°
C028	Grocka	N 44°38′	E 20°46′	43.5°	101.1°
C034	Memambetsu	N 43°55′	E 144°12′	33.8°	208.4°
C050	Alma Ata	N 43°16′	E 77°23′	33.2°	151.2°
C051	Vladivostock	N 43°47′	E 132°02′	32.7°	198.0°
〃	〃	N 43°41′	E 132°10′	32.7°	198.1°
C059	Panaguritsch	N 42°31′	E 24°11′	40.8°	103.5°
C076	Tashkent	N 41°25′	E 69°12′	32.3°	143.7°
C086	Tortosa	N 40°49′	E 00°30′	43.9°	79.8°
C098	Toledo	N 39°53′	W 04°03′	43.9°	74.9°
C117	Onagawa	N 38°26′	E 141°28′	28.1°	206.8°
C126	Ashkabad	N 37°57′	E 58°06′	30.4°	133.2°
C137	Almeria	N 36°51′	W 02°28′	40.7°	75.5°
C143	San Farnando	N 36°28′	W 06°12′	41.0°	71.5°
C147	Kakioka	N 36°14′	E 140°11′	25.8°	205.9°
C163	Tehran	N 35°44′	E 51°23′	29.2°	126.6°
C214	Shimosato	N 33°34′	E 135°56′	22.8°	202.4°
C215	Surlari	N 44°41′	E 26°15′	42.5°	106.2°
C223	Aso	N 32°53′	E 131°01′	21.8°	198.0°
C236	Tucson	N 32°15′	W 110°50′	40.6°	312.1°
C239	Misallat	N 29°45′	E 30°54′	27.0°	106.7°
C245	Kanoya	N 31°25′	E 130°53′	20.3°	198.0°
C251	Quetta	N 30°11′	E 66°57′	21.5°	139.8°
C256	Helwan	N 29°52′	E 31°20′	27.1°	106.5°
C264	Tenerife	N 28°29′	W 16°17′	35.1°	58.7°
C273	Tamanrasset	N 22°48′	W 05°31′	27.6°	68.5°
C277	Honolulu	N 21°18′	E 201°54′	21.0°	266.4°
〃	〃	N 21°19′	W 158°00′	21.1°	266.4°
C287	Teoloyucan	N 19°45′	W 99°11′	29.7°	327.0°
C300	San Juan	N 18°23′	E 293°53′	30.1°	3.2°
〃	〃	N 18°07′	W 66°09′		
C311	M'Bour	N 14°24′	W 16°58′	21.4°	55.1°
C339	Nitsanim	N 31°44′	E 34°36′	28.3°	110.1°
C360	Alushta-Simferopol	N 44°41′	E 34°25′	40.9°	113.7°
C362	Irkutsk (Patrony)	N 52°10′	E 104°27′	40.5°	174.7°
〃	Irkutsk (Zuy)	N 52°28′	E 104°02′	40.8°	174.4°
〃	Irkutsk	N 53°04′	E 105°32′	41.4°	175.6°
C364	Tbilisi	N 42°05′	E 44°42′	36.6°	122.2°

TABLE 1.1 (continued)

CSAGI No.	Station	Geographic		Geomagnetic	
		Lat.	Long.	Lat.	Long.
C401	Burlington	N 35°49′	W 106°04′	44.8°	316.5°
C405	Dallas	N 32°59′	W 96°45′	43.1°	327.6°
C406	Espanola	N 36°00′	E 253°57′	45.0°	316.5°
C425	Centro Geofisico	N 22°58′	W 82°09′	34.3°	345.3°
	Pendelli	N 38°02′	E 23°51′	36.4°	102.3°
	Hatizyozima	N 33°07′	E 139°48′	23.0°	206.7°
	Novo-Kazalinsk	N 45°48′	E 62°07′	37.6°	139.1°
	Karaganda	N 49°49′	E 78°05′	40.3°	149.3°
	Novosibirsk	N 54°51′	E 83°15′	44.8°	158.6°
	Khabarovsk	N 48°31′	E 135°10′	38.0°	200.7°
E528	Cha-Pa	N 22°21′	E 103°50′	10.7°	173.3°
E538	Alibag	N 18°38′	E 72°52′	9.3°	143.6°
E542	Hyderabad	N 17°25′	E 78°33′	7.6°	149.6°
E547	Baguio	N 16°25′	E 120°35′	4.9°	189.2°
E553	Muntinlupa	N 14°22′	E 121°01′	2.8°	189.7°
E556	Guam	N 13°27′	E 144°45′	3.7°	212.8°
″	″	N 13°35′	E 144°52′	·3.8°	212.9°
E562	Annamalainagar	N 11°24′	E 79°41′	1.4°	149.4°
E556	Kodaikanal	N 10°14′	E 77°28′	0.4°	147.1°
E568	Addis Ababa	N 09°02′	E 38°46′	5.3°	109.2°
E571	Ibadan	N 07°27′	E 03°54′	10.7°	74.7°
E574	Palmyra Is.	N 05°53′	W 162°50′	5.1°	265.8°
E575	Paramaribo	N 05°49′	W 55°13′	17.2°	14.4°
E578	Fuquene	N 05°48′	W 73°44′	17.1°	355.1°
E583	Bangui	N 04°26′	E 18°34′	4.8°	88.5°
E585	Fanning Is.	N 03°55′	W 159°23′	3.8°	268.8°
E586	Moca	N 03°21′	E 08°40′	5.8°	78.6°
E590	Tatuoka	S 01°12′	W 48°31′	9.8°	20.8°
E593	Talara	S 04°38′	W 81°18′	7.0°	347.7°
E595	Chiclayo	S 06°48′	W 79°46′	4.7°	349.3°
E596	Chimbote	S 09°06′	W 78°22′	2.4°	350.7°
E603	Trivandrum	N 08°29′	E 76°57′	−1.2°	146.4°
E606	Koror	N 07°16′	E 134°32′	−3.5°	203.4°
E618	Jarvis Is.	S 00°23′	W 160°03′	−0.6°	269.0°
E621	Nairobi	S 01°17′	E 36°48′	−4.4°	105.2°
E624	Lwiro	S 02°12′	E 28°48′	−3.7°	97.2°
E625	Hollandia	S 02°35′	E 140°31′	−12.7°	210.3°
E629	Vassouras	S 22°24′	W 43°39′	−11.8°	23.8°
E631	Leopoldville	S 04°16′	E 15°22′	−3.0°	83.6°
E634	Kuyper	S 06°02′	E 106°44′	−17.7°	175.5°
E640	Luanda	S 08°55′	E 13°10′	−7.2°	80.5°
E642	Port Moresby	S 09°24′	E 147°09′	−18.8°	217.9°
E644	Elizabethville	S 11°40′	E 27°28′	−12.7°	93.9°

TABLE 1.1 (continued)

CSAGI No.	Station	Geographic		Geomagnetic	
		Lat.	Long.	Lat.	Long.
E646	Huancayo	S 12°03′	W 75°20′	−0.4°	353.8°
E653	Apia	S 13°48′	W 171°47′	−16.1°	260.3°
E672	Tahiti	S 17°33′	W 149°37′	−15.3°	282.8°
E702	La Quiaca	S 22°06′	W 65°36′	−10.4°	3.2°
E712	Easter Is.	S 27°10′	W 109°25′	−18.0°	322.7°
E740	Yauca	S 15°32′	W 74°40′	−3.9°	354.5°
E743	Midway	N 28°13′	W 177°22′	24.2°	247.1°
	Davao	N 07°05′	E 125°35′	−4.1°	194.5°
	Tulsa	N 35°54′	E 264°13′	45.9°	328.3°
	Arequipa	S 16°28′	W 71°29′	4.9°	357.6°
	Tangerang	S 06°10′	E 106°38′	−17.6°	175.4°
	Lunping	N 25°11′	E 121°10′	13.8°	189.5°
	Dar Es Salaam	S 06°31′	E 39°11′	−10.1°	107.4°
	Tsumeb	S 19°13′	E 17°42′	−18.5°	84.3°
C901	Tananarive	S 18°55′	E 47°33′	−23.8°	112.4°
C914	Lourenco Marques	S 25°55′	E 32°35′	−27.7°	95.7°
C918	Brisbane	S 27°32′	E 152°55′	−35.9°	227.0°
C925	Watheroo	S 30°19′	E 115°53′	−42.0°	185.7°
C932	Pilar	S 31°40′	W 63°53′	−20.0°	4.6°
C933	Ganangara	S 31°47′	E 115°57′	−43.4°	185.8°
C957	Hermanus	S 34°26′	E 19°14′	−33.3°	80.4°
C976	Las Acacias	S 35°00′	W 57°41′	−23.5°	10.1°
C988	Trelew	S 43°15′	W 65°19′	−31.6°	3.1°
B966	Toolangi	S 37°32′	E 145°28′	−46.8°	221.0°
B979	Amberley	S 43°09′	E 172°43′	−47.7°	252.7°
B998	Port aux Français	S 49°21′	E 70°12′	−57.5°	127.7°
A951	Novolazarevskaya	S 70°46′	E 11°49′	−66.1°	53.2°
A959	Pionerskaya	S 69°44′	E 95°30′	−80.5°	145.9°
A961	Macquarie Is.	S 54°30′	E 158°57′	−61.2°	243.4°
A962	Orcadas del Sur.	S 60°44′	W 44°47′	−50.1°	18.8°
A973	Argentine Is.	S 65°15′	W 64°16′	−53.6°	3.3°
A976	Oasis	S 66°16′	E 100°43′	−77.6°	160.4°
A977	Wilkes	S 66°16′	E 110°31′	−78.0°	179.1°
A978	Mirny	S 66°33′	E 93°01′	−77.2°	146.4°
A979	Dumont d'Urville	S 66°40′	E 140°01′	−75.8°	231.5°
A980	Mawson	S 67°35′	E 62°55′	−73.1°	102.3°
A984	Syowa Station	S 69°02′	E 39°36′	−69.6°	77.1°
A986	Roi Baudouin	S 70°26′	E 24°19′	−67.9°	62.8°
A988	Cape Hallett	S 72°19′	E 170°13′	−74.7°	278.9°
A989	Halley Bay	S 75°31′	W 26°36′	−65.6°	24.1°
A991	Scott Base	S 77°51′	E 166°47′	−78.9°	295.3°
A995	Little America	S 78°11′	E 197°48′	−73.9°	312.5°
A996	Vostok	S 78°27′	E 106°52′	−89.2°	77.8°

TABLE 1.1 (continued)

CSAGI No.	Station	Geographic		Geomagnetic	
		Lat.	Long.	Lat.	Long.
A997	Marie Byrd	S 79°57′	W 119°50′	−70.4°	336.3°
〃	Byrd Station	S 79°59′	W 120°00′	−70.4°	336.2°
A999	South Pole (Amundsen-Scott)	S 90°00′	—	−78.3°	0.0°
	Lazarev	S 69°58′	E 12°54′	−65.5°	55.0°
	Plateau	S 79°15′	E 40°30′	−77.2°	52.5°
	Belyj Is.	N 73°03′	W 70°00′	63.7°	155.5°
	Amderwa	N 69°46′	E 61°41′	61.3°	148.0°
	Tambej	N 71°05′	E 71°08′	61.7°	155.0°
	Kamennyj Mys	N 68°05′	E 73°06′	58.6°	154.8°
	Seyakha	N 70°01′	E 72°05′	60.6°	155.1°
	Kzyle Agatch	N 45°22′	E 78°44′	35.8°	153.3°
	Zhelamia Cape	N 76°58′	E 68°35′	67.5°	157.6°
	Kharasovoy	N 71°08′	E 66°50′	61.8°	152.0°
	Numto	N 63°31′	E 71°22′	54.0°	151.6°

FIG. 1.9. The hut for absolute measurement of the geomagnetic field at the Yatsugatake Magnetic Observatory, University of Tokyo, Japan. The photo was taken by one of the authors (Honkura).

FIG. 1.10. The interior of the absolute measurement hut of the Peking Standard Seismological Observatory. The photo was taken by one of the authors (Rikitake) in 1978.

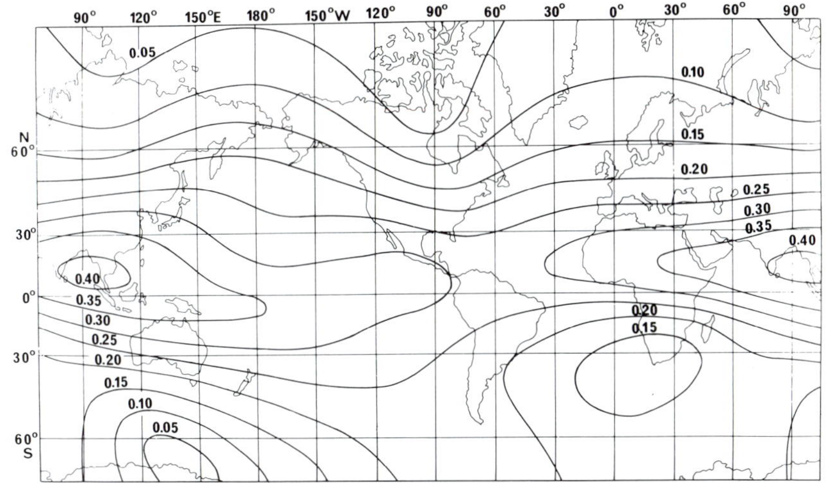

FIG. 1.11. The magnetic chart for the horizontal intensity for 1975 in units of Γ.

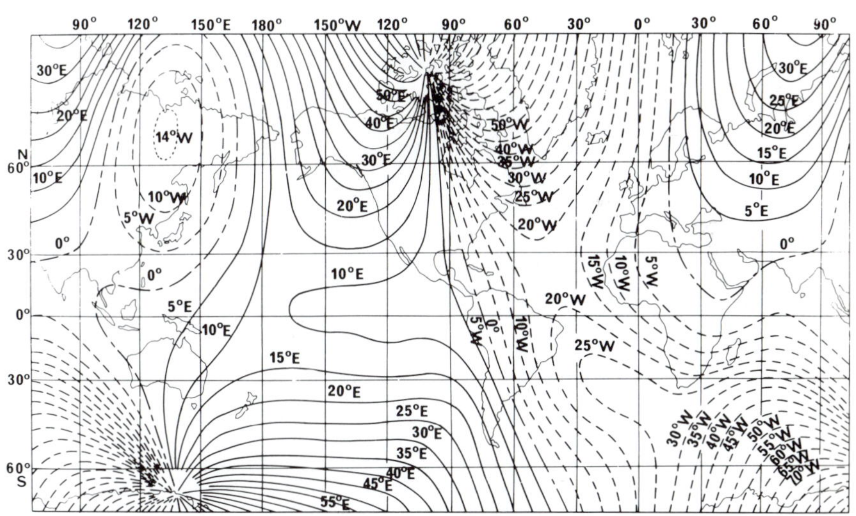

FIG. 1.12. The magnetic chart for the declination for 1975.

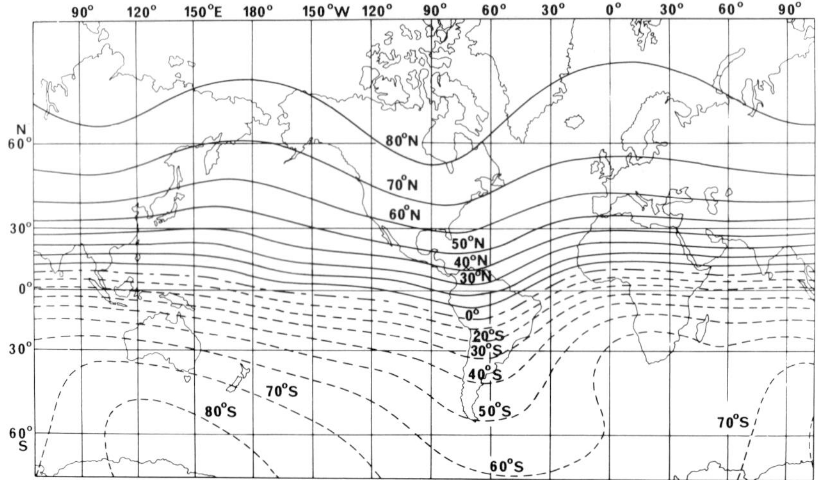

FIG. 1.13. The magnetic chart for the inclination for 1975.

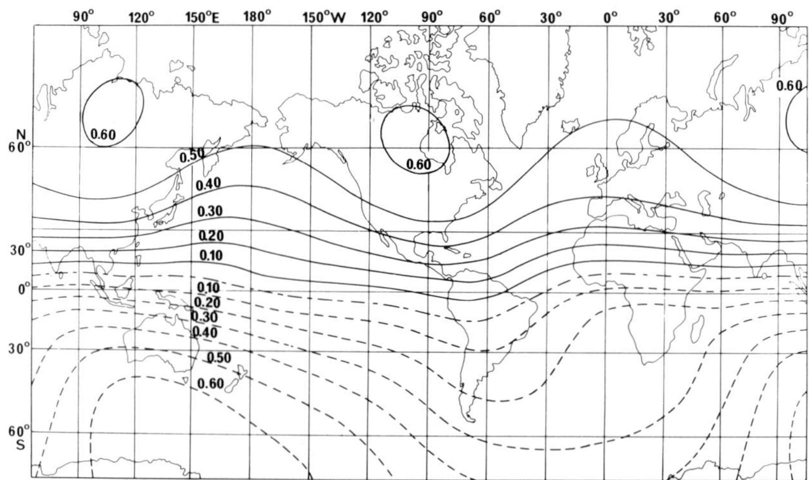

FIG. 1.14. The magnetic chart for the vertical intensity for 1975 in units of Γ.

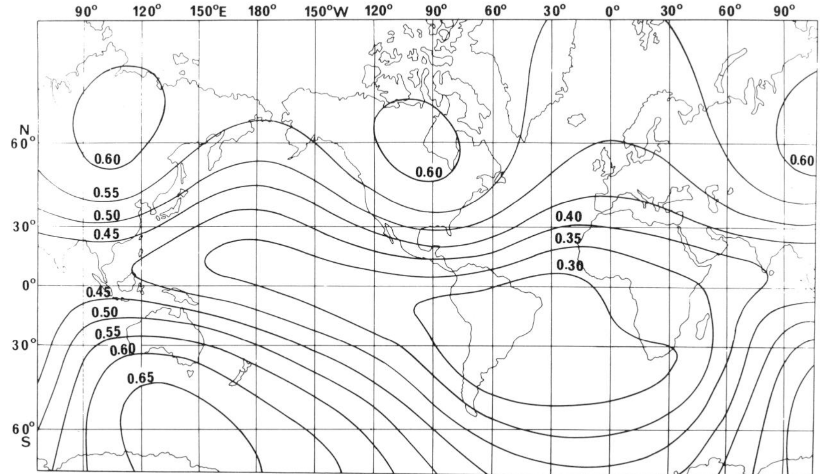

FIG. 1.15. The magnetic chart for the total intensity for 1975 in units of Γ.

measurement building of the Peking Standard Seismological Observatory. We can see many classical magnetometers as well as the proton precession magnetometer in the building.

Usually, a standard magnetic observatory is equipped with a rapid-run variograph, earth-current meters and other instruments in addition to those for absolute measurements and normal-run variometers.

1.2.5 Magnetic chart

The distribution of the elements of the geomagnetic field is customarily shown on a map. Such a representation is called a magnetic chart. World magnetic charts are issued by agencies in the U.S.A., the U.S.S.R. and England every 5–10 years. Figures 1.11, 1.12, 1.13, 1.14 and 1.15 are the simplified world magnetic charts for H, D, I, Z and F for the 1975 epoch.

In addition to the world magnetic charts, many countries have magnetic charts for their own countries. For instance, it is clear from the D chart in Japan that the declination amounts to about 10° west in the northernmost part of Japan, while it amounts to only 5° west in the southernmost part of the country.

1.2.6 Magnetic pole

Looking at Fig. 1.12, one sees that there are two points where all the equal-declination lines converge. At these points, the declination becomes indeterminate. In other words, the horizontal intensity vanishes. These points are called the north and south magnetic or magnetic-dip poles, respectively. In

1975, the north magnetic pole was located at a point in the Bathurst Is. in Arctic Canada, at a location 76.2°N and 100.0°W. Meanwhile, the south magnetic pole is located at 65.8°S and 139.3°E, a little off the Adelie coast in Antarctica.

It is obvious that a straight line connecting the two poles does not pass through the center of the earth, so that, strictly speaking, the geomagnetic field cannot be represented by that of a bar magnet placed at the center of the earth.

1.3 Spherical Harmonic Analysis of the Geomagnetic Field

It has been customary for the distributions of the geomagnetic field, as revealed by magnetic charts, to be subjected to a spherical harmonic analysis. This allows us to express the geomagnetic field in terms of a series of spherical functions.

1.3.1 Spherical harmonic function

Spherical harmonic functions are important for geophysics because any quantity which has a distribution over a spherical surface can be expressed by a superposition of spherical surface harmonics. The situation is much the same as that for a Fourier series.

Spherical harmonics are usually described by functions of spherical coordinates (r, θ, ϕ) which are related to Cartesian coordinates (x, y, z) by

$$x = r \sin\theta \cos\phi, \quad y = r \sin\theta \sin\phi, \quad z = r\cos\theta. \tag{1.3}$$

A function defined by

$$P_n(\cos\theta) = \sum_{s=0}^{\leqq n/2} (-1)^s \frac{(2n-2s)!}{2^n s!(n-s)!(n-2s)!} \cos^{n-2s}\theta \tag{1.4}$$

is called the zonal harmonic or Legendre function of degree n where n is a positive integer. P_n is a polynomial in $\cos\theta$. In a region which does not contain $\theta = 0$ or π, it is possible to define a zonal harmonic of the second kind $Q_n(\cos\theta)$. But no actual application of Q_n is considered in this book.

Equation (1.4) can be rewritten as

$$P_n(\cos\theta) = \frac{(2n)!}{2^{2n}(n!)^2}\left\{2\cos n\theta + 2\frac{1\cdot n}{1\cdot(2n-1)}\cos(n-2)\theta \right. \\ \left. + 2\frac{1\cdot 3\cdot n(n-1)}{1\cdot 2\cdot(2n-1)(2n-3)}\cos(n-4)\theta + \cdots\right\}. \tag{1.5}$$

If n is odd, the series ends with a multiple of $\cos\theta$ and, if n is even, the last term is constant, and does not contain the factor 2.

The $P_n(\cos\theta)$ for n of low degree are given as

$$\begin{aligned} P_0(\cos\theta) &= 1 \\ P_1(\cos\theta) &= \cos\theta \\ P_2(\cos\theta) &= (3\cos 2\theta + 1)/4 \\ P_3(\cos\theta) &= (5\cos 3\theta + 3\cos\theta)/8 \\ P_4(\cos\theta) &= (35\cos 4\theta + 20\cos 2\theta + 9)/64 \\ &\cdots\cdots \end{aligned} \tag{1.6}$$

A function $P_{n,m}(\cos\theta)$ defined by

$$P_{n,m}(\cos\theta) = \sin^m\theta \, \mathrm{d}^m P_n(\cos\theta)/\mathrm{d}(\cos\theta)^m \tag{1.7}$$

is called the associated Legendre function. n and m are positive integers and $m \leqq n$. When $m = 0$, we have $P_{n,0} = P_n$. $P_{n,m}$ satisfies

$$\frac{\mathrm{d}^2 P}{\mathrm{d}\theta^2} + \cot\theta \frac{\mathrm{d}P}{\mathrm{d}\theta} + \left\{ n(n+1) - \frac{m^2}{\sin^2\theta} \right\} P = 0. \tag{1.8}$$

The $P_{n,m}(\cos\theta)$ for n and m of low degree and order are given as

$$\begin{aligned} P_{1,1}(\cos\theta) &= \sin\theta \\ P_{2,1}(\cos\theta) &= (3/2)\sin 2\theta \\ P_{2,2}(\cos\theta) &= (3/2)(1 - \cos 2\theta) \\ P_{3,1}(\cos\theta) &= (3/8)(\sin\theta + 5\sin 3\theta) \\ P_{3,2}(\cos\theta) &= (15/4)(\cos\theta - \cos 3\theta) \\ P_{3,3}(\cos\theta) &= (15/4)(3\sin\theta - \sin 3\theta) \\ P_{4,1}(\cos\theta) &= (5/16)(2\sin 2\theta + 7\sin 4\theta) \\ P_{4,2}(\cos\theta) &= (15/16)(3 + 4\cos 2\theta - 7\cos 4\theta) \\ P_{4,3}(\cos\theta) &= (105/8)(2\sin 2\theta - \sin 4\theta) \\ P_{4,4}(\cos\theta) &= (105/8)(3 - 4\cos 2\theta + \cos 4\theta) \\ &\cdots\cdots \end{aligned} \tag{1.9}$$

When n and m take on larger values, numerical values of $P_{n,m}$ become extremely large. In order to avoid such inconvenience, SCHMIDT (1935) proposed the use of the partly normalized function $P_n^m(\cos\theta)$ which is called the Schmidt spherical surface harmonic. $P_n^m(\cos\theta)$ is defined by

TABLE 1.2. $P_n^m/P_{n,m}$ for $n, m \leqq 4$.

n	m = 1	2	3	4
1	1			
2	$\frac{1}{3}\sqrt{3} = 0.5774$	$\frac{1}{6}\sqrt{3} = 0.2887$		
3	$\frac{1}{6}\sqrt{6} = 0.4082$	$\frac{1}{30}\sqrt{15} = 0.1291$	$\frac{1}{60}\sqrt{10} = 0.0527$	
4	$\frac{1}{10}\sqrt{10} = 0.3162$	$\frac{1}{30}\sqrt{5} = 0.0745$	$\frac{1}{420}\sqrt{70} = 0.0199$	$\frac{1}{840}\sqrt{35} = 0.0070$

$$P_n^m(\cos\theta) = \begin{cases} P_{n,m}(\cos\theta) & m = 0 \\ \left\{2\dfrac{(n-m)!}{(n+m)!}\right\}^{1/2} P_{n,m}(\cos\theta) & m > 0. \end{cases} \tag{1.10}$$

Table 1.2 shows ratios $P_n^m/P_{n,m}$ for $n, m \leqq 4$. It is customary to use P_n^m in handling problems in geomagnetism although $P_{n,m}$ is also used for theoretical problems.

1.3.2 Spherical harmonic analysis

It is in general possible to express a function $f(\theta, \phi)$, which is defined on a spherical surface, in the form

$$f(\theta, \phi) = \sum_{n=0}^{\infty} \sum_{m=0}^{n} (A_n^m \cos m\phi + B_n^m \sin m\phi) P_n^m(\cos\theta). \tag{1.11}$$

The procedure by which the coefficients A_n^m and B_n^m are determined is called spherical harmonic analysis.

It had been a difficult task to determine the coefficients up to large n and m until the use of high-speed computers became commonplace. In order to make numerical work easy, for instance, $f(\theta, \phi)$ at equally spaced values of θ and ϕ are first obtained by interpolation in many classical analyses. The coefficients are then determined using the orthogonality property of spherical harmonic functions. Many of the classical analyses of the geomagnetic field can be found in CHAPMAN and BARTELS (1940, pp. 631–638).

It has in recent years become fashionable to determine a set of the coefficients, which are truncated at certain n and m, by some least-squares procedure. In the light of the development of the high-speed computer, it is not particularly difficult to determine the coefficients up to n and $m=12$.

In an actual analysis of the geomagnetic field, however, there are many points to be considered seriously. First of all, the coefficients are affected by errors due to uneven distribution of data, geomagnetic transient fluctuations and local anomalies. In addition to these difficulties, mutual dependency of the coefficients should be taken into account. In an actual analysis of the geomagnetic field, however, truncation at $n = 8$ or so would not be unreasonable.

1.3.3 Separation of the external field from the internal one

In free space, it is possible to define a magnetic potential W which satisfies

$$\nabla^2 W = 0. \tag{1.12}$$

Equation (1.12) can be written in spherical polar coordinates (r, θ, ϕ) as

$$\frac{1}{r^2}\frac{\partial}{\partial r}\left(r^2\frac{\partial W}{\partial r}\right) + \frac{1}{r^2 \sin\theta}\frac{\partial}{\partial\theta}\left(\sin\theta\frac{\partial W}{\partial\theta}\right) + \frac{1}{r^2\sin^2\theta}\frac{\partial^2 W}{\partial\phi^2} = 0. \tag{1.13}$$

A general solution of (1.13) is written as

$$\begin{aligned} W = a \sum_{n=1}^{\infty}\sum_{m=0}^{n} \{&(c_n^m \cos m\phi + s_n^m \sin m\phi) P_n^m(\cos\theta)(a/r)^{n+1} \\ &+ (\bar{c}_n^m \cos m\phi + \bar{s}_n^m \sin m\phi) P_n^m(\cos\theta)(r/a)^n\} \end{aligned} \tag{1.14}$$

where $P_n^m(\cos\theta)$ is a spherical function (see Subsection 1.3.1).

In the case of a geomagnetic problem, it is customary to take the origin of the coordinates at the center of the earth, so that a denotes the earth's radius.

On differentiating (1.14), the northward component of the magnetic field at the earth's surface ($r = a$) is obtained as

$$\begin{aligned} X &= \frac{1}{a}\frac{\partial W}{\partial\theta} \\ &= \sum_{n=1}^{\infty}\sum_{m=0}^{n} \{(c_n^m + \bar{c}_n^m)\cos m\phi + (s_n^m + \bar{s}_n^m)\sin m\phi\} \mathrm{d}\, P_n^m/\mathrm{d}\theta \end{aligned} \tag{1.15}$$

and the downward component as

$$\begin{aligned} Z &= \partial W/\partial r \\ &= \sum_{n=1}^{\infty}\sum_{m=0}^{n} \{(n\bar{c}_n^m - \overline{n+1}\, c_n^m)\cos m\phi + (n\bar{s}_n^m - \overline{n+1}\, s_n^m)\sin m\phi\}\, P_n^m. \end{aligned} \tag{1.16}$$

On the other hand, spherical harmonic analyses (see Subsection 1.3.2) of the observed data lead to

$$
\begin{aligned}
X &= \sum_{n=1}^{\infty} \sum_{m=0}^{n} (a_n^m \cos m\phi + b_n^m \sin m\phi) \mathrm{d} P_n^m/\mathrm{d}\theta \\
Z &= \sum_{n=1}^{\infty} \sum_{m=0}^{n} (\bar{a}_n^m \cos m\phi + \bar{b}_n^m \sin m\phi) P_n^m .
\end{aligned}
\tag{1.17}
$$

On comparing (1.15) and (1.16) to (1.17), we obtain

$$
\begin{aligned}
c_n^m + \bar{c}_n^m &= a_n^m & \quad s_n^m + \bar{s}_n^m &= b_n^m \\
-\overline{n+1}\, c_n^m + n\bar{c}_n^m &= \bar{a}_n^m & \quad -\overline{n+1}\, s_n^m + n\bar{s}_n^m &= \bar{b}_n^m
\end{aligned}
\tag{1.18}
$$

and so the coefficients of the magnetic potential are obtained as

$$
\begin{aligned}
c_n^m &= \frac{n a_n^m - \bar{a}_n^m}{2n+1} & \quad s_n^m &= \frac{n b_n^m - \bar{b}_n^m}{2n+1} \\
\bar{c}_n^m &= \frac{\overline{n+1}\, a_n^m + \bar{a}_n^m}{2n+1} & \quad \bar{s}_n^m &= \frac{\overline{n+1}\, b_n^m + \bar{b}_n^m}{2n+1} .
\end{aligned}
\tag{1.19}
$$

Similar results can also be deduced from the combination of Y (eastward component) or $-\partial W/a \sin\theta\partial\phi$ and Z.

It is therefore possible to separate c_n^m from $\bar{c}_n^m$ and s_n^m from $\bar{s}_n^m$. When there is no magnetic matter outside the earth, the magnetic potential tends to vanish as $r \to \infty$, so that a typical term of W is proportional to $(a/r)^{n+1}$. It is therefore evident that c_n^m and s_n^m are the coefficients of magnetic potential arising from sources within the earth. In contrast, a typical term in W becomes proportional to $(r/a)^n$ when the magnetic matter exists only outside the earth. $\bar{c}_n^m$ and $\bar{s}_n^m$ are hence the coefficients of magnetic potential arising from sources outside the earth.

It has been known since the time of GAUSS (1839) that, for the main geomagnetic field, $|c_n^m|, |s_n^m| \gg |\bar{c}_n^m|, |\bar{s}_n^m|$, so that the part of the geomagnetic field of external origin can be practically ignored.

1.3.4 Gauss coefficient

When the part of external origin is ignored, the potential of the geomagnetic field can be expressed as

$$
W = a \sum_{n=1}^{\infty} \sum_{m=0}^{n} (a/r)^{n+1} (g_n^m \cos m\phi + h_n^m \sin m\phi)\, P_n^m(\cos\theta) \tag{1.20}
$$

where g_n^m and h_n^m are called the Gauss coefficients.

A number of reviews (e.g. MAUERSBERGER, 1959; RIKITAKE, 1966a; VESTINE, 1967) on hitherto-made analyses have been presented. According to BARRACLOUGH (1978), a total of 264 spherical harmonic models

TABLE 1.3. Spherical harmonic coefficients for the main field in units of 10^{-4} e.m.u. (10^{-8}T).

Source	Year	g_1^0	g_1^1	h_1^1	g_2^0	g_2^1	h_2^1	g_2^2	h_2^2
Erman and Petersen (1874)	1829	−3201	−284	601	−8	257	−4	−14	146
Gauss (1839)	1835	−3235	−311	625	51	292	12	−2	157
Adams, J. C. (1900); Adams, W. G. (1898)	1845	−3219	−278	578	9	284	−10	4	135
Adams, J. C. (1900); Adams, W. G. (1898)	1880	−3168	−243	603	−49	297	−75	61	149
Schmidt (1898)	1885	−3168	−222	595	−50	278	−71	65	149
Fritsche (1897)	1885	−3164	−241	591	−35	286	−75	68	142
Neumayer and Petersen (1889)	1885	−3157	−248	603	−53	288	−75	65	146
Dyson and Furner (1923)	1922	−3095	−226	592	−89	299	−124	144	84
Jones and Melotte (1953)	1942	−3039	−218	555	−117	294	−150	157	51
Afanasieva (1946)	1945	−3032	−229	590	−125	288	−146	150	48
Vestine *et al.* (1947)	1945	−3057	−211	581	−127	296	−166	164	54
Finch and Leaton (1957)	1955	−3055	−277	590	−152	303	−190	158	24
Cain *et al.* (1967)	1965	−3040.1	−216.4	577.8	−154.0	299.8	−193.2	159.0	20.3
Leaton *et al.* (1965)	1965	−3037.5	−208.7	576.9	−164.8	295.4	−199.5	157.9	11.6
Peddie and Fabiano (1976)	1975	−3005.57	−201.70	567.05	−193.20	300.13	−204.44	161.97	−6.92
Langel *et al.*(1980b)	1979.85	−2998.96	−195.86	560.81	−199.48	302.72	−212.73	166.16	−19.61

of the geomagnetic field can be found in the literature. Some of the sets of Gauss coefficients of low degree, for typical analyses, are given in Table 1.3. The last set of coefficients in the table is that for the MAGSAT data (see Section 1.6).

1.4 Dipole and Non-Dipole Fields

When a magnetic dipole is placed at the origin of Cartesian coordinates, the potential W of the magnetic field at a point $P(x, y, z)$, arising from the dipole, is given by

$$W = \frac{M_x x + M_y y + M_z z}{r^3} \tag{1.21}$$

where M_x, M_y and M_z are the components of magnetic moment of the dipole in the x, y and z directions, respectively. The relationship between spherical polar coordinates (r, θ, ϕ) and Cartesian coordinates has already been given by (1.3).

(1.21) can be rewritten as

$$W = \frac{M_x}{r^2} \sin\theta\cos\phi + \frac{M_y}{r^2}\sin\theta\sin\phi + \frac{M_z}{r^2}\cos\theta$$

which is further rewritten as

$$W = \frac{M_x}{r^2} P_1^1(\cos\theta)\cos\phi + \frac{M_y}{r^2} P_1^1(\cos\theta)\sin\phi + \frac{M_z}{r^2} P_1(\cos\theta). \tag{1.22}$$

Comparing (1.22) to (1.20), we obtain

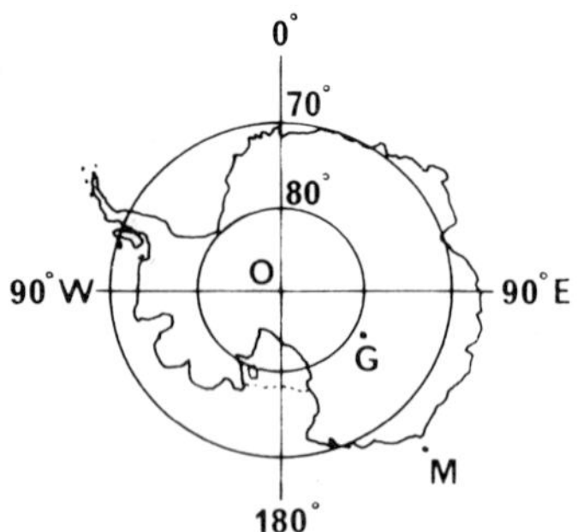

FIG. 1.16. The three poles in Antarctica for 1975. O, Geographical pole; G, Geomagnetic pole; M, Magnetic pole.

$$
\begin{aligned}
g_1^0 &= M_z/a^3 \\
g_1^1 &= M_x/a^3 \\
h_1^1 &= M_y/a^3.
\end{aligned}
\tag{1.23}
$$

The physical meaning of Gauss coefficients of the 1st degree thus becomes clear. The magnetic moment $M(=\sqrt{M_x^2 + M_y^2 + M_z^2})$ of the dipole is then given by

$$M = a^3\sqrt{(g_1^0)^2 + (g_1^1)^2 + (h_1^1)^2}. \tag{1.24}$$

The direction of the dipole is specified by

$$
\begin{aligned}
\tan\theta_0 &= \sqrt{(g_1^1)^2 + (h_1^1)^2}/g_1^0 \\
\tan\phi_0 &= h_1^1/g_1^1.
\end{aligned}
\tag{1.25}
$$

The colatitudes and longitudes of the points at which the extension of the axis of the dipole intersects the earth's surface are given by two sets of (θ_0, ϕ_0) in the northern and southern hemispheres, respectively. These points are called the geomagnetic poles.

In 1975, the north and south geomagnetic poles were located at (78.6°N, 70.5°W) in Greenland and (78.6°S, 109.5°E) in the interior of the antarctic

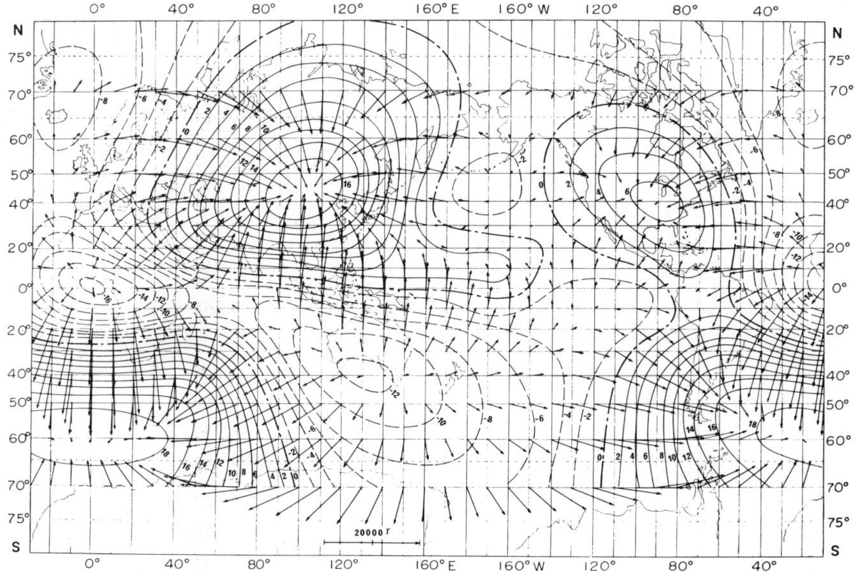

FIG. 1.17. The non-dipole field for 1965. The contours of vertical intensity are drawn every 1,000 γ, while the scale for the arrows representing the horizontal field is shown in the figure (Yukutake, personal communication, 1971).

continent, respectively. It is therefore evident that the axis of the dipole is inclined to the earth's rotation axis by 11.4°. There are three poles in the polar regions. Figure 1.16 shows the geographic, geomagnetic and magnetic poles in the antarctic region.

The geomagnetic field expressed by the spherical harmonics, with coefficients g_n^m and h_n^m of degree higher than 1, is called the non-dipole field. Figure 1.17 shows the world distribution of horizontal and vertical components of the non-dipole field for 1965 (T. Yukutake, personal communication, 1971). It can be seen from the figure that the intensity of the non-dipole field amounts to 20% or so of that of the dipole field. Two of the most remarkable features of Fig. 1.17 are the intense foci in Mongolia and the south of Africa. It also appears that the non-dipole field is low over the Pacific area.

1.5 International Geomagnetic Reference Field

Magnetic charts differ from one another, even for the same epoch, depending on the data used. As geomagnetic anomalies are defined by the deviation from the standard geomagnetic field, such discrepancies between magnetic charts give rise to much inconvenience.

In order to avoid such a difficulty, the International Association of Geomagnetism and Aeronomy (IAGA), adopted the International Geomagnetic Reference Field (IGRF) for 1965 (IAGA Commission II, Working Group 4, 1969). Eighty Gauss coefficients up to degree 8 and their time-derivatives were decided. These coefficients are essentially weighted means of those obtained by a number of spherical harmonic analyses. It is planned that an IGRF should fit not only the geomagnetic field on the earth's surface but also the field in the space surrounding the earth, as revealed by satellite magnetic surveys (see the following section).

In practice, the difference between the actual geomagnetic field and the IGRF becomes noticeable as time goes on. It is also noticeable that the geomagnetic field around Japan, Australia and so on is considerably different from the IGRF for 1965 and its candidate for 1975 (e.g. IGRF Evaluation Group of Japan, 1975).

A new IGRF for 1975 was adopted in 1975 (IAGA Division I, Study Group, Geomagnetic Reference Fields, 1975). The IGRF was further revised in 1981. The Gauss coefficients and their time-derivatives for 1980 are reproduced in Table 1.4.

1.6 Satellite Magnetic Survey

Measurements of the earth's magnetic field by satellites have recently

TABLE 1.4. IGRF 1980 coefficients.

		Main field (nT)		Secular change (nT/year)	
n	m	g	h	$\dot{g}$	$\dot{h}$
1	0	−29988		22.4	
1	1	−1957	5606	11.3	−15.9
2	0	−1997		−18.3	
2	1	3028	−2129	3.2	−12.7
2	2	1662	−199	7.0	−25.2
3	0	1279		0.0	
3	1	−2181	−335	−6.5	0.2
3	2	1251	271	−0.7	2.7
3	3	833	−252	1.0	−7.9
4	0	938		−1.4	
4	1	783	212	−1.4	4.6
4	2	398	−257	−8.2	1.6
4	3	−419	53	−1.8	2.9
4	4	199	−298	−5.0	0.4
5	0	−219		1.5	
5	1	357	46	0.4	1.8
5	2	261	149	−0.8	−0.4
5	3	−74	−150	−3.3	0.0
5	4	−162	−78	0.2	1.3
5	5	−48	92	1.4	2.1
6	0	49		0.4	
6	1	65	−15	0.0	−0.5
6	2	42	93	3.4	−1.4
6	3	−192	71	0.8	0.0
6	4	4	−43	0.8	−1.6
6	5	14	−2	0.3	0.5
6	6	−108	17	−0.1	0.0
7	0	70		−1.0	
7	1	−59	−83	−0.8	−0.4
7	2	2	−28	0.4	0.4
7	3	20	−5	0.5	0.2
7	4	−13	16	1.6	1.4
7	5	1	18	0.1	−0.5
7	6	11	−23	0.1	−0.1
7	7	−2	−10	0.0	1.1
8	0	20		0.8	
8	1	7	7	−0.2	−0.1
8	2	1	−18	−0.3	−0.7
8	3	−11	4	0.3	0.0
8	4	−7	−22	−0.8	−0.8
8	5	4	9	−0.2	0.2
8	6	3	16	0.7	0.2
8	7	7	−13	−0.3	−1.1
8	8	−1	−15	1.2	0.8

TABLE 1.5. MGST (6/80) coefficients in nT.

Internal Coefficients											
n	m	g_n^m	h_n^m	n	m	g_n^m	h_n^m	n	m	g_n^m	h_n^m
1	0	−29989.6		8	0	18.4		11	5	−0.4	0.6
1	1	−1958.6	5608.1	8	1	6.8	6.9	11	6	−0.3	−0.1
2	0	−1994.8		8	2	−0.1	−17.9	11	7	1.7	−2.4
2	1	3027.2	−2127.3	8	3	−10.8	4.0	11	8	1.8	−0.3
2	2	1661.6	−196.1	8	4	−7.0	−22.3	11	9	−0.6	−1.4
3	0	1279.9		8	5	4.3	9.2	11	10	2.1	−1.6
3	1	−2179.8	−334.4	8	6	2.7	16.1	11	11	3.5	0.6
3	2	1251.4	270.7	8	7	6.3	−13.1	12	0	−1.6	
3	3	833.0	−251.1	8	8	−1.2	−14.8	12	1	0.4	0.6
4	0	938.3		9	0	5.6		12	2	−0.1	0.6
4	1	782.5	211.6	9	1	10.4	−21.1	12	3	−0.1	2.3
4	2	398.4	−256.7	9	2	1.1	15.2	12	4	0.6	−1.5
4	3	−419.2	52.0	9	3	−12.6	8.9	12	5	0.5	0.5
4	4	199.3	−297.6	9	4	9.5	−4.8	12	6	−0.6	0.2
5	0	−217.4		9	5	−3.3	−6.5	12	7	−0.4	−0.4
5	1	357.6	45.2	9	6	−1.3	9.0	12	8	0.1	0.0
5	2	261.0	149.4	9	7	6.8	9.5	12	9	−0.4	0.0
5	3	−73.9	−150.3	9	8	1.4	−5.9	12	10	−0.2	−1.5
5	4	−162.0	−78.1	9	9	−5.1	2.1	12	11	0.7	0.3
5	5	−48.3	91.8	10	0	−3.3		12	12	0.0	0.7
6	0	48.3		10	1	−3.5	1.4	13	0	0.0	
6	1	65.2	−14.5	10	2	2.5	0.4	13	1	−0.5	−0.4
6	2	41.4	93.4	10	3	−5.3	2.6	13	2	0.3	0.4
6	3	−192.2	70.6	10	4	−2.1	5.6	13	3	−0.7	1.6
6	4	3.5	−42.9	10	5	4.6	−4.2	13	4	0.0	0.0
6	5	13.7	−2.4	10	6	3.1	−0.4	13	5	1.2	−0.6
6	6	−107.6	16.9	10	7	0.6	−1.3	13	6	−0.4	−0.1
7	0	71.7		10	8	1.8	3.5	13	7	0.4	0.8
7	1	−59.0	−82.4	10	9	2.8	−0.5	13	8	−0.6	0.2
7	2	1.6	−27.5	10	10	−0.5	−6.2	13	9	0.2	0.8
7	3	20.5	−4.9	11	0	2.4		13	10	0.1	0.5
7	4	−12.6	16.1	11	1	−1.3	0.7	13	11	0.4	−0.1
7	5	0.6	18.1	11	2	−1.9	1.7	13	12	−0.4	0.0
7	6	10.6	−22.9	11	3	2.2	−1.1	13	13	0.0	−0.1
7	7	−2.0	−9.9	11	4	0.1	−2.7				
External Coefficients											
1	0	20.4									
1	1	−0.6	−0.4								

become possible. Among the several satellites which so far have measured the geomagnetic field, the three Pogo (Polar Orbiting Geophysical Observatory) satellites, Ogo 2, 4 and 6, provided a large amount of data covering the whole orbit during 1965–71 (LANGEL, 1979). Their altitude ranged from about 400 km to about 1,500 km. The on-board magnetometers were of rubidium-vapor type and their accuracy was approximately 6 gammas.

The data obtained by Pogo satellite measurements have been used for producing maps of regional magnetic anomalies (REGAN *et al.*, 1975) as well as for deriving a global magnetic field model (LANGEL *et al.*, 1980a). In determining regional magnetic anomalies, a global model consisting of Gauss coefficients with $n \leqq 13$ was first determined and then values calculated from the model were subtracted from the observed ones, thus yielding residuals which can be regarded as magnetic anomalies. However, at high altitude the magnetic field intensity of internal origin is reduced to a considerable extent and, therefore, the effect of the external magnetic field is expected to be significant, particularly in the auroral zones where strong field-aligned electric currents are believed to flow on magnetically disturbed days. Hence, areas for which magnetic anomalies are to be investigated were restricted to latitudes between 50°N and 50°S. Further data reduction techniques were used to produce an anomaly map from the observed data, as described in detail by BHATTACHARYYA (1977), REGAN (1979), and MAYHEW (1979).

The magnetic anomaly map derived from the Pogo data disclosed some remarkable anomalies. A typical one is the Bangui anomaly found over the central part of Africa (REGAN *et al.*, 1975). The Bangui anomaly is characterized by a magnetic low of 12 gammas at the satellite altitude. By making use of the technique of upward continuation, aeromagnetic data over Central Africa were reduced to the satellite altitude in order to compare different kinds of data at the same altitude. Then it was found that they were in good agreement, which confirms the validity of an anomaly map derived from satellite data.

Magnetic anomalies thus obtained are likely to originate in the lithosphere. In fact, the distribution of magnetic anomalies in Africa is well correlated with geological and tectonic features there (REGAN *et al.*, 1975). Similar correlations were brought to light in other regions. Thus, satellite data are expected to provide valuable information on the structure of lithosphere.

In the meantime, the launch of a satellite called Magsat was planned. This new satellite can measure, at low altitude, the three components of the earth's magnetic field as well as its total intensity (LANGEL, 1979). It was launched on October 30, 1979, into a near-polar orbit; the orbital plane, close to the twilight meridian, was chosen to avoid the ionospheric currents on the day-side which cause a magnetic variation called *Sq*. Cesium vapor and fluxgate magnetometers were used to measure the total intensity and the three components, respectively. The initial altitude was between 561.2 km and 352.4 km and this decreased gradually until the reentry into the earth's atmosphere on June 11, 1980. The lowest limit of altitude for which magnetic field data are available is about 190 km (LANGEL *et al.*, 1980b).

There are two main objectives in solid earth geomagnetism; a study of the global distribution of the geomagnetic field and a study of magnetic anomalies of lithospheric origin. In the case of the former, the Gauss coefficients could be determined from data on magnetically quiet days, such as on November 5 and 6, 1979, as shown in Table 1.5. This particular model is called the MGST (6/80) field model for the epoch 1979. 85. Analyses of Magsat data are being carried out extensively in order to investigate magnetic anomalies of lithospheric origin.

CHAPTER 2

CHAPTER 2

SECULAR VARIATION OF THE GEOMAGNETIC FIELD

2.1 *Secular Variation in Historical Times*

It is said that Henry Gellibrand (1597–1636), an English astronomer, was the first to discover the secular variation of the geomagnetic field (CHAPMAN and BARTELS, 1940, p. 910). Many observations of geomagnetic secular variation in historical times at various observatories are summarized in CHAPMAN and BARTELS (1940). It is reported that the declination in London, which was 11°15′E in 1580, decreased gradually, becoming zero in 1657. After that, it moved further westward reaching a maximum value amounting to 24°08′W in 1805. The declination then tended to shift eastward from the westward maximum reaching a value of 11°30′W in 1935. The inclination changed over a range approximately from 66° to 74° during the period and the horizontal intensity in London also varied over a wide range. In 1853, this amounted to 0.1734 Γ. It increased gradually, reaching 0.1853 Γ in 1913 and after that tended to decrease reaching 0.1835 Γ in 1933.

The secular variation in the declination in Japan has been known since 1600, mostly based on measurements by magnetic compass on board occidental ships in the early years. Figure 2.1 shows the secular variation in declination in Japan as summarized by IMAMITI (1956). The measured values at various locations in Japan are reduced to those at Kakioka, where we have a standard magnetic observatory. The present distribution of the declination is assumed for the reduction procedure. It is seen in Fig. 2.1 that the declination underwent a secular change covering a range of 15° in Japan within about 400 years.

SMITH and NEEDHAM (1967) presented a similar variation in the declination in China as shown in Fig. 2.2. As the data are taken from various parts of the Chinese territory, the errors in the curve in Fig. 2.2 must be fairly large.

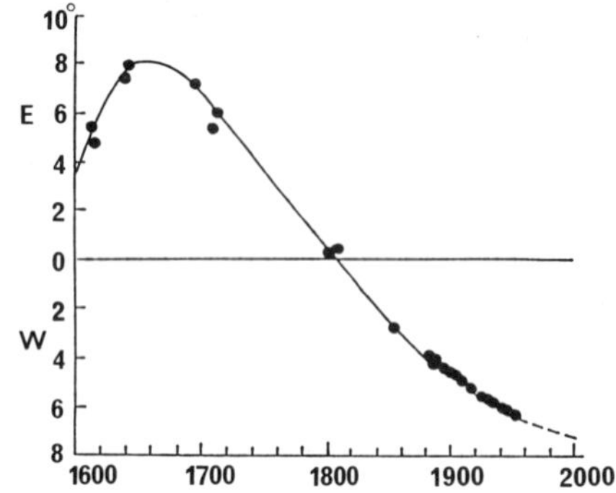

FIG. 2.1. Secular variation in declination in Japan (IMAMITI, 1956).

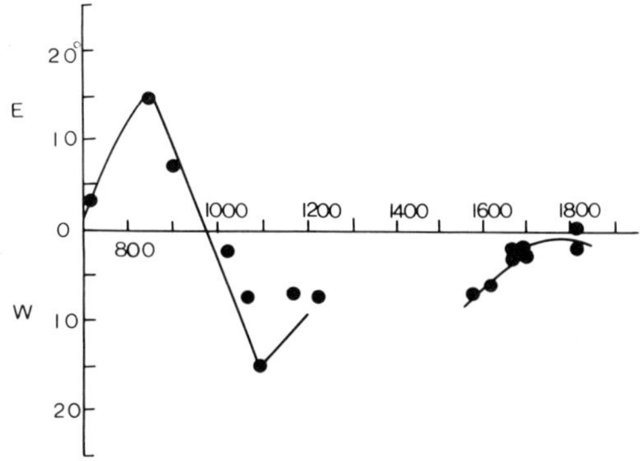

FIG. 2.2. Secular variation in declination in China (SMITH and NEEDHAM, 1967).

2.2 World Distribution of Secular Variation

Equal-variation lines of any element of the geomagnetic field can be shown on a world map. Such a representation of geomagnetic secular variation is called an isoporic chart. Figures 2.3 and 2.4 show the 1975 world distributions of secular variation for the total intensity and declination, respectively. It can be seen in the figures that the distribution of geomagnetic secular variation is fairly complicated.

2.3 Decrease in the Earth's Dipole Moment

Looking at Table 1.3, we clearly observe that g_1^0 has been decreasing steadily since the middle of the last century. This is also the case for g_1^1 and h_1^1 although we also see some fluctuations. It is then interesting to examine how the magnetic moment of the earth's dipole has decreased during the past 150

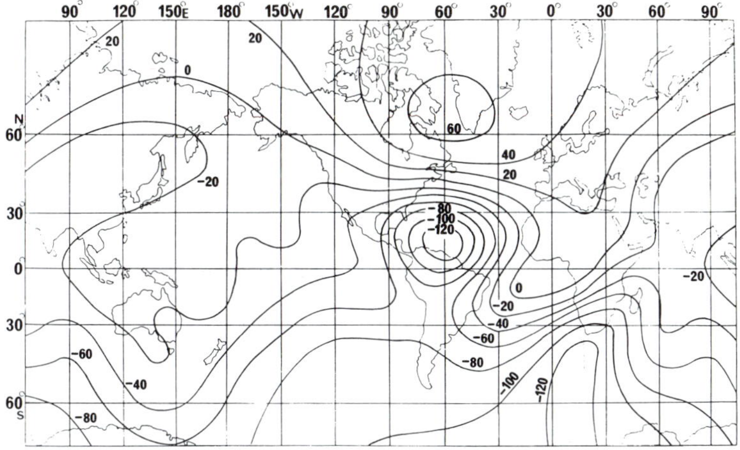

FIG. 2.3. Magnetic chart for the secular variation in total intensity for 1975 in units of γ/yr.

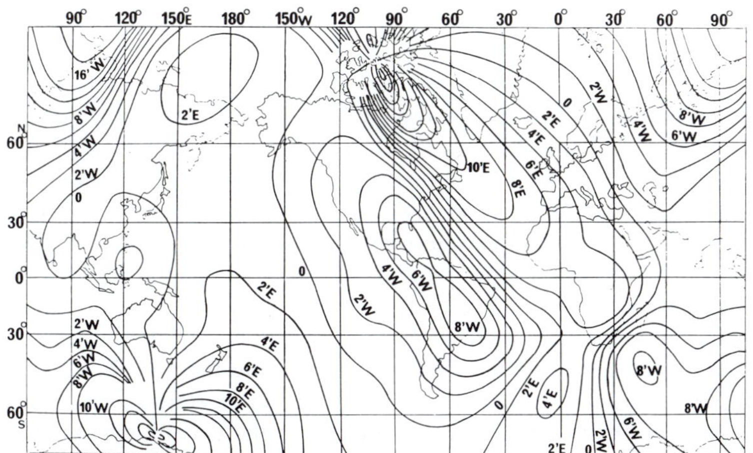

FIG. 2.4. Magnetic chart for the secular variation in declination for 1975 in units of min/yr.

years.

Figure 2.5 shows the dipole moments for various epochs plotted against time. In the figure we can see an almost monotonic decrease in the moment. The figure is drawn on the basis of Table 1.3 and other auxiliary data. It has been said that the rate of decrease amounts to about 5% per 100 years. According to the data collected in recent years, the rate seems to have increased, so that we would have no dipole field in a further 1,200 years,

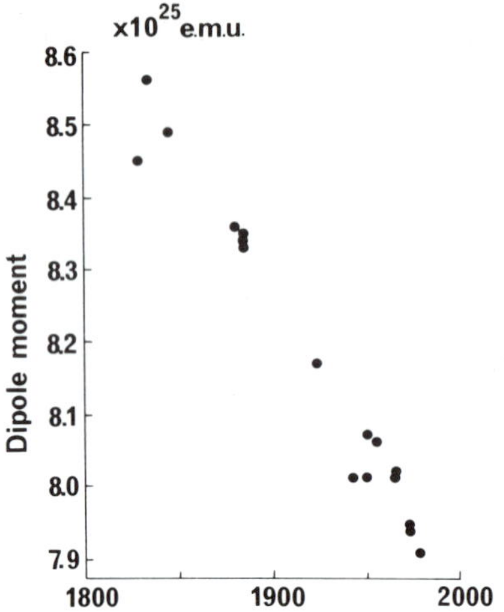

FIG. 2.5. Secular variation in the earth's dipole moment (1e.m.u. = 10^3 Am^2).

should the present rate of decrease continue.

2.4 Westward Drift of the Non-Dipole Field

It has been noticed that the intersection point between the equator and one of the zero-declination lines, which is at the moment running through the North and South American continents, has been steadily drifting since the 16th century. Navigators reported that the point was found somewhere around 30°E in logitude at the beginning of the 16th century and it is now found at about 75°W, so that the amount of drift exceeds 90° during the period of four and a half centuries. The drift velocity amounts to about 0.2°/yr on the average. Figure 2.6 shows the movement of the intersection point. Such a figure was first presented by ELSASSER (1950) and is revised using the new data.

It has been noticed that it is not only the intersection point mentioned above, but also the broad pattern of the geomagnetic field, which has tended to move westwards very slowly since the 17th century. A number of researchers have tried to determine the velocity of the westward drift (BULLARD *et al.*, 1950; WHITHAM, 1958; YUKUTAKE, 1962, 1968a, 1968b, 1970, 1971; NAGATA, 1962; YUKUTAKE and TACHINAKA, 1968a, 1968b, 1969). According to them, the velocity of the westward drift is in the range 0.2–0.4 degrees of longitude per year.

2.5 Northward Shift of the Dipole

Close examination of an isoporic chart such as Fig. 2.3 enables us to

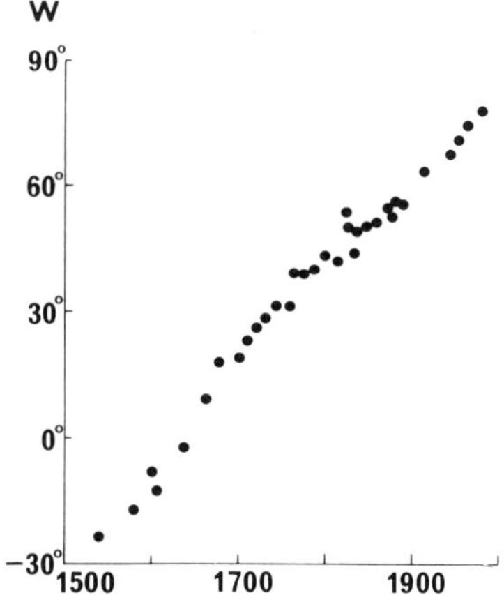

FIG. 2.6. The secular movement of the intersection point between the equator and one of the zero-declination lines.

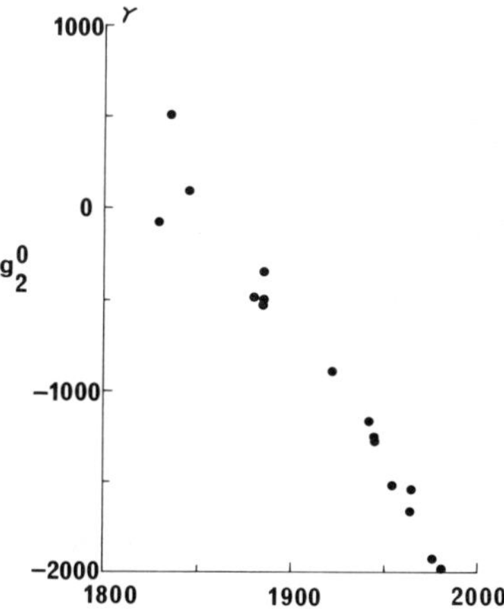

FIG. 2.7. Secular variation in Gauss coefficient g_2^0.

recognize the general increase, in the northern hemisphere, and decrease, in the southern hemisphere, of the geomagnetic total field intensity. Such a secular variation is well reflected by the time-variation of g_2^0. The g_2^0 versus time plot, made on the basis of Table 1.3, in fact indicates a marked decrease in g_2^0 during these 150 years, as can be seen in Fig. 2.7.

If an axial dipole having a magnetic moment M is placed at a point on the earth's rotation axis, the magnetic potential due to the dipole is given by

$$W = M \sum_n n z_0^{n-1} r^{-n-1} P_n(\cos\theta) \tag{2.1}$$

where z_0 is the distance between the dipole and the earth's center. If $z_0 \ll a$ (the radius of the earth), (2.1) leads to

$$W = (M/a^2)P_1 + (2Mz_0/a^3)P_2 \tag{2.2}$$

at $r = a$ provided higher power terms of z_0/a are ignored.

It is therefore clear from (2.2) that, if the dipole proceeds toward the north pole, or if z_0 increases, an increase in the absolute value of g_2^0, which takes a minus sign, takes place. This shifting of the dipole towards the north during these 150 years was pointed out by NAGATA (1962).

2.6 *Drifting and Standing Non-Dipole Fields*

Close examination of the non-dipole field as shown in Fig. 1.17, for successive epochs, disclosed the fact that the drift velocities of the non-dipole highs and lows are not always the same. It seems likely that the Mongolian focus of the vertical intensity has hardly moved although its intensity has increased considerably since the 17th century. In contrast, the center of the African focus has moved from the Indian Ocean to the Gulf of Guinea, that is, from 100° to 0°E during the same period. The positive and negative foci in the Pacific area do not seem to have moved, their intensities, also, being almost unchanged. It is therefore certain that only a few of the non-dipole highs and lows undergo a marked drift.

These characteristics of the non-dipole field led YUKUTAKE and TACHINAKA (1969) to a study in which they tried to separate the non-dipole field into standing and drifting parts.

They assumed that the geomagnetic potential can be expressed as

$$W = \sum_n \sum_m W_n^m P_n^m(\cos\theta) \tag{2.3}$$

where

$$W_n^m = F_n^m \cos(m\phi + \varphi_n^m) + K_n^m \cos m\{\phi + v_n^m(t - \tau_n^m)\} \tag{2.4}$$

in which the spherical harmonic coefficients of the standing part are denoted by F_n^m, while the phase angle of that part is given by φ_n^m. K_n^m and τ_n^m represent the amplitude of the drifting component and its phase measured in units of time. v_n^m is the drift velocity.

As for v_n^m, YUKUTAKE and TACHINAKA (1969) considered the following three cases:

(A) Rigid rotating field model, (B) dispersive velocities for Fourier components when the potential along parallels are expanded into Fourier series and (C) dispersive velocities for individual spherical harmonics. It turns out, however, that the (B) model results in the best fit between the synthesized and actual fields although the (A) and (C) models also lead to fairly similar results. For this reason, discussion was based mostly on the (B) model. v_n^m's for the (B) model are shown in Table 2.1. In all the models adopted, drift velocities have been obtained by analyzing the secular variation. It is believed that the velocity obtained from a secular variation analysis provides an approximation to that of the drifting field which is better than that obtained from an analysis of the non-dipole field itself. This is because the former velocity seems to be affected less by the non-drifting field.

$$F_n^m \cos \varphi_n^m,\ F_n^m \sin \varphi_n^m,\ K_n^m \cos m v_n^m \tau_n^m \text{ and } K_n^m \sin m v_n^m \tau_n^m$$

are first determined from a condition

$$\sum_i [\{F_n^m \cos \varphi_n^m + K_n^m \cos m v_n^m (t_i - \tau_n^m) - g_n^m(t_i)\}^2 + \{F_n^m \sin \varphi_n^m + K_n^m \sin m v_n^m (t_i - \tau_n^m) + h_n^m(t_i)\}^2] = \min. \tag{2.5}$$

where g_n^m and h_n^m are the Gauss coefficients and the summation is taken over the data for different epochs. Mathematically speaking, the above procedure for determining four unknowns is not quite logical because they are not independent, but it seems likely that there are no serious defects for practical purposes. F_n^m, φ_n^m, K_n^m and τ_n^m are then calculated as reproduced in Table 2.2, t in (2.4) being time, with the origin at year 1800, v_n^m being measured in units of 0.01°/yr, and a special time unit being used for measuring τ_n^m.

Figures 2.8 and 2.9 show the vertical components of standing and drifting fields synthesized from the Gauss coefficients as given in Table 2.2. The non-dipole field, formed by the superposition of both the fields, agrees so well with the field actually observed for all epochs, that the rms differences in the vertical

TABLE 2.1. Velocity model (B) for the drifting field (YUKUTAKE and TACHINAKA, 1969).

m	v_n^m
1	0.166°/yr
2	0.339
3	0.269
4	0.189
$m \geq 5$	0.3

TABLE 2.2. Gauss coefficients for the standing and drifting parts of the non-dipole field (YUKUTAKE and TACHINAKA, 1969).

n	m	F_n^m	φ_n^m	K_n^m	τ_n^m
1	1	3503γ	−78.4°	3896γ	9.70
2	1	1571	112.9	3570	1.31
2	2	399	139.8	1879	1.68
3	1	1607	−124.7	1907	−5.47
3	2	1076	−7.9	225	1.14
3	3	342	−29.8	540	1.36
4	1	920	25.9	556	7.26
4	2	638	34.1	121	2.42
4	3	231	118.4	252	−1.33
4	4	224	68.4	161	1.71
5	1	479	27.9	179	8.73
5	2	291	−4.0	54	−1.69
5	3	54	167.0	63	0.71
5	4	113	127.3	35	−0.65
5	5	101	−60.1	81	0.64
6	1	58	45.8	48	4.83
6	2	49	−148.5	67	2.22
6	3	154	172.1	45	−1.09
6	4	20	170.9	5	0.24
6	5	104	−95.3	55	1.03
6	6	40	−176.8	34	0.57

component, averaged over the harmonic terms up to degree 6, amount to only 200 γ or so in the cases of the analyses in recent years.

Comparing Fig. 2.9 to Fig. 2.8, we see that the drifting field is somewhat simpler in its distribution than the standing one. The maximum intensities of the two fields are not greatly different from one another. It is interesting to note that the tremendous increase in the vertical component of the Mongolian focus, i.e. from nearly zero to 20,000 γ within a period of 400 years, can be reproduced for all epochs and that a very small displacement of the focal center is also demonstrated. Similarly good fits are also found for the African non-dipole focus, the Pacific lows and so on.

The westward drift rate of the spherical harmonic coefficients can also be calculated with the coefficients given in Table 2.2. It should be pointed out that small rates of westward drift for P_1^1 as well as P_3^2 are then obtained.

The origin of the non-dipole field and its secular variation are currently believed to lie in the earth's core where magnetohydrodynamic processes prevail. According to the theory of the origin of the geomagnetic field (see Chapter 4), however, the non-dipole field should be subject to a westward drift. The finding of a standing field of appreciable intensity leads us to

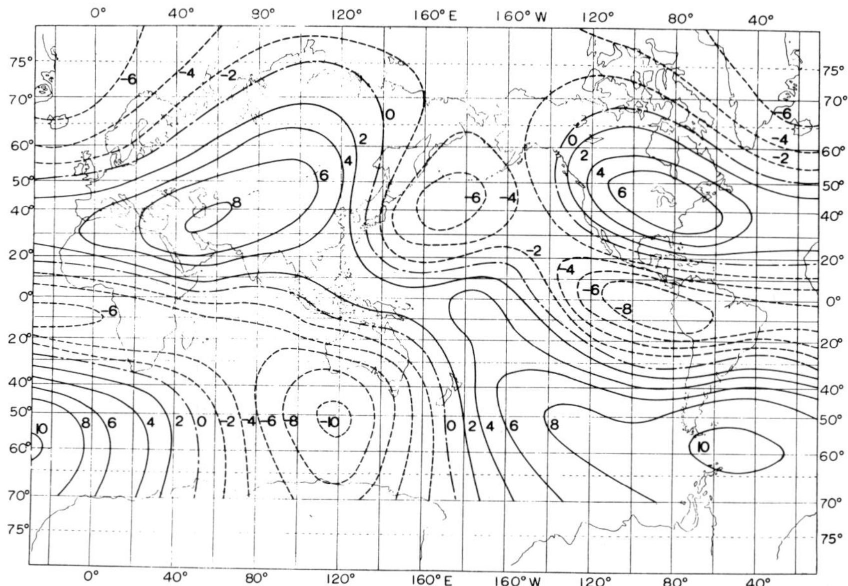

FIG. 2.8. The vertical component of standing non-dipole field in units of 1,000 γ (YUKUTAKE and TACHINAKA, 1969).

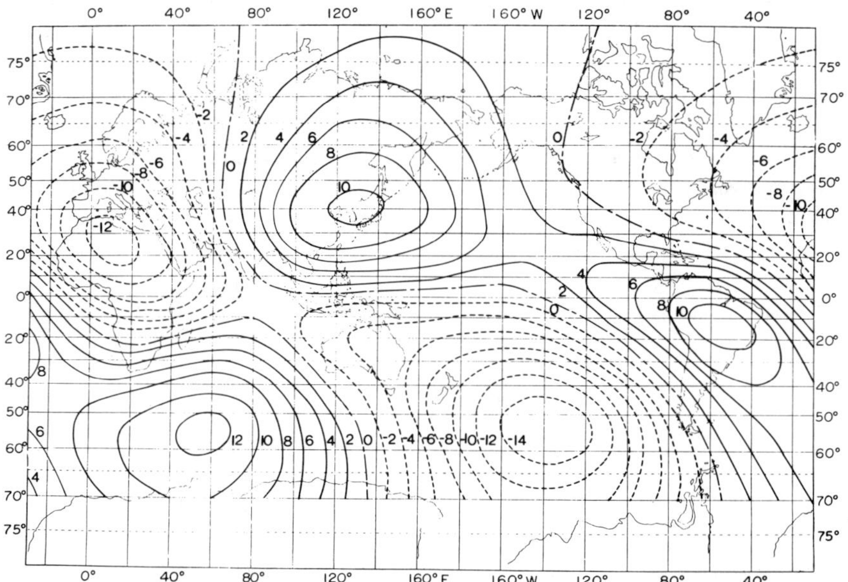

FIG. 2.9. The vertical component of drifting non-dipole field in units of 1,000 γ (YUKUTAKE and TACHINAKA, 1969).

reconsider the possibility of the crust-mantle origin of a part of the main field, though the high temperature in the interior of the earth makes it difficult to accept an explanation by permanent magnetization. It is possible that the life-time of the apparently standing part is very long, so that no change of the field could be traced within a period of a few hundred years.

2.7 *Change in the Rate of the Earth's Rotation and Geomagnetic Secular Variation*

In an attempt to search for a cause of slight fluctuations in earth's rotation, VESTINE (1953) examined a possible relationship between the observed change in the rate of rotation and the drift velocity of the eccentric dipole (a kind of approximate representation of geomagnetic field distribution in terms of a dipole located at an arbitrary place, e.g. BARTELS, 1936). In association with a minimum in the westward drift velocity of the eccentric dipole, which appeared around 1910, retardation in the rate of earth's rotation was observed, as pointed out by VESTINE and KAHLE (1968). Figure 2.10 shows the relationship between the westward drift velocity of the eccentric dipole and deviation of the length of the day from its average (KAHLE *et al.*, 1969). It is found from the figure that a change in the rate of rotation precedes, by about 7 years, the corresponding change in the westward drift velocity.

YUKUTAKE (1973) claimed, however, that movement of the geomagnetic field is not appropriately estimated in terms of the eccentric dipole, and recommended the use of secular variation data, particularly the spherical harmonics term of $(n, m) = (2,2)$ for secular variations, because this term is dominant in representation of drifting fields. In addition to secular variations, as characterized by the above dominant term, YUKUTAKE (1973) used data of

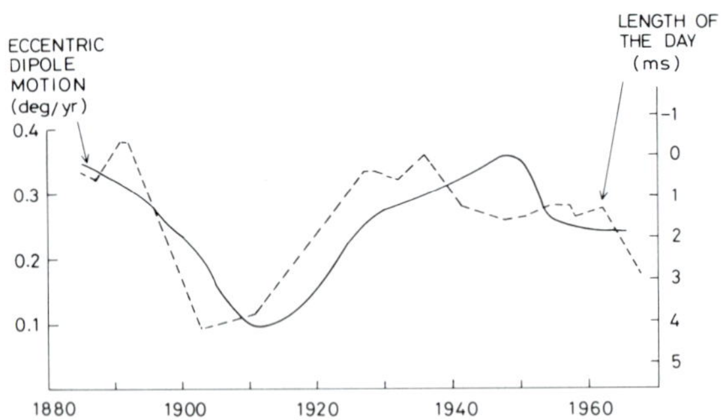

FIG. 2.10. Secular variation of the westward drift velocity of the eccentric dipole (solid line) and fluctuation of the day from its average (broken line) (KAHLE *et al.*, 1969).

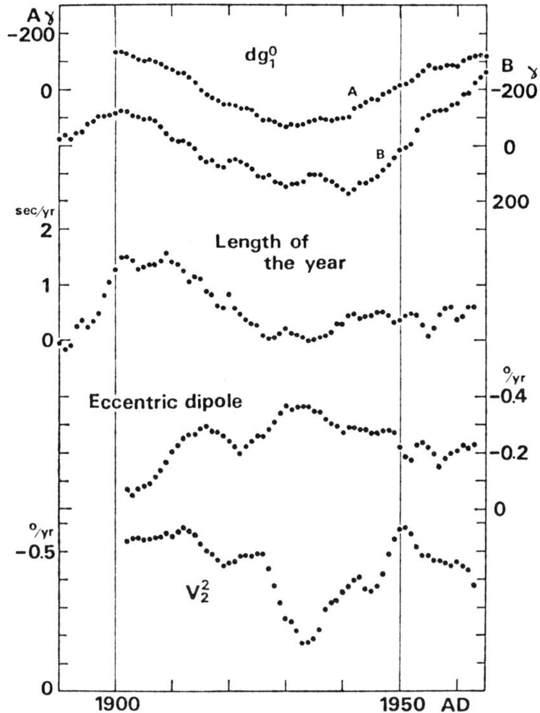

FIG. 2.11. Secular variations having an apparent period of 65 years. dg_1^0 indicates the change in Gauss coefficient g_1^0 as determined from the analysis of data at 21 (A) and 6 (B) observatories, respectively. V_2^2 denotes the potential of geomagnetic secular variation for the spherical harmonic component $n = m = 2$ (YUKUTAKE, 1973).

dipole moment changes in his attempt to examine a possible relationship between geomagnetic secular variations and changes in the rate of the earth's rotation. As theoretically estimated by YUKUTAKE (1972), based on electromagnetic core-mantle coupling, a change in the toroidal field in the core and the lower mantle arising from a change in the dipole field would result in a change in the state of electromagnetic coupling and hence a change in rotation speed of the mantle and the core (see Section 4.8).

Along the lines mentioned above, YUKUTAKE (1973) examined apparently periodic variations with periods of roughly 8,000, 400 and 65 years, respectively. The result for the apparent period of 65 years is shown in Fig. 2.11. The change in the length of the year seems to be in-phase with the change in the dipole moment represented by dg_1^0. The result for the period of about

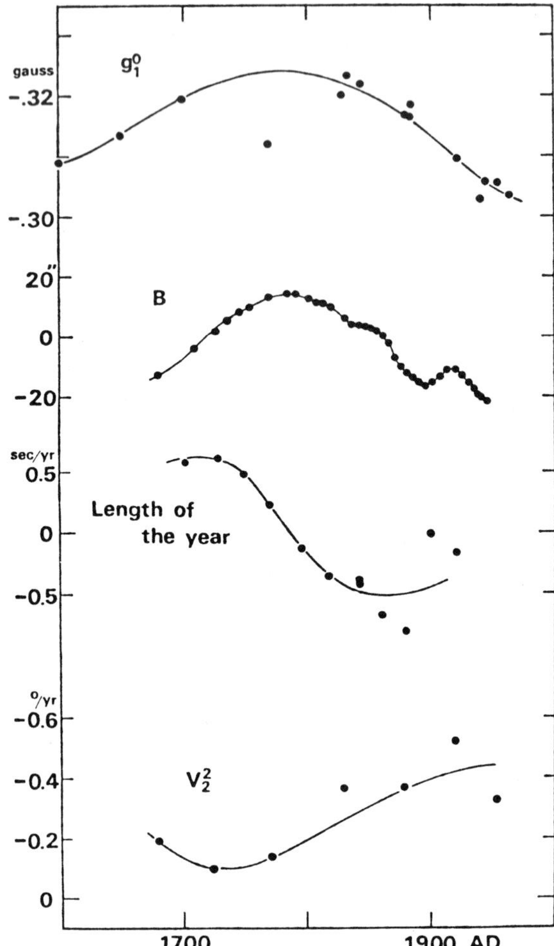

FIG. 2.12. Secular variations having an apparent period of 400 years. g_1^0 is the Gauss coefficient representing the geomagnetic dipole term. B shows a change in the moon's longitude which is ascribed to a change in the earth's rotation rate. V_2^2 denotes the potential of geomagnetic secular variation for the spherical harmonic component $n = m = 2$ (YUKUTAKE, 1973).

400 years is shown in Fig. 2.12. In this case, some phase shifts between the length of the year and other quantities are observed. Such a relationship can also be recognized for the period of 8,000 years. It is of interest to note that the change in the length of the year per unit change in the dipole moment tends to become more marked as the period becomes shorter.

The above facts provide strong evidence supporting the proposition that electromagnetic phenomena within the earth are closely related to the earth's rotation. Some theoretical aspects will be described in Section 4.8.

CHAPTER 3

CHAPTER 3

PALEOMAGNETISM

3.1 Natural Remanent Magnetization

Igneous rocks are generally magnetized and the intensity of magnetization is controlled by magnetic minerals contained in such rocks. In particular, basaltic and andesitic rocks usually have strong magnetization, since they contain more magnetic minerals such as magnetite (Fe_3O_4), hematite (αFe_2O_3), ilmenite ($FeTiO_3$), and so on than other kinds of rocks. Such magnetic minerals are Fe–Ti–oxides represented by FeO–Fe_2O_3–TiO_2 ternary system. For more details of magnetic properties of rocks, see a textbook on rock magnetism (e.g. NAGATA, 1961; NAGATA and OZIMA, 1967).

Magnetization which rocks naturally hold has been called the natural remanent magnetization (NRM). NRM of a rock sample has been expected to provide some information on the geomagnetic field at the time when the sample was formed. If this is the case, the history of an ancient geomagnetic field can be examined by measuring NRM of rocks of various ages. In fact, studies of rock magnetism revealed that some kinds of remanent magnetization are stable, as will be shown in the following, and they gave a physical basis for paleomagnetism, a study of an ancient geomagnetic field which is preserved in NRM.

3.1.1 Thermoremanent magnetization

It is well known that a ferromagnetic material loses its magnetization at temperatures higher than the so-called Curie point. On the other hand, magnetization is acquired in the presence of a magnetic field during cooling from temperatures higher than the Curie point. This kind of magnetization is called the thermoremanent magnetization (TRM). Strong NRM found for basaltic lava is evidently of TRM origin. The nature of TRM has been extensively studied (e.g. DUNLOP and WEST, 1969) and the following characteristics have been discovered:

1) The intensity of TRM acquired in a weak magnetic field is proportional to the intensity of the magnetic field and the direction of TRM is parallel to that of the magnetic field.

2) If the partial thermoremanent magnetization (PTRM), that is TRM acquired during a certain temperature range, is integrated from room temperature to the Curie point, the resultant magnetization corresponds to TRM.

Because of its high stability, in addition to the above characteristics, TRM provides the most reliable information on the past geomagnetic field.

3.1.2 Chemical remanent magnetization

HAIGH (1958) and NAGATA and KOBAYASHI (1958) demonstrated that in association with a chemical reaction $\alpha Fe_2O_3 \rightarrow Fe_3O_4$ magnetization is acquired at temperatures lower than the Curie point of magnetite (Fe_3O_4). Such a remanent magnetization has been called the chemical remanent magnetization (CRM). CRM is also stable and has a nature similar to (1) in Subsection 3.1.1. Stable magnetization of some sedimentary rocks are supposed to be of CRM origin. However, careful attention must be paid to CRM when it is used for paleomagnetic work, since in general the time of CRM acquisition does not necessarily coincide with the time of rock formation.

3.1.3 Detrital remanent magnetization

When fine magnetic grains in oceans and large lakes gradually fall down to the bottom, the directions of their magnetization, which were initially random, tend statistically to become parallel to the geomagnetic field. Thus sedimentary rocks containing such grains are likely to have a remanent magnetization called the detrital (or depositional) remanent magnetization (DRM). Such a mechanism for DRM acquisition has been well explained by theoretical considerations (e.g. NAGATA *et al.*, 1943). However, the inclination of DRM tends to become smaller, by about 10–20°, than that of the magnetic field in which DRM was acquired. This effect, called the inclination error, has been accounted for as rotation of grains during settling at the bottom.

The importance of DRM acquisition during the process of sediment consolidation has also been considered as clearly demonstrated by HAMANO (1980). Although no systematic error in the inclination of DRM was observed in his experiment, HAMANO (1980) claimed that his result concerning the DRM inclination might not be considered as a generally acceptable conclusion because the direction of DRM depends strongly on a rather complicated acquisition process. The mechanism of DRM acquisition during consolidation of sediments was also investigated experimentally by OTOFUJI and SASAJIMA (1981) on the basis of a consolidation process due to artificially applied centrifugal force, and the importance of magnetization acquisition during consolidation was confirmed.

Although the detailed mechanism of DRM acquisition is still under investigation, measurements of DRM of deep-sea sediments have provided important data concerning geomagnetic reversals, as will be described later.

3.1.4 Stability of NRM and magnetic cleaning

Paleomagnetic studies are based on the presumption that NRM acquired at the time of rock formation would have been stably preserved. Although NRM of TRM, CRM or DRM origin is known to be stable, NRM of rocks also includes other kinds of remanent magnetization such as isothermal remanent magnetization (IRM) and viscous remanent magnetization (VRM). IRM denotes magnetization acquired in a magnetic field at a constant temperature, while VRM is magnetization acquired in a magnetic field during a certain time interval t and its intensity is proportional to $\log t$. Therefore, NRM cannot be used for paleomagnetic work unless such secondary components are removed in some way or other.

Secondary NRM is generally unstable and therefore it can be removed using the a.c. demagnetization method; that is, if an alternating magnetic field is applied to a sample placed in a null magnetic field and the intensity of the applied a.c. field is reduced gradually to zero, unstable components disappear more effectively than TRM and CRM. Such a demagnetization process is well demonstrated in Fig. 3.1. As implied in the

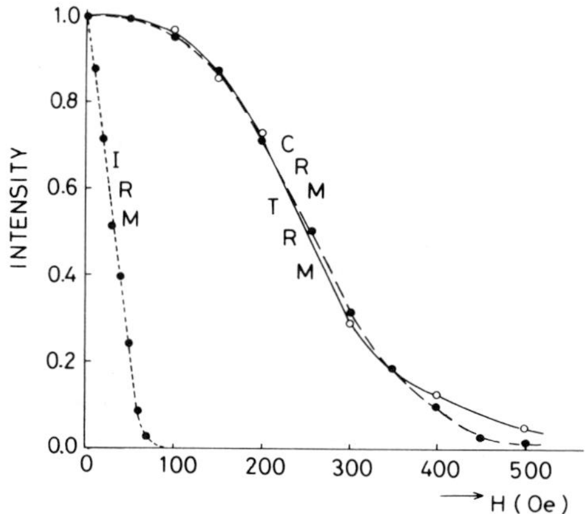

FIG. 3.1. Reduction in the intensity of TRM, CRM, and IRM, respectively, of magnetite during the a.c. demagnetization process (KOBAYASHI, 1959).

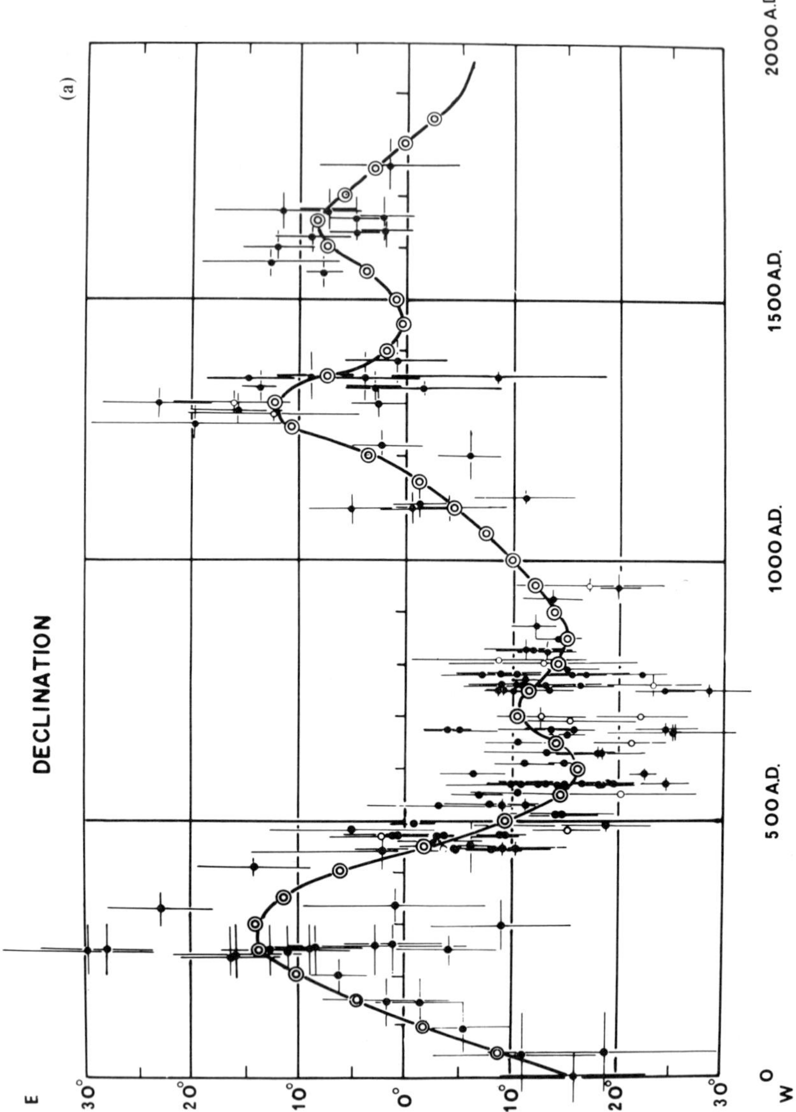
DECLINATION
(a)
E
30°
20°
10°
0°
10°
20°
30°
W
0
500 A.D.
1000 A.D.
1500 A.D.
2000 A.D.

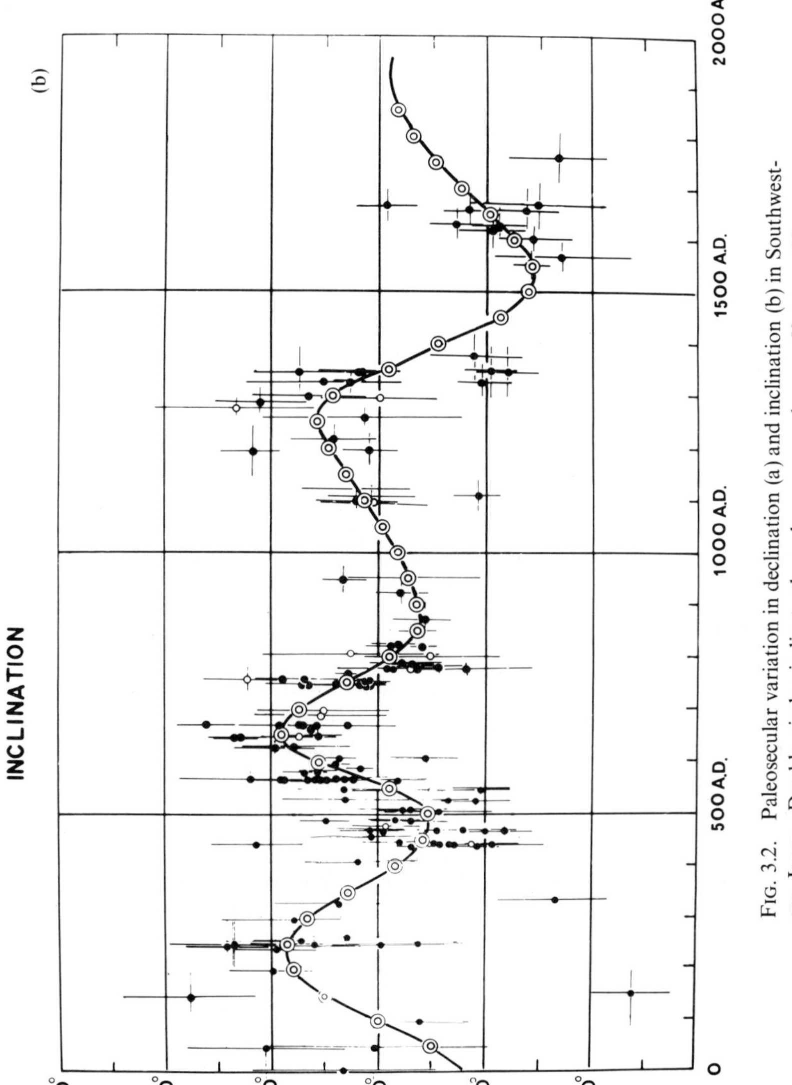

FIG. 3.2. Paleosecular variation in declination (a) and inclination (b) in Southwestern Japan. Double circles indicate the values averaged every 50 years (HIROOKA, 1971).

figure, if demagnetization is made at an appropriate level of magnetic field intensity, unstable components such as IRM and VRM can be removed, whereas stable components are demagnetized little. An alternative method of demagnetization has also been used for removing secondary components: the thermal demagnetization method, in which a sample is heated once and then cooled in a null magnetic field (see e.g. NAGATA, 1961). Demagnetization of unstable components is called magnetic cleaning and it is one of the necessary procedures in paleomagnetic studies.

3.2 Paleosecular Variation

The geomagnetic field undergoes secular changes as described in Chapter 2. In particular, the dipole moment is decreasing at a rate of 5% in 100 years. This tendency may be a transient one and nothing is clear until we examine a more long-term trend. Fortunately TRM-bearing rock samples of known age are available; for instance, those taken from lava flows associated with historic volcanic eruptions, ancient furnaces and potteries, and so on. If paleomagnetic work is applied to those samples, a paleosecular variation may be disclosed.

In fact, WATANABE (1958, 1959) measured NRM of samples taken from ancient furnaces and discovered a secular variation of inclination and declination in Japan, while YUKUTAKE (1961) and YUKUTAKE *et al.* (1964) used lava and ash-fall tuff for the same purpose. The study of paleomagnetism in historic and archeological times is referred to as archeomagnetism.

Secular variations in declination and inclination based on archeomagnetism have been extensively studied for various regions over the world. For example, for the southwestern part of Japan, HIROOKA (1971) disclosed secular variations in declination and inclination until about 2,000 years before present, as shown in Fig. 3.2. It is evident that variations as much as ten degrees or more are not unusual.

The intensity of the past geomagnetic field has usually been determined using the so-called Thellier method proposed by THELLIER and THELLIER (1959). The method is based on the characteristics of PTRM as described in Subsection 3.1.1.

Suppose that J_N is the intensity of NRM measured for a sample. The same sample will acquire TRM if it is heated to a temperature higher than the Curie point T_c and then cooled to the room temperature T_0 in the presence of a magnetic field with intensity F_0. Here we denote TRM thus obtained by $J_T(T_c, T_0; F_0)$. Then the intensity F of the geomagnetic field at the time of NRM acquisition can be inferred as

$$\frac{F}{F_0} = \frac{J_N}{J_T}. \tag{3.1}$$

Similarly, $J_T(T_{j+1}, T_j; F_0)$ indicates PTRM acquired during cooling from T_{j+1} to T_j in F_0. Then $J_T(T_i, T_0; F_0)$ can be written as

$$J_T(T_i, T_0; F_0) = \sum_{j=0}^{i-1} J_T(T_{j+1}, T_j; F_0) \tag{3.2}$$

because of the additive law of PTRM. Now we heat the sample from T_0 to T_i and then cool it to T_0 in the null magnetic field. NRM measured after this procedure, denoted by $I(T_i, T_0; 0)$, should satisfy the following relation:

$$I(T_i, T_0; 0) = J_N - \frac{F}{F_0} J_T(T_i, T_0; F_0). \tag{3.3}$$

Therefore, F can be determined more reliably by measuring $I(T_i, T_0; 0)$ and $J_T(T_i, T_0; F_0)$ for various T_i between T_0 and T_c.

One of the examples of the Thellier method is shown in Fig. 3.3. The sample was taken from lava which flowed down from the crater at the 1471 eruption of Sakurajima Volcano, Japan (TANAKA, 1980). In this case, the total intensity was determined as 0.533 gauss (53.3 μT) from the slope of the straight line in the figure.

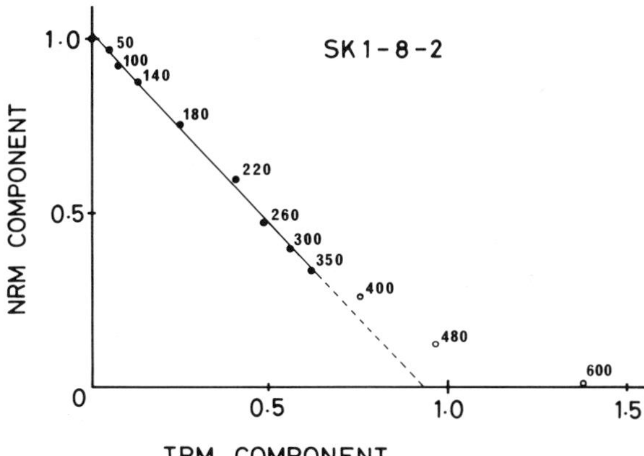

FIG. 3.3. An example of the Thellier method applied to a lava sample. Numerals indicate the temperature in degrees centigrade. In this case, the data at temperatures higher than 400°C deviate from the tendency at low temperature. The total intensity was estimated as 0.533 gauss (53.3 μT) from the gradient of the straight line (TANAKA, 1980).

Studies based on the above method have been undertaken for samples from various regions of the world (e.g. NAGATA and OZIMA, 1967; SMITH, 1967; KITAZAWA and KOBAYASHI, 1968; BUCHA, 1969; BUCHA *et al.*, 1970; KITAZAWA, 1970) and the paleosecular variation of geomagnetic intensity for the past 10,000 years or so before present has been disclosed. TANAKA (1980) extended this kind of study to an age of 30,000 years before present on the basis of samples in the Japanese regions.

In order to study any geomagnetic intensity variation of a global nature, it is necessary to convert the geomagnetic intensity at a certain location into the moment of a magnetic dipole located at the earth's center. On the assumption that the geomagnetic field is purely of dipole origin, the intensity F observed at a location where the geomagnetic colatitude and inclination are θ and I, respectively, yields the virtual dipole moment M as

$$M = Fa^3(4 - 3\sin^2\theta)^{-\frac{1}{2}} = \frac{1}{2}Fa^3(1 + 3\cos^2 I)^{\frac{1}{2}} \tag{3.4}$$

where a denotes the radius of the earth. Figure 3.4 shows a paleosecular variation of the virtual dipole moment as given by TANAKA (1980). Data, with error bars, for the period before 10,000 years are less reliable than the other data, without error bars. This is partly because of the poor capability of the ^{14}C method in determining ages much older than the half life of ^{14}C (estimated as about 5,568 years).

The data for the period from 10,000 years to the present show that the dipole moment underwent a periodic variation with a period of several thousand years. The model of geomagnetic reversal proposed by COX (1969) is partly based on the above evidence of a periodic variation of the dipole moment. Such a periodic tendency seems to be traced back to about 15,000 years before present. At the present stage, little stress will be put on the continuation of that tendency further into the past, although a rather weak moment seems to have persisted during the period from 25,000 to 15,000 years before present.

3.3 Reversal of the Geomagnetic Field

MATUYAMA (1929) showed that many samples taken from Japan, Korea and eastern China possessed NRM having an orientation completely opposite to the direction of the present geomagnetic field. In view of the fact that such samples are systematically older than others showing the same (normal) orientation as the present field, it was thought that the polarity of the geomagnetic field must have been reversed around the beginning of the Quaternary period.

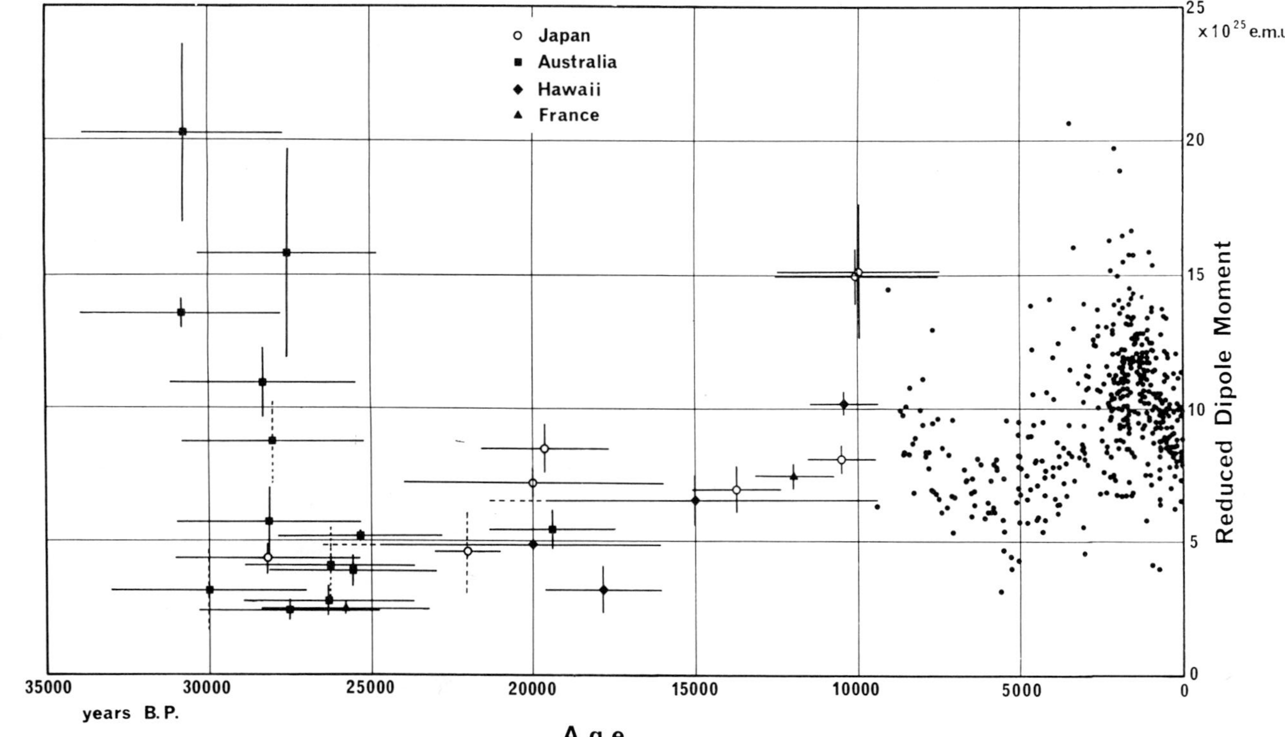

FIG. 3.4. Paleosecular variation of the virtual dipole moment (TANAKA, 1980). (1 e.m.u. = 10^3 Am^2)

Meanwhile, NAGATA *et al.* (1951, 1952) discovered the existence of reverse thermoremanent magnetization (hereafter referred to as RTRM); that is, TRM acquired in a direction opposite to that of the applied magnetic field (often called self-reversal). If this phenomenon occurs generally, one of the most basic assumptions in paleomagnetism is no longer valid. Further studies of RTRM, however, imply that RTRM is a special property related to the ilmenite-hematite solid solution with a certain range of titanium content (UYEDA, 1958). Theoretically, the possibility of self-reversal was first suggested by NÉEL (1951) and the physical mechanism of self-reversal was discussed in detail by ISHIKAWA and SHONO (1963). In any case, RTRM has seldom been found for titanomagnetite which is the main magnetic mineral of igneous rocks

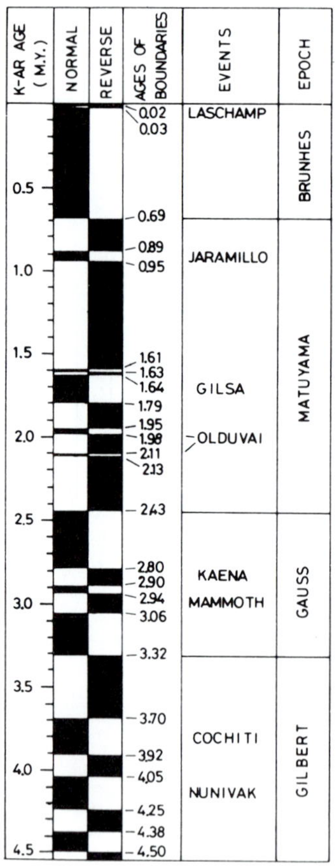

FIG. 3.5. Polarity time scale for the period from 4.5 million years ago to the present. Black portions indicate respective polarities (COX, 1969).

and, therefore, RTRM is unlikely to occur generally in natural circumstances. On the other hand, on many occasions NRM showing a reversed direction has been found and thus the geomagnetic field reversal has been considered to be a real phenomenon.

In the meantime, the K-Ar method made it possible to determine the ages of fairly old rocks. Paleomagnetic work on rock samples with ages determined in this way disclosed a history of geomagnetic field reversal, represented by a polarity time scale (Cox *et al.*, 1964). Figure 3.5 shows one of the most representative polarity time scales for the period from about 4.5 million years ago to the present (Cox, 1969). The polarity has been the same as the present

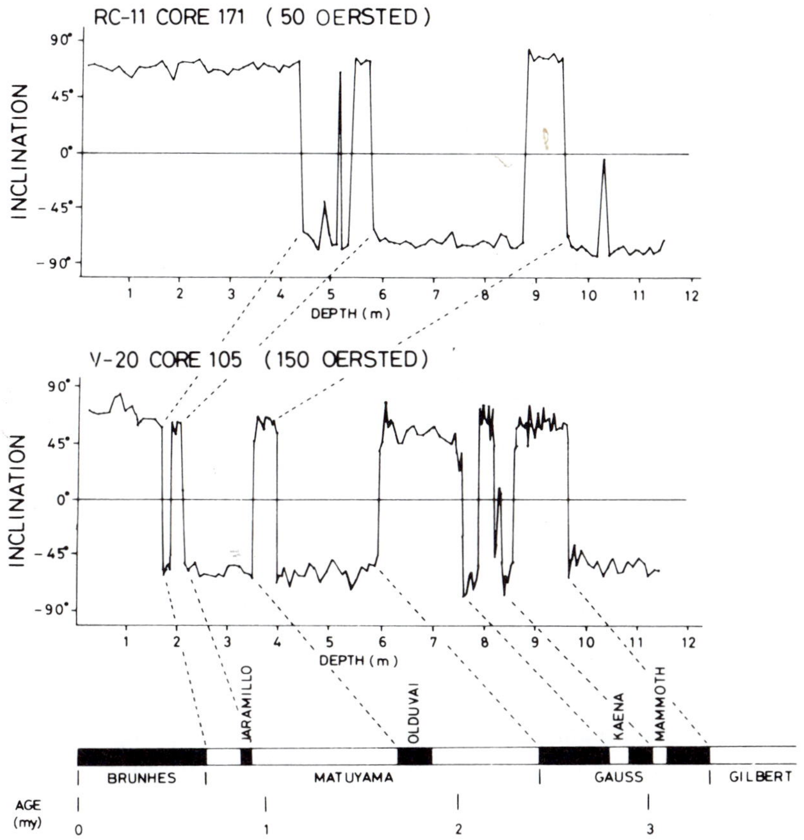

FIG. 3.6. Inclination data derived from core samples at two different sites in the north Pacific. The values shown were obtained after demagnetization at levels of 50 (R-11 core 171) and 150 (V-20 core 105) oersted, respectively (OPDYKE, 1972).

geomagnetic field since 0.69 million years ago, except for some polarity reversals of short duration called 'events'. This period of predominantly normal polarity is called the Brunhes normal epoch. Three other epochs have been proposed: the Matuyama reverse, Gauss normal, and Gilbert reverse epochs. However, as shown in Fig. 3.5, the polarity has changed quite often, even during such epochs, characterized by events of short duration.

Other evidence of geomagnetic field reversal has been obtained from measurements of DRM for core samples taken from deep-ocean floor sediments. Figure 3.6 shows inclination data obtained for sediment cores from the north Pacific (OPDYKE, 1972). Some polarity reversals can be clearly recognized and moreover, for each polarity reversal, corresponding reversals are identified in the polarity time scale shown in Fig. 3.6. Similar studies have been made for core samples taken from other regions and the pattern in a sequence of reversals for one region is found to be in good agreement with that in another region.

It is expected that a higher sedimentation rate would yield DRM providing higher resolution in inclination variation. In view of this, sediment cores from large lakes where the sedimentation rate is usually high have been used for a study of the detailed history of the geomagnetic field. In the case of a 197.2 meter long core taken from Lake Biwa, Japan, a few geomagnetic excursions of very short duration were found, during which the geomagnetic pole position moved markedly away from the previous position, yet the field did not undergo a complete reversal (YASKAWA *et al.*, 1973). Recently, even wet sediments have been used for studies of paleosecular variation (e.g. LUND and BANERJEE, 1979).

3.4 Geomagnetic Polarity Stratigraphy

Marine magnetic surveys of the geomagnetic total intensity have been so extensively carried out that many magnetic anomalies have been found over various oceans. A very prominent pattern of magnetic anomaly, found over the ridge areas, was interpreted by VINE and MATTHEWS (1963) as direct evidence for sea-floor spreading (see Chapter 6). According to the interpretation, a sequence of normal and reverse magnetization has been preserved in the oceanic curst since its creation at the ridge, providing information on the polarity time scale. The importance of studies of marine magnetic anomalies is increased by their ability to provide a continuous sequence of normal and reverse polarities. Thus the polarity time scale, as determined back to about 4 million years ago, from measurements of NRM of continental rocks, could be extended to 80 million years by HEIRTZLER *et al.* (1968) and later to 160 million years by LARSON and PITMAN (1972) and

LARSON and HILDE (1975) as shown in Fig. 3.7.

Revisions of polarity time scale have been in progress as accuracy in age determination based on the K-Ar and other methods increases for continental rock samples. Dating of core samples has recently become possible as a result of the Deep Sea Drilling Project (DSDP) and the time scale for marine magnetic anomalies has also been revised. Recent revisions are reviewed by BUTLER and OPDYKE (1979), although most of them are rather minor.

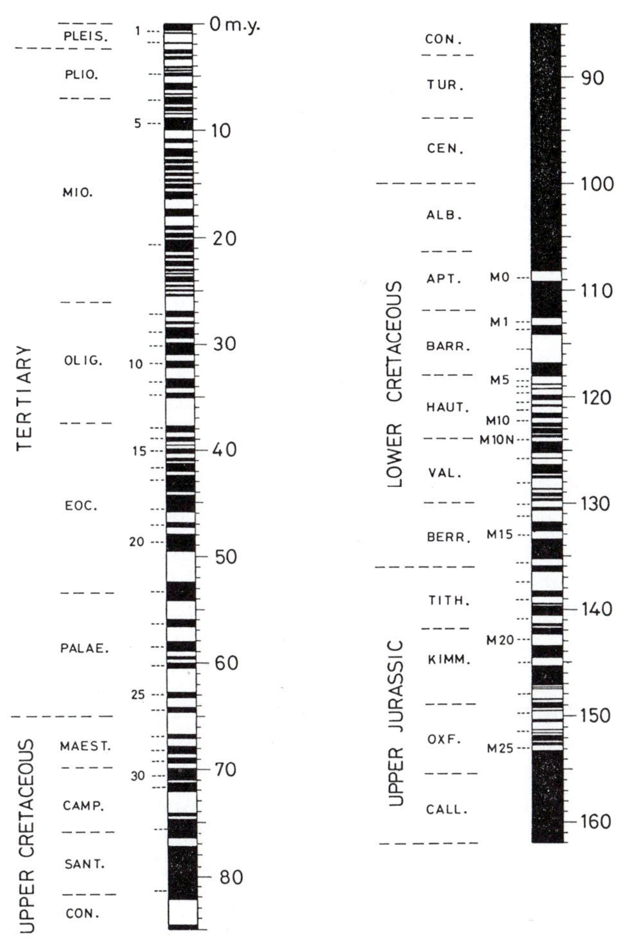

FIG. 3.7. Polarity time scale for the period from 160 million years ago to the present as derived from marine magnetic anomalies. Black portions indicate normal polarity (LARSON and PITMAN, 1972; LARSON and HILDE, 1975).

Attempts are also being made to determine geomagnetic stratigraphy for the Paleozoic and Precambrian periods on the basis of paleomagnetic work for sedimentary sequences of these ages (e.g. IRVING and PULLAIAH, 1976; KIRSCHVINK, 1978).

3.5 Morphology of Transitional Field

Numerous occurrences of geomagnetic field reversal impose severe constraints on models of geomagnetic field generation mechanism. In this respect, a study of the process of field reversal must be of great interest. Nonetheless, features of the transitional field during a reversal process are not well understood, because the typical time duration, of some thousand years, for a field reversal process is by no means long enough for paleomagnetic studies.

It seems possible, at least in principle, to examine transitional fields by measuring DRM in detail for a sequence of sediments which includes a field reversal and where the sedimentation rate is so high that high resolution is attainable. Alternatively, an intrusive rock body can be used for this purpose, since TRM acquisition proceeds gradually deeper into the rock body and consequently a transition of the geomagnetic field will be recorded continuously during a field reversal.

A growing number of studies of the transitional field have been accumulated and FULLER *et al.* (1979) reviewed some of the most reliable data. On the basis of nine sets of data on the transitional field, FULLER *et al.* (1979) examined VGP (virtual geomagnetic pole; see Subsection 3.6.1) paths for transitional fields, field intensity variations during reversals, and time duration required for a reversal process to be completed. Although not all of the data contributed to all of the above aspects and some discrepancies exist between different data sets, FULLER *et al.* (1979) derived the following conclusions:

VGP path

During a field reversal the VGP path tends to move along a certain meridian line depending on the location where a sample was taken, although some fluctuations are superimposed on such a general trend. However, if the longitude of the meridian line is measured with respect to the longitude of a sample site, it turns out that the path tends to follow the 0° meridian line during a reversal from reverse to normal polarities, while during a reversal of opposite sense the tendency is completely opposite; the path generally traces the 180° meridian line. Such a striking tendency was first pointed out by HOFFMAN (1977).

Field intensity

The intensity seems to decrease substantially during a reversal; probably it is reduced to 10% or so of the intensity which existed before the reversal process. On some occasions, however, a very strong field seems to appear but such a stage lasts only for a short time.

Time duration required for a reversal process

The estimated time duration ranges from 1,000 to 10,000 years; most probably 4,000–5,000 years as derived from oceanic sediment records. The time during which the intensity first decreases from the initial level and then recovers may be longer than the time required for a sequence of directional change.

Figure 3.8 shows an example of a transitional field; in this case, changes in intensity and direction were examined for samples from the Tatoosh intrusion in Washington, U.S.A. (DODSON *et al.*, 1978; FULLER *et al.*, 1979).

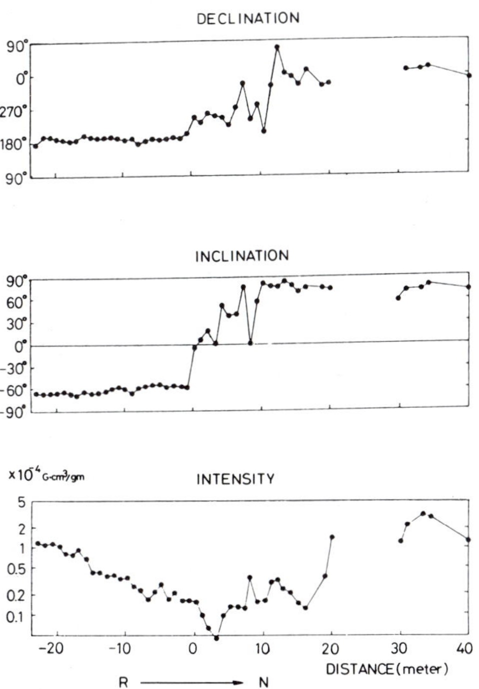

FIG. 3.8. An example of changes in declination, inclination, and intensity, respectively, during a geomagnetic field reversal from reverse to normal polarities (FULLER *et al.*, 1979).

Two simple yet typical models can be considered for the mechanism of field reversal: a dipole rotating from one polarity to the other without changing its intensity (e.g. STEINHAUSER and VINCENZ, 1973) and a dipole which first decays and then grows in the opposite direction (e.g. SMITH, 1967; WATKINS, 1969; LARSON *et al.*, 1971). In the latter case, the non-dipole fields control the geometry of transitional field. Comparing the VGP paths derived from samples at California and Japan, HILLHOUSE and COX (1976) claimed that the latter was the mechanism for the Matuyama-Brunhes transition. The latter model also seems to be supported by the conclusions deduced by FULLER *et al.* (1979).

Meanwhile, HOFFMAN and FULLER (1978) proposed a sophisticated reversal mechanism. According to the mechanism, a new quadrupole or octupole field is supposed to begin to grow at the initial stage of a reversal process. As it continues to grow, the initial dipole field is gradually modified and then a new reversed polarity tends to prevail, resulting finally in its dominance over the entire earth.

3.6 *Polar Wandering and Continental Drift*

3.6.1 *Virtual geomagnetic pole*

Suppose that the declination and inclination of the geomagnetic field are known at a point $S(\varphi_0, \lambda_0)$ on the earth's surface, where φ_0 and λ_0 denote the geographical latitude and longitude at S. If the geomagnetic field is due to a magnetic dipole placed at the earth's center, the geomagnetic pole P should be located in the direction of declination D at S. Now we consider a meridian line which passes through both S and P as shown in Fig. 3.9. Here θ is the angle between OP and OS where O denotes the earth's center. For simplicity, we consider the case in which the dipole direction coincides with the earth's

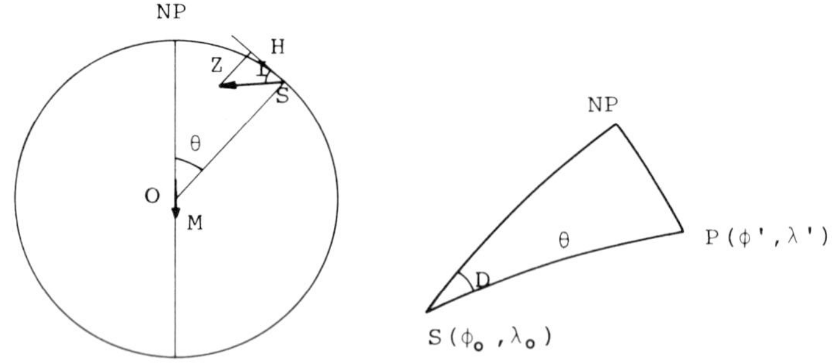

FIG. 3.9. Interrelation between a measuring point S, the direction of the geomagnetic field at S, and the virtual geomagnetic pole P.

rotation axis, as shown in Fig. 3.9. Then the horizontal (H) and vertical (Z) components at S are derived from (1.22) as

$$
\begin{aligned}
H &= \frac{1}{a}\left[\frac{\partial W}{\partial \theta}\right]_{r=a} = -\frac{M}{a^3}\sin\theta \\
Z &= \left[\frac{\partial W}{\partial r}\right]_{r=a} = -\frac{2M}{a^3}\cos\theta.
\end{aligned}
\tag{3.5}
$$

(in this case, $M_x = M_y = 0$, $M_z = M$).
From the relation $\tan I = Z/H$, where I denotes the inclination at S, we obtain

$$\cos\theta = \frac{1}{2}\tan I. \tag{3.6}$$

This relation holds good for any dipole directions. In a spherical triangle shown in Fig. 3.9, where φ' and λ' denote the geographical latitude and longitude respectively, the following relations hold:

$$
\begin{aligned}
&\sin\varphi' = \sin\varphi_0\cos\theta + \cos\varphi_0\sin\theta\cos D \\
&\sin\beta = \frac{\sin\theta\sin D}{\cos\varphi'} \\
&\beta = \begin{cases} \lambda' - \lambda & (\cos\theta \geqq \sin\varphi\sin\varphi') \\ 180^\circ - (\lambda' - \lambda) & (\cos\theta < \sin\varphi\sin\varphi'). \end{cases}
\end{aligned}
\tag{3.7}
$$

Thus from a data set (D, I) at $S(\varphi_0, \lambda_0)$ φ' and λ' at P can be determined. The point P is called the virtual geomagnetic pole (VGP).

The VGP determined from NRM of rock samples at a particular location scatters to some extent because of secular variations of the non-dipole fields. However, such an effect will be cancelled out by the averaging procedure over a time span of 10,000 years or so and the averaged pole position is likely to represent the pole of the dipole field. The pole thus determined is called the paleomagnetic pole.

3.6.2 Polar wandering and continental drift

The paleomagnetic pole is usually considered to be roughly identical to the pole of the earth's rotation, since the discrepancy between them is expected to be small compared with the accuracy in paleomagnetic pole determination. It was shown that the pole determined from NRM measurements for old rock samples taken in Europe deviates significantly from the present pole position. Moreover, time dependence of the pole position for one land mass turned out to be so systematic that a smooth curve could be drawn connecting pole positions in chronological order (RUNCORN, 1965). Such a curve has been called the apparent polar wandering curve.

It would be reasonable to speculate that the polar wandering curve determined for one land block should coincide with that for another land block. However, the polar wandering curve determined for Europe was found to deviate systematically from that for North America (RUNCORN, 1965). Such a discrepancy could be interpreted as implying a motion of one block relative to another. Thus paleomagnetic data happened to provide physical evidence supporting the hypothesis of continental drift. Since then, extensive studies of continental drift based on polar wandering curves have been undertaken by many workers; for details, see a textbook on paleomagnetism (e.g. IRVING, 1964; MCELHINNY, 1973). Many recent results relating to continental drift can be found in a review paper by VAN DER VOO (1979).

CHAPTER 4

CHAPTER 4

THEORIES OF ORIGIN OF THE GEOMAGNETIC FIELD

Paleomagnetic evidence showing the existence of the earth's magnetic field for more than 10^9 years has been accumulated and it has been found that its intensity has not differed much from the present intensity, although numerous polarity reversals have occurred (see Chapter 3). In addition, the non-stationary features of the geomagnetic field, such as secular changes, have also been brought to light. Such features of the geomagnetic field evidently cannot be interpreted simply in terms of permanent magnetization of rocks in the earth's interior. Moreover, the temperature is so high in the earth's interior that most of the material comprising the earth can exibit no permanent magnetization.

Attempts have been made to interpret the existence of the earth's magnetic field in terms of physically plausible mechanisms such as the gyromagnetic effect, rotating electric charges, electromagnetic induction by magnetic storms, the thermoelectric effect, the Hall effect, rotation of a massive body, and so on (RIKITAKE, 1966a). However, none of these hypotheses has survived until today as a likely mechanism.

In the meantime, it was revealed that the Sun and some stars also have their own magnetic fields. Since then much attention has been focused on rotation and the fluid state, at least partly, as common features of these celestial bodies. Thus, magnetic field generation through an interaction between the magnetic field and fluid motion within a celestial body has been speculated upon as the most likely origin of the magnetic field, and eventually such a hypothesis was formulated as a theory called the "dynamo theory."

4.1 Basic Concept of the Dynamo

The earth's outer core consists of an electrically conducting fluid, in which electric currents of some origin would flow, resulting in generation of the magnetic field. However, such electric currents must be subject to free-decay with a time constant governed by the electromagnetic properties of the core. Therefore, there must be a mechanism of electric current regeneration for the magnetic field to be maintained against the free-decay. Such a regeneration mechanism is well demonstrated by a simple disk dynamo model as shown in Fig. 4.1 (BULLARD, 1955).

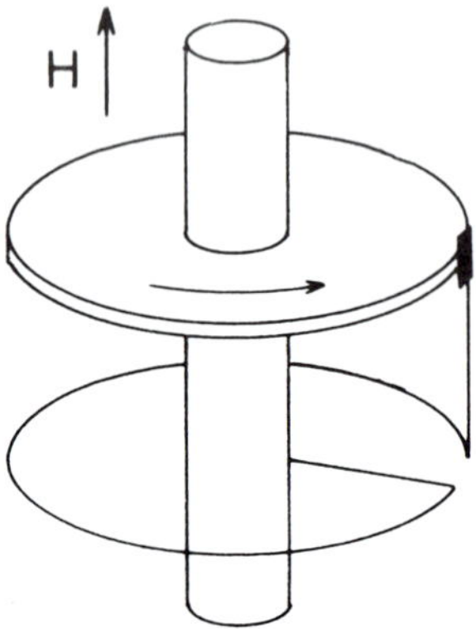

FIG. 4.1. A disk dynamo model (BULLARD, 1955).

The conducting disk rotates about the axle while the magnetic field is applied parallel to the axle. Electromotive force is then generated within the disk in the radial direction. The potential difference between the axle and the disk periphery can be written as

$$V = \omega \int_a^b Hr \, \mathrm{d}r \tag{4.1}$$

where a and b are the radii of the axle and the disk, respectively, and ω denotes the angular velocity. Suppose that the magnetic field H is generated by the electric current I flowing in the circular coil. Then

$$\int_a^b Hr \, \mathrm{d}r = MI \tag{4.2}$$

where $2\pi M$ is the mutual inductance between the coil and the periphery of the disk. Therefore, the following equation holds:

$$L \frac{\mathrm{d}I}{\mathrm{d}t} + RI = M\omega I \tag{4.3}$$

where L and R denote the self-inductance and resistance of the coil, respectively. If ω is constant, the solution of (4.3) is derived as

$$I = I_0 \exp\{(M\omega - R)t/L\} \tag{4.4}$$

where I_0 is the electric current at $t = 0$. However, the Lorentz force whose integrated moment about the axle is given by

$$F = I\int_a^b rH\,\mathrm{d}r = MI^2 \tag{4.5}$$

acts on the system and consequently we have

$$C\frac{\mathrm{d}\omega}{\mathrm{d}t} = G - MI^2 \tag{4.6}$$

where C indicates the moment of inertia of the disk around the axle and G is the torque, which is taken as constant, applied to the disk.

In the case of $GM > 0$, we define τ and y as

$$\begin{aligned} \tau &= \left(\frac{2GM}{CL}\right)^{\frac{1}{2}} t \\ y &= \ln(MI^2/G) \end{aligned} \tag{4.7}$$

then from (4.3) and (4.6)

$$\frac{\mathrm{d}^2 y}{\mathrm{d}\tau^2} = 1 - \mathrm{e}^y. \tag{4.8}$$

Expressing as $x = \mathrm{d}y/\mathrm{d}\tau$, we obtain

$$\begin{aligned} x^2 &= A + 2(y - \mathrm{e}^y) \\ A &= x_0^2 - 2(y_0 - \mathrm{e}^{y_0}) \end{aligned} \tag{4.9}$$

where x_0 and y_0 are initial values and given by

$$\begin{aligned} x_0 &= \left(\frac{2CM}{GL}\right)^{\frac{1}{2}}\left(\omega_0 - \frac{R}{M}\right) \\ y_0 &= \ln\left(\frac{MI_0^2}{G}\right). \end{aligned} \tag{4.10}$$

For $A > 2$, (4.9) yields a closed circle in the x-y plane as shown in Fig. 4.2 and consequently any solutions are periodic. BULLARD (1955) conducted numerical integration for $A = 3$, 10 and obtained variations of the electric current and angular velocity as shown in Fig. 4.3. The inductive relationship between the electric current and the angular velocity is well demonstrated in the figure.

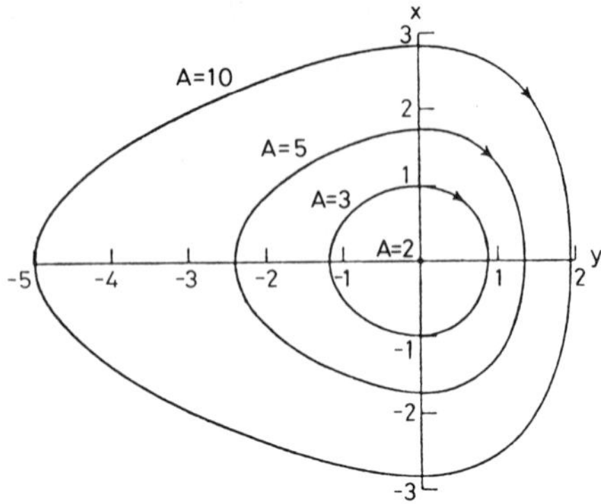

FIG. 4.2. Trajectories in the x-y plane for various values of A (BULLARD, 1955).

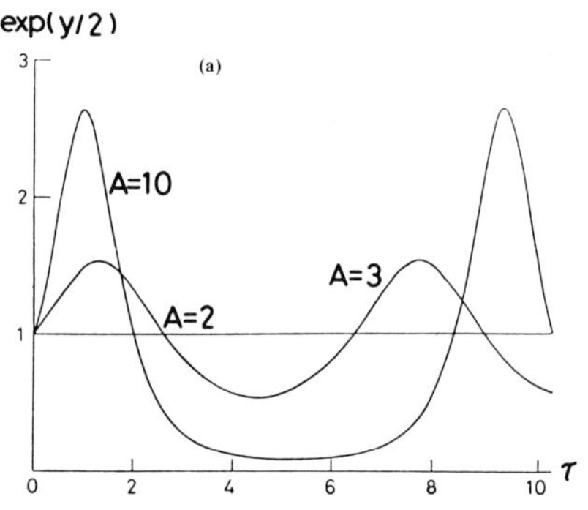

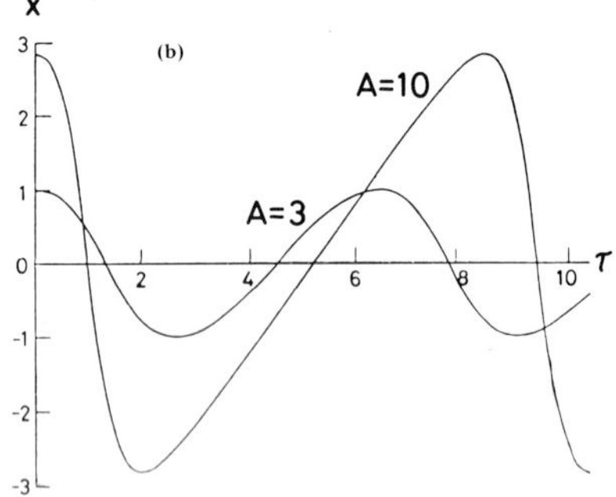

FIG. 4.3. Variations of the electric current as represented by y (a) and the angular velocity as represented by x (b), respectively, for $A = 3$, 10. The case $A = 2$ corresponds to the steady state (BULLARD, 1955).

A steady state can be obtained when $\omega = R/M$ and $I = \pm\sqrt{G/M}$. It turns out, therefore, that in this system the intensity of the magnetic field fluctuates about the steady state value. It is concluded that this system acts as a self-exciting dynamo, although no polarity reversal occurs.

In the case of $GM < 0$, we define τ and y as

$$\tau = \left(-\frac{2GM}{CL}\right)^{\frac{1}{2}} t$$
$$y = \ln\left(-\frac{MI^2}{G}\right). \tag{4.11}$$

Then

$$\begin{aligned} x^2 &= A_1 - 2(\mathrm{e}^y + y) \\ A_1 &= x_0^2 + 2(\mathrm{e}^{y_0} + y_0). \end{aligned} \tag{4.12}$$

In this case, however, it can be shown that the system never acts as a self-exciting dynamo.

The basic concepts of dynamo action can be deduced from this simple model. First of all, there seems to exist a self-regenerative mechanism through the interaction between the magnetic field and fluid motion. The model also suggests the conditions necessary for the regenerative mechanism; that is, differential rotation and the lack of reflexional symmetry (MOFFATT, 1978). The former is indicated by the combination of the rotating disk and the fixed coil with which the disk is connected by brushes. The latter is implied in the condition that GM must be positive for the regeneration mechanism to be possible. Finally, it appears to be essential to take into account the effect of the electric current (or the magnetic field) on fluid motion, controlled by the so-called Lorentz force, when one tries to examine a non-steady state, particularly polarity reversal.

4.2 *Hydromagnetic Equations for the Dynamo*

In hydromagnetism, the interaction between the magnetic field and fluid motion plays an important role and we cannot treat equations describing fluid motion independently from those for electromagnetic fields. The velocity field satisfies the Navier-Stokes equation,

$$\frac{\mathrm{d}\boldsymbol{v}}{\mathrm{d}t} = -\frac{1}{\rho}\nabla p - \nabla U + 2\boldsymbol{v}\times\boldsymbol{\omega} + \nu\nabla^2\boldsymbol{v} + \frac{1}{\rho}\boldsymbol{i}\times\boldsymbol{B} \tag{4.13}$$

where $\boldsymbol{v}$, $\boldsymbol{i}$, and $\boldsymbol{B}$ are the velocity, the electric current density, and the magnetic induction, respectively. ρ, p, U, and ν denote the density, the pressure, the gravitational potential, and the kinematic viscosity, respectively. The third term on the right indicates the Coriolis force. Electromagnetic fields can be expressed by the Maxwell equations as

$$\begin{aligned} \operatorname{curl} \boldsymbol{E} &= -\frac{\partial \boldsymbol{B}}{\partial t} \\ \operatorname{curl} \boldsymbol{H} &= 4\pi \boldsymbol{i} \\ \operatorname{div} \boldsymbol{B} &= 0 \\ \boldsymbol{i} &= \sigma(\boldsymbol{E} + \boldsymbol{v} \times \boldsymbol{H}) \end{aligned} \tag{4.14}$$

in electromagnetic units, where $\boldsymbol{E}$ is the electric field, and σ incicates the electrical conductivity. Suppose that the medium is homogeneous, then taking the curl of the second equation of (4.14), we obtain

$$\frac{\partial \boldsymbol{H}}{\partial t} = \operatorname{curl} (\boldsymbol{v} \times \boldsymbol{H}) + \nu_m \nabla^2 \boldsymbol{H} \tag{4.15}$$

where

$$\nu_m = \frac{1}{4\pi\sigma}. \tag{4.16}$$

ν_m is called the magnetic viscosity. (4.13) and (4.15) are the fundamental equations for a hydromagnetic dynamo theory. It is a mathematically complicated problem to treat these equations generally and, therefore, only (4.15) is usually treated for a given velocity field. Then the problem is reduced to the so-called kinematic dynamo problem. In a kinematic dynamo theory, (4.15) is the fundamental equation and is usually called the induction equation.

Now we introduce the magnetic Reynolds number as defined by

$$R_m = \frac{LV}{\nu_m} \tag{4.17}$$

where L and V denote the characteristic length and velocity, respectively. Then the last term of (4.15), that is the diffusion term, is of the order of $1/R_m$ compared with the other terms. In the following we consider the physical meaning of R_m. Suppose that $\boldsymbol{v} = 0$ in (4.15), then $\boldsymbol{H}$ is subject to free-decay with a decay time given by

$$\tau = \frac{L^2}{\nu_m}. \tag{4.18}$$

Then we have

$$R_m = \frac{\tau V}{L} = \frac{\tau}{T} \tag{4.19}$$

where T is the characteristic time for mechanical motion. Thus R_m provides information on how long the decay time is, compared with the time required for fluid to move over a distance L. In the case of the earth, $\nu_m \sim 3 \times 10^5$ (see Section 10.6), $V \sim 0.01$ cm/sec, $L \sim 3 \times 10^8$ cm and, therefore, $R_m \sim 10$.

If R_m is infinitely large in the following equation,

$$\frac{\mathrm{d}}{\mathrm{d}t}\int B_n \mathrm{d}S = \nu_m \int \nabla^2 B_n \mathrm{d}S \tag{4.20}$$

where S is any material surface and B_n is the component of $\boldsymbol{B}$ normal to the surface (e.g. Roberts, 1967; p. 44), we have

$$\frac{\mathrm{d}}{\mathrm{d}t}\int B_n \mathrm{d}S = 0 \tag{4.21}$$

which implies that the magnetic flux moves together with the fluid. If R_m is a finite number, the magnetic lines of force will be distorted to a certain extent by fluid motion. Such a distortion process is important in the dynamo mechanism. Hagiwara (1967) examined how a pair of convection cells influences the magnetic field which is initially uniform. As shown in Fig. 4.4, the magnetic lines of force are carried up or down by an upward or downward fluid stream.

4.3 Physical Conditions Necessary for a Dynamo Process

The axisymmetric case would naturally be attempted first, for mathematical simplicity in solving a hydromagnetic problem for a rotating body. However, Cowling (1934) showed that a steady axisymmetric field cannot be maintained by axisymmetric fluid motions. Such an "anti-dynamo" theorem was also confirmed by Backus and Chandrasekhar (1956), Cowling (1957), and Backus (1957).

One of the necessary conditions for dynamo action seems to be a differential rotation as suggested from the simple disk dynamo model. If convective motion prevails in the earth's core, differential rotation can be generated as will be described below. In a rotating sphere, the Coriolis force acts on the fluid convecting in the radial direction so as to conserve the angular momentum. Consequently, inward fluid motion accelerates the rotation, while outward motion gives rise to retardation of rotation. Because of the

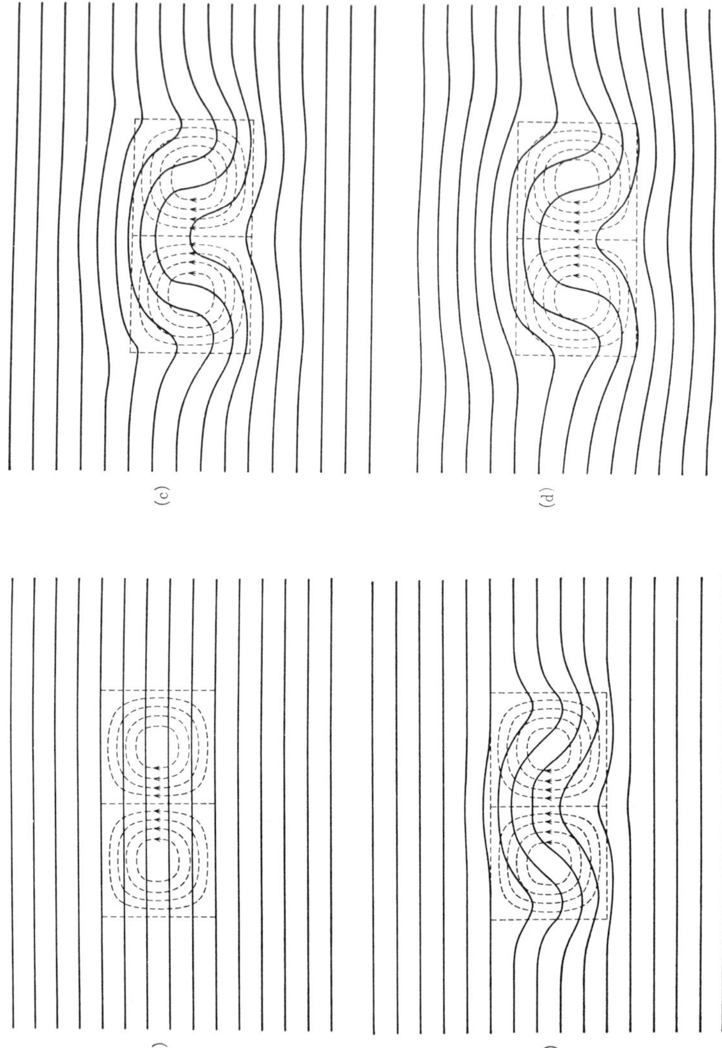

FIG. 4.4. Distortion of magnetic lines of force caused by fluid convection at (a) the initial stage ($t = 0$), (b) $t = 20\Delta t$, (c) $t = 60\Delta t$, and (d) $t = 200\Delta t$, where Δt is taken as about 210 years. R_m is assumed to be 10 (HAGIWARA, 1967).

differential rotation thus created, the magnetic lines of force representing a dipole-like field (called the poloidal magnetic field) are pulled toward the accelerated rotation and the so-called toroidal magnetic field is created as shown schematically in Fig. 4.5. The toroidal field has no radial component and, therefore, it is confined within the sphere. MOFFATT (1978) treated theoretically the generation of a toroidal magnetic field from a poloidal magnetic field through a differential rotation and pointed out that the generation of the toroidal field is a result of diffusion expressed by the last term of (4.15).

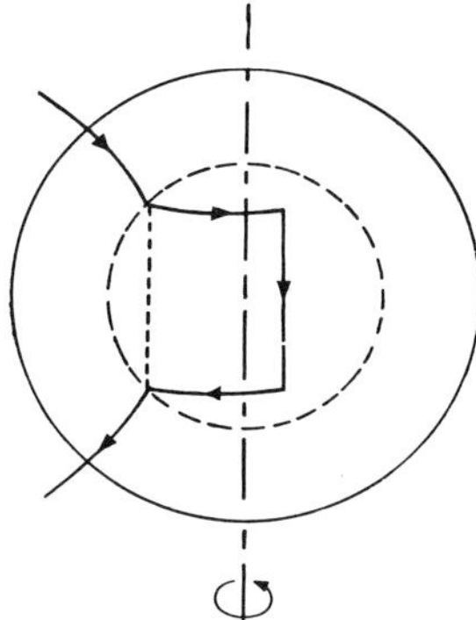

FIG. 4.5. Schematic representation of the creation of a toroidal magnetic field from a poloidal one by differential rotation.

However strong a toroidal field may be, it must eventually decay or dissipate unless the initial dipole-type field, from which the toroidal field is generated, is maintained. In other words, there must be a kind of feedback mechanism from the toroidal field to the poloidal field. Based on a rather simple consideration, ELSASSER (1955) pointed out that a poloidal field cannot be amplified by rotationally symmetric motion and for the dynamo action, three-dimensional motion is necessary. ELSASSER (1955) suggested a radially convecting motion, which will be distorted by the Coriolis force, resulting in a spiral motion. Then spiral magnetic lines of force would be created from the toroidal field as well. Thus magnetic loops are created in a plane perpendicular to the direction of the initial toroidal field, as shown schematically in Fig. 4.6. Suppose that a number of loops of this kind are created in the sphere, then statistically they may form a stable poloidal field even if individual loops are unstable. PARKER (1955) treated this problem theoretically on the basis of a local coordinate system and showed fairly strictly that a number of loops can be created in the meridian plane. It now appears that the combination of differential rotation and radial convection acts as a dynamo mechanism.

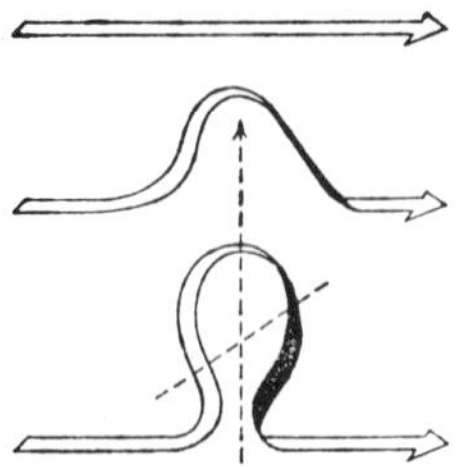

FIG. 4.6. Schematic representation of the creation of a poloidal magnetic field from a toroidal one by convective motion affected by the Coriolis force (ELSASSER, 1955).

4.4 *Kinematic Dynamo Models*

4.4.1 *Bullard process*

In contrast to the Elsasser-Parker process, in which eddies of rather small scale are important, a dynamo action suggested by BULLARD (1949) is generated through large-scale motions in the sphere; the process suggested has been called the Bullard process.

In general, the magnetic and velocity fields can be expanded in a series of spherical harmonics, and then large-scale fields are expressed by spherical harmonics of low degree. In a manner similar to the treatment of ELSASSER (1946a, b), the magnetic field can be separated into toroidal ($\boldsymbol{T}$) and poloidal ($\boldsymbol{S}$) fields as

$$\begin{aligned} \boldsymbol{T} &= \text{curl}\ (\boldsymbol{r}\psi) \\ \boldsymbol{S} &= C\ \text{curl curl}\ (\boldsymbol{r}\psi) \end{aligned} \tag{4.22}$$

where C is a constant and ψ is a scalar function which must satisfy conditions derived from equations governing the magnetic field. The characteristics of toroidal and poloidal fields are well recognized in the case of rotational symmetry, since in this case $\partial/\partial\varphi = 0$ and, therefore, $T_r = T_\theta = 0$, $S_\varphi = 0$. The toroidal field has T_φ component only and consequently magnetic lines of force are confined in circles around the rotation axis, while the poloidal field represents magnetic fields in meridional planes such as a field due to an axial dipole placed at the center of sphere.

Bullard expressed ψ as

$$\psi = \sum_{n,m} (\psi_n^{mc} + \psi_n^{ms}) \tag{4.23}$$

where

$$\psi_n^{mc} = R(r)\,P_n^m(\cos\theta)\cos m\varphi$$
$$\psi_n^{ms} = R(r)\,P_n^m(\cos\theta)\sin m\varphi. \quad (4.24)$$

P_n^m denotes the associated Legendre function of degree n and order m. Then toroidal and poloidal fields are expressed correspondingly as T_n^{mc}, T_n^{ms}, S_n^{mc}, and S_n^{ms}, respectively. In the induction equation (4.15), $\boldsymbol{H}$ appearing on the left can be regarded as the induced field due to the coupling (indicated by the first term on the right) between the inducing field $\boldsymbol{H}$ and the velocity $\boldsymbol{v}$.

The process that BULLARD (1949) put forward as a possible dynamo action is shown in Table 4.1 and also in Fig. 4.7. In this process S_2^{2c} motion can be reversed, yet the system still works as a dynamo. Therefore, the magnetic field expressed by S_1 can grow in either direction. In addition to T_1-S_2^{2c} motion selected by BULLARD (1949), T_1-S_2^c motion and T_1-S_1^c motion also yield a closed chain. However, these systems are more complicated than the Bullard process. In the case of T_1-S_1 motion or T_1-S_2 motion, no closed chain can be attained and, therefore, such a combination of motions cannot generate a dynamo process.

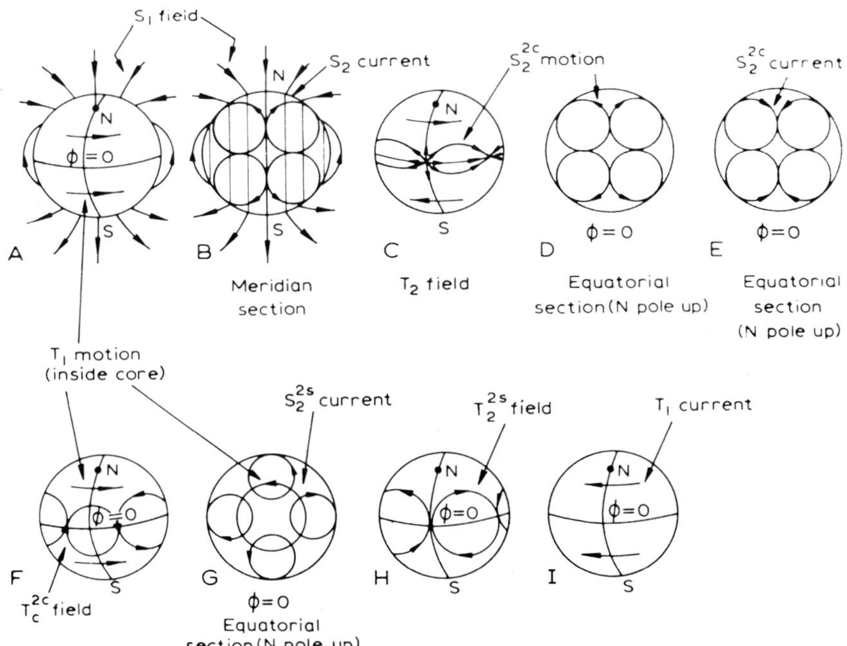

FIG. 4.7. The Bullard process for dynamo action (BULLARD, 1949).

TABLE 4.1. Interactions between magnetic fields and fluid motions.

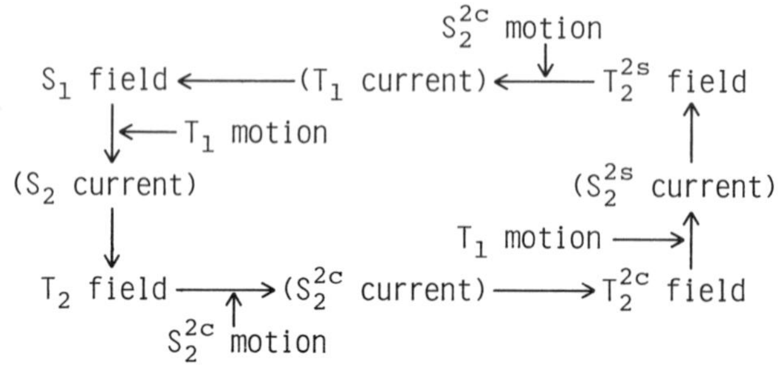

```
                            S2^2c motion
                                 ↓
S1 field ←———(T1 current) ←———— T2^2s field
   ↓← T1 motion                       ↑
(S2 current)                   (S2^2s current)
   ↓                                  ↑
                        T1 motion ———→
T2 field ———→ (S2^2c current) ———→ T2^2c field
          ↑
     S2^2c motion
```

4.4.2 Bullard-Gellman dynamo model

Through the following transformation,

$$t \to \frac{a^2}{\nu_m} t$$

$$r \to ar \tag{4.25}$$

$$\boldsymbol{v} \to \frac{\nu_m}{a} \boldsymbol{v}.$$

t and $\boldsymbol{v}$ are now dimensionless parameters. Then (4.15) is transformed to

$$\frac{\partial \boldsymbol{H}}{\partial t} = \nabla^2 \boldsymbol{H} + V \operatorname{curl}(\boldsymbol{v} \times \boldsymbol{H}) \tag{4.26}$$

where we replaced $\boldsymbol{v}$ with $V\boldsymbol{v}$ so that V becomes a parameter for the magnitude of the velocity. (4.26) is now an eigenvalue problem.

TAKEUCHI and SHIMAZU (1952a, b, 1953, 1954) examined quantitatively whether or not the Bullard process actually works as a dynamo. BULLARD and GELLMAN (1954) also investigated the Bullard process in detail by making use of a computer which was then being brought into practical use. The model which they treated has been called the Bullard-Gellman (B-G) model.

In the B-G model, the (r, θ, ϕ) components of toroidal and poloidal magnetic fields are given, respectively, as

$$T_r = 0, \quad T_\theta = \frac{T(r)}{r \sin\theta} \frac{\partial Y}{\partial \phi}, \quad T_\phi = -\frac{T(r)}{r} \frac{\partial Y}{\partial \theta}$$

$$S_r = \frac{n(n+1)}{r^2} S(r) Y, \quad S_\theta = \frac{1}{r} \frac{\partial S(r)}{\partial r} \frac{\partial Y}{\partial \theta}, \quad S_\phi = \frac{1}{r \sin\theta} \frac{\partial S(r)}{\partial r} \frac{\partial Y}{\partial \phi}. \tag{4.27}$$

where Y denotes spherical harmonics $P_n^m(\cos\theta)\begin{matrix}\sin\\ \cos\end{matrix} m\phi$. $T(r)$ and $S(r)$ are scalar functions of r. We expand $\boldsymbol{H}$ as

$$\boldsymbol{H} = \sum_{\beta} (\boldsymbol{S}_\beta + \boldsymbol{T}_\beta) \tag{4.28}$$

where β denotes (n, m). Substituting (4.28) into (4.26), we obtain

$$\sum_{\beta}\left[\operatorname{curl}^2(\boldsymbol{S}_\beta + \boldsymbol{T}_\beta) + \frac{\partial \boldsymbol{S}_\beta}{\partial t} + \frac{\partial \boldsymbol{T}_\beta}{\partial t} - V\operatorname{curl}(\boldsymbol{v}\times\boldsymbol{S}_\beta) - V\operatorname{curl}(\boldsymbol{v}\times\boldsymbol{T}_\beta)\right] = 0. \tag{4.29}$$

By applying operations $\boldsymbol{S}'_\gamma\cdot$ (scalar product) and $\boldsymbol{T}'_\gamma\cdot$ to (4.29), in which $S'(r) = T'(r) = 1$, and integrating at $r = r$ with respect to θ and ϕ, we have

$$\begin{aligned}
\frac{\partial S_\gamma}{\partial t} &= \frac{\partial^2 S_\gamma}{\partial r^2} - \gamma(\gamma+1)r^{-2}S_\gamma \\
&\quad + \frac{Vr^4}{\gamma(\gamma+1)N_\gamma}\sum_{\beta}\iint \boldsymbol{S}'_\gamma\cdot\operatorname{curl}(\boldsymbol{v}\times\boldsymbol{S}_\beta + \boldsymbol{v}\times\boldsymbol{T}_\beta)\sin\theta\,\mathrm{d}\theta\,\mathrm{d}\phi \\
\frac{\partial T_\gamma}{\partial t} &= \frac{\partial^2 T_\gamma}{\partial r^2} - \gamma(\gamma+1)r^{-2}T_\gamma \\
&\quad + \frac{Vr^2}{N_\gamma}\sum_{\beta}\iint \boldsymbol{T}'_\gamma\cdot\operatorname{curl}(\boldsymbol{v}\times\boldsymbol{S}_\beta + \boldsymbol{v}\times\boldsymbol{T}_\beta)\sin\theta\,\mathrm{d}\theta\,\mathrm{d}\phi.
\end{aligned} \tag{4.30}$$

Now we expand the velocity field as

$$\boldsymbol{v} = \sum_{\alpha} (\boldsymbol{S}_\alpha + \boldsymbol{T}_\alpha) \tag{4.31}$$

$\boldsymbol{v}$ must satisfy the boundary conditions: $v_r = 0$ at $r = 1$. Thus $S_\alpha = 0$ at $r = 1$. In the case of a viscous fluid, $v_\theta = v_\phi = 0$ at $r = 1$. However, it is considered that the effect of the boundary extends little beyond a thin boundary layer, and hence we treat as $v_\theta \neq 0$, $v_\phi \neq 0$ at $r = 1$. Substitution of (4.31) into (4.30) yields

$$\begin{aligned}
r^2\frac{\partial S_\gamma}{\partial t} &= r^2\frac{\partial^2 S_\gamma}{\partial r^2} - \gamma(\gamma+1)S_\gamma - V\sum_{\alpha\beta}[(S_\alpha S_\beta S_\gamma) + (T_\alpha S_\beta S_\gamma) + (S_\alpha T_\beta S_\gamma)] \\
r^2\frac{\partial T_\gamma}{\partial t} &= r^2\frac{\partial^2 T_\gamma}{\partial r^2} - \gamma(\gamma+1)T_\gamma \\
&\quad - V\sum_{\alpha\beta}[(S_\alpha S_\beta T_\gamma) + (T_\alpha S_\beta T_\gamma) + (S_\alpha T_\beta T_\gamma) + (T_\alpha T_\beta T_\gamma)]
\end{aligned} \tag{4.32}$$

where

$$(S_\alpha S_\beta S_\gamma) = -\frac{K}{2N_\gamma}\left[\alpha(\alpha+1)\{\alpha(\alpha+1)-\beta(\beta+1)-\gamma(\gamma+1)\}S_\alpha\frac{\partial S_\beta}{\partial r}\right.$$

$$\left.+\beta(\beta+1)\{\alpha(\alpha+1)-\beta(\beta+1)+\gamma(\gamma+1)\}\frac{\partial S_\alpha}{\partial r}S_\beta\right]$$

$$(T_\alpha S_\beta S_\gamma) = -L\beta(\beta+1)T_\alpha S_\beta/N_\gamma$$

$$(S_\alpha T_\beta S_\gamma) = -L\alpha(\alpha+1)S_\alpha T_\beta/N_\gamma$$

$$(S_\alpha S_\beta T_\gamma) = \frac{L}{N_\gamma}\left[\alpha(\alpha+1)S_\alpha\frac{\partial^2 S_\beta}{\partial r^2}\right.$$

$$+\left\{[\alpha(\alpha+1)+\beta(\beta+1)-\gamma(\gamma+1)]\frac{\partial S_\alpha}{\partial r}-\alpha(\alpha+1)\frac{2S_\alpha}{r}\right\}\frac{\partial S_\beta}{\partial r}$$

$$\left.+\beta(\beta+1)\left(\frac{\partial^2 S_\alpha}{\partial r^2}-\frac{2}{r}\frac{\partial S_\alpha}{\partial r}\right)S_\beta\right]$$

$$(T_\alpha S_\beta T_\gamma) = -\frac{K}{2N_\gamma}\left[\{\beta(\beta+1)[\alpha(\alpha+1)-\beta(\beta+1)+\gamma(\gamma+1)]\right.$$

$$+\gamma(\gamma+1)[\alpha(\alpha+1)+\beta(\beta+1)-\gamma(\gamma+1)]\}T_a\frac{\partial S_\beta}{\partial r}$$

$$\left.+\beta(\beta+1)\{\alpha(\alpha+1)-\beta(\beta+1)+\gamma(\gamma+1)\}\left(\frac{\partial T_\alpha}{\partial r}-\frac{2T_\alpha}{r}\right)S_\beta\right]$$

$$(S_\alpha T_\beta T_\gamma) = \frac{K}{2N_\gamma}\left[\alpha(\alpha+1)\{-\alpha(\alpha+1)+\beta(\beta+1)+\gamma(\gamma+1)\}S_\alpha\frac{\partial T_\beta}{\partial r}\right.$$

$$+\left\{\alpha(\alpha+1)[-\alpha(\alpha+1)+\beta(\beta+1)+\gamma(\gamma+1)]\left(\frac{\partial S_\alpha}{\partial r}-\frac{2S_\alpha}{r}\right)\right.$$

$$\left.\left.+\gamma(\gamma+1)[\alpha(\alpha+1)+\beta(\beta+1)-\gamma(\gamma+1)]\frac{\partial S_\alpha}{\partial r}\right\}T_\beta\right]$$

$$(T_\alpha T_\beta T_\gamma) = -L\gamma(\gamma+1)T_\alpha T_\beta/N_\gamma.$$

K and L denote the Adams-Gaunt and Elsasser integrals, respectively, written as

$$K_{\alpha\beta\gamma} = \iint Y_\alpha Y_\beta Y_\gamma \sin\theta \mathrm{d}\theta \mathrm{d}\phi, \tag{4.34}$$

$$L_{\alpha\beta\gamma} = \iint Y_\alpha \left(\frac{\partial Y_\beta}{\partial \theta}\frac{\partial Y_\gamma}{\partial \phi} - \frac{\partial Y_\beta}{\partial \phi}\frac{\partial Y_\gamma}{\partial \theta} \right) \mathrm{d}\theta \mathrm{d}\phi.$$

A list of numerical results of these integrals is shown in GIBSON and ROBERTS (1969).

WINCH (1974) pointed out that the Adams-Gaunt and Elsasser integrals can be expressed in terms of the Wigner 3-j coefficients. For the Adams-Gaunt integral,

$$K_{j_1 j_2 j_3}^{m_1 m_2 m_3} = \frac{1}{4\pi}\int_0^{2\pi}\int_0^{\pi} Y_{j_1}^{m_1} Y_{j_2}^{m_2} Y_{j_3}^{m_3} \sin\theta \mathrm{d}\theta \mathrm{d}\phi$$
$$= [(2j_1+1)(2j_2+1)(2j_3+1)]^{1/2} \begin{pmatrix} j_1 & j_2 & j_3 \\ m_1 & m_2 & m_3 \end{pmatrix} \begin{pmatrix} j_1 & j_2 & j_3 \\ 0 & 0 & 0 \end{pmatrix} \tag{4.35}$$

where Y_j^m is defined by

$$Y_j^m = (-1)^m \frac{2(2j+1)(j-m)!}{2-\delta_m^0 (j+m)!} P_j^m(\cos\theta) \mathrm{e}^{im\phi} \tag{4.36}$$

and the Wigner 3-j coefficient is given as

$$\begin{pmatrix} j_1 & j_2 & j_3 \\ m_1 & m_2 & m_3 \end{pmatrix} = (-1)^{j_1 - j_2 - m_3}$$
$$\times \left\{ \frac{(j_1+j_2-j_3)!(j_1-j_2+j_3)!(-j_1+j_2+j_3)!}{(j_1+j_2+j_3+1)!} \right\}^{1/2}$$
$$\times [(j_1+m_1)!(j_1-m_1)!(j_2+m_2)!(j_2-m_2)!(j_3+m_3)!(j_3-m_3)!]^{1/2}$$
$$\times \sum_t (-1)^t [(j_1-m_1-t)!(j_3-j_2+m_1+t)!(j_2+m_2-t)!$$
$$\times (j_3-j_1-m_2+t)!(j_1+j_2-j_3-t)!t!]^{-1}. \tag{4.37}$$

In the above expression, the coefficient is zero unless arguments in all of the factorial expressions preceding the summation symbol are non-negative and the summation is carried out only for values of t which make arguments of the factorial expressions non-negative. The orthogonality conditions are as follows:

$$\sum_{j_3} (2j_3 + 1)\begin{pmatrix} j_1 & j_2 & j_3 \\ m_1 & m_2 & m_3 \end{pmatrix}\begin{pmatrix} j_1 & j_2 & j_3 \\ m'_1 & m'_2 & m_3 \end{pmatrix} = \delta_{m_1}^{m'_1}\delta_{m_2}^{m'_2}$$

$$\sum_{m_1 m_2} \begin{pmatrix} j_1 & j_2 & j_3 \\ m_1 & m_2 & m_3 \end{pmatrix}\begin{pmatrix} j_1 & j_2 & j'_3 \\ m_1 & m_2 & m_3 \end{pmatrix} = \frac{1}{2j_3 + 1}\delta_{j_3}^{j'_3}. \tag{4.38}$$

Similarly, the Elsasser integral is written as

$$L_{j_1 j_2 j_3}^{m_1 m_2 m_3} = \frac{i}{2}[(2j_1 + 1)(2j_2 + 1)(2j_3 + 1)]^{1/2}$$
$$\times \begin{pmatrix} j_1 - 1 & j_2 - 1 & j_3 - 1 \\ 0 & 0 & 0 \end{pmatrix}\begin{pmatrix} j_1 & j_2 & j_3 \\ m_1 & m_2 & m_3 \end{pmatrix}$$
$$\times \left\{\frac{(-j_1 + j_2 + j_3)(j_1 - j_2 + j_3)(j_1 + j_2 - j_3)}{(j_1 + j_2 + j_3)}\right\}^{1/2}$$
$$\times [(j_1 + j_2 + j_3 - 1)(j_1 + j_2 + j_3 + 1)]^{1/2}. \tag{4.39}$$

In the case of the steady state, for which BULLARD and GELLMAN (1954) solved the equations numerically, (4.32) become

$$r^2\frac{d^2 S_\gamma}{dr^2} - \gamma(\gamma + 1)S_\gamma = V\sum_{\alpha\beta} [(S_\alpha S_\beta S_\gamma) + (T_\alpha S_\beta S_\gamma) + (S_\alpha T_\beta S_\gamma)]$$

$$r^2\frac{d^2 T_\gamma}{dr^2} - \gamma(\gamma + 1)T_\gamma = V\sum_{\alpha\beta} [(S_\alpha S_\beta T_\gamma) + (T_\alpha S_\beta T_\gamma) + (S_\alpha T_\beta T_\gamma) + (T_\alpha T_\beta T_\gamma)]. \tag{4.40}$$

The boundary conditions which have to be satisfied are that the normal component of $\boldsymbol{B}$ and the tangential components of $\boldsymbol{H}$ must be continuous at the surface of sphere. Outside the sphere, $\boldsymbol{H}$ satisfies

$$\nabla^2 \boldsymbol{H} = 0. \tag{4.41}$$

A solution of (4.41) can be expressed as

$$\boldsymbol{H} = \sum_{\beta} c_\beta(t) \operatorname{grad} (Y_\beta / r^{\beta + 1}). \tag{4.42}$$

Continuity conditions derived from (4.42) and (4.28) yield

$$c_\beta = -\beta S_\beta$$
$$T_\beta = 0 \qquad \text{at } r = 1.$$
$$\frac{\partial S_\beta}{\partial r} + \beta S_\beta = 0 \tag{4.43}$$

The velocity field assumed in the B-G model is a combination of T_1 and S_2^{2c} motions and their radial functions are given as

$$Q_T = \varepsilon r^3 \quad \text{or} \quad \varepsilon r^2(1-r) \qquad (4.44)$$
$$Q_S = r^3(1-r)^2$$

where ε is a constant which controls the ratio of the T_1 to S_2^{2c} motions. For the above velocity field, equations for toroidal and poloidal magnetic fields were derived up to the fourth degree of spherical harmonics. In numerical calculations, the radial distance between the center and the surface was divided into 10 equal portions and the differential equations were transformed into a set of finite difference equations. As a result of elaborate numerical calculations for very complicated equations, BULLARD and GELLMAN (1954) showed that an eigenvalue exists and consequently the B-G model works as a steady dynamo.

4.4.3 *Braginskii's dynamo and constraints on fluid motion*

In the Cowling theorem, nothing is suggested about the possibility of dynamo action in the case of *nearly* axisymmetric fluid motion. In such a system, the effectiveness of dynamo action should be higher, the larger the magnetic Reynolds number. With such a presumption in mind, BRAGINSKII (1965) treated a nearly axisymmetric dynamo for the condition $R_m \to \infty$. The fluid motion that Braginskii assumed is described as

$$\boldsymbol{v} = \boldsymbol{v}_0 + \boldsymbol{v}_1' + \boldsymbol{v}_2 \qquad (4.45)$$

where $\boldsymbol{v}_0$ is azimuthal and axisymmetric, $\boldsymbol{v}_2$ is meridional and axisymmetric, and $\boldsymbol{v}_1'$ is non-axisymmetric. Their magnitudes are assumed to be

$$v_1' \sim v_0 R_m^{-1/2}, \quad v_2 \sim v_0 R_m^{-1} \qquad (4.46)$$

where v_0 is referred to in defining R_m. Substituting (4.45) into (4.15), that is the induction equation, we would have the asymptotic form for $\boldsymbol{H}$:

$$\boldsymbol{H} = \sum_{n=0}^{\infty} \boldsymbol{H}_n \qquad (4.47)$$

where

$$H_n = \mathrm{O}(R_m^{-n/2} H_0). \qquad (4.48)$$

$\boldsymbol{H}$ can be separated into axisymmetric and non-axisymmetric parts. Thus, the dynamo process is controlled by the term $\boldsymbol{S}$ defined by

$$\boldsymbol{S} = \langle \boldsymbol{v}_1' \times \boldsymbol{H}' \rangle \qquad (4.49)$$

where $\boldsymbol{H}'$ denotes the non-axisymmetric part and the bracket indicates the averaging procedure with respect to ϕ. $\boldsymbol{S}$ can also be written as

$$\boldsymbol{S} = \sum_{n=2}^{\infty} \boldsymbol{S}_n = \left\langle \boldsymbol{v}_1' \times \left(\sum_{k=1}^{\infty} \boldsymbol{H}_k \right) \right\rangle \tag{4.50}$$

where

$$S_n = \mathrm{O}(R_m^{-n/2} H_0). \tag{4.51}$$

For the condition that the highest order of S_ϕ (the ϕ component of $\boldsymbol{S}$) is $\mathrm{O}(R_m^{-2} H_0)$, BRAGINSKII (1965) showed that dynamo action is possible only if $\boldsymbol{v}_1'$ contains both cos $m\phi$ and sin $m\phi$ terms. In other words, no fluid motion having a ϕ dependence of only cos $m\phi$ or only sin $m\phi$ works as a dynamo.

This kind of study was extended further to the next highest order, $\mathrm{O}(R_m^{-5/2} H_0)$, by TOUGH (1967) and TOUGH and GIBSON (1969), and it was shown that the above conclusion holds valid. However, it has not been proved yet whether the above conclusion is generally true or not.

Another constraint on the type of fluid motion was derived as a by-product of the theoretical treatment of the B-G dynamo model. BULLARD and GELLMAN (1954) proved that any purely toroidal motion cannot maintain a poloidal magnetic field. The same conclusion was also obtained by BACKUS (1958).

4.4.4 Bullard-Gellman-Lilley dynamo model

In the B-G model, spherical harmonics of which the degree is higher than four were truncated, yet it has been uncertain whether the effect of such truncation is negligibly small or not. In the meantime, GIBSON and ROBERTS (1969) showed that the eigenvalue tends to become unstable as the truncation level is upgraded. This result casts severe doubt on the possibility of a dynamo

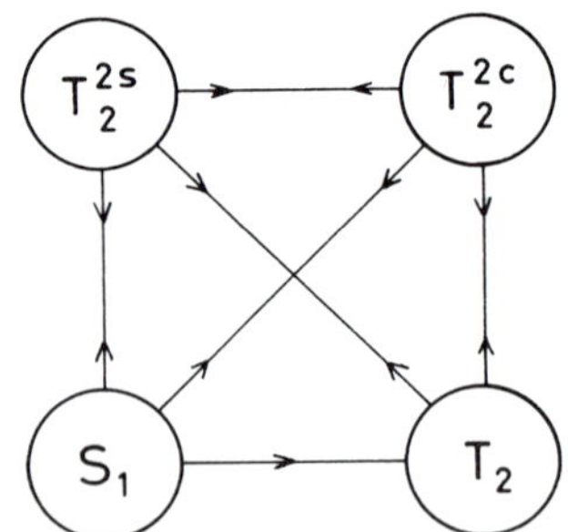

FIG. 4.8. Interaction diagram for $n \leqq 2$ in the B-G-L dynamo model. New interrelations between S_1 and T_2^{2c} and also between T_2 and T_2^{2s} are introduced by the addition of S_2^{2s} motion (LILLEY, 1970).

action in the B-G model. Moreover, the type of fluid motion in the B-G model, that is T_1 and S_2^{2c}, is against the constraint imposed by Braginskii and described in the preceding section.

In view of these, LILLEY (1970) added another type of motion expressed by S_2^{2s} to the B-G model. In this case, the fluid motion does not violate Braginskii's condition. This model has been called the Bullard-Gellman-Lilley dynamo model. A diagram of the interaction between the magnetic field and fluid motion is shown in Fig. 4.8 for $n \leqq 2$. Lilley assumed radial functions

$$\begin{aligned} T_1 \ (r) &= 10r^2(1 - r^2) & & \\ S_2^{2c}(r) &= r^3(1 - r^2)^2 & & \\ S_2^{2s}(r) &= 1.6r^3(1 - 4r^2)^2 & & (0 \leqq r \leqq 0.5) \\ S_2^{2s}(r) &= 0 & & (0.5 < r \leqq 1). \end{aligned} \tag{4.52}$$

Lilley made numerical computations for some cases; the radial distance was divided variously and spherical harmonics were truncated at various levels. The sixth degree was the highest. In the case of 16 divisions of radial distance, the eigenvalue became 24.92, 22.28, 17.76, 21.28, 20.73 for $n = 2, 3, 4, 5, 6$, respectively, as the truncation level. It appears, therefore, that the eigenvalue is stable.

Eigenfunctions representing various types of magnetic field were also examined with special reference to the stability. Lilley found that the T_2, T_2^{2c}, and T_2^{2s} fields are dominant among various types of magnetic field and they are also stable, while the S_1 field, which represents a dipole field, is unstable and even its polarity changes as the truncation level is upgraded. In the B-G-L dynamo model, the system consisting of the T_2, T_2^{2c}, and T_2^{2s} magnetic fields seems to be a dominant feature. In any case, the B-G-L model was shown to work as a dynamo at this stage of research.

GUBBINS (1973a) also made numerical calculations for the B-G-L dynamo model only to find that for $n = 5$ no eigenvalue exists between 0 and 600 when the radial distance was divided by 32 equal grid points. Next he made an attempt to find an eigenvalue for the fluid motion consisting of T_1, S_2^{2c}, and S_2^{2s} with radial functions somewhat different from those in the B-G-L model. However, no evidence of convergence of the eigenvalue was obtained for any fluid motions that he examined.

GUBBINS (1973a) also considered the fluid motion expressed by

$$\boldsymbol{v} = \boldsymbol{T}_1 + \boldsymbol{S}_2^{2c} + \boldsymbol{T}_3^{2c} \tag{4.53}$$

where the radial functions are given by

$$
\begin{aligned}
T_1(r) &= 10r^4 \\
S_2^{2c}(r) &= r^5 - r^6 \\
T_3^{2c}(r) &= 0.07 \sin^5 (\pi r).
\end{aligned}
\tag{4.54}
$$

This model includes a rising and twisting motion which seems necessary for the process of meridional field generation from the toroidal field, as pointed out by ELSASSER (1955) and PARKER (1955). It turned out, however, that this model does not work as a dynamo, either.

4.4.5 *Gubbins dynamo model*

GUBBINS (1973b) pointed out that the constraint imposed on the fluid motion by BRAGINSKII (1965) is not so strong as has been supposed, since generalization of the expansion of the power series in $R_m^{-1/2}$ has not been verified (TOUGH and GIBSON, 1969; SOWARD, 1972). GUBBINS (1973b) further presumed that Braginskii's anti-dynamo restriction might be removed if the magnitude of non-symmetric motion is sufficiently large.

In the meantime, the possibility of dynamo action due to turbulent motion has been widely believed, as will be described in the next section. In this connection, GUBBINS (1973a) speculated that small-scale motions might maintain a large-scale magnetic field, and investigated dynamo action for a velocity field:

$$\boldsymbol{v} = \varepsilon \boldsymbol{S}_n + \boldsymbol{T}_n \tag{4.55}$$

where the radial function for poloidal and toroidal fields is assumed to be

$$-r^2 \sin (n\pi r) \tanh n\pi(1 - r). \tag{4.56}$$

In this case there are two independent solutions; one is expressed in terms of $\cos m\phi$ only and the other $\sin m\phi$ only. If $m = 0$, the magnetic field decays according to the Cowling theorem. However, there is no reason to rule out the possibility of maintaining a magnetic field for $m \neq 0$.

As in the turbulent dynamo model, we write $\boldsymbol{H}$ as

$$\boldsymbol{H} = \boldsymbol{H}_0 + \boldsymbol{H}' \tag{4.57}$$

where $\boldsymbol{H}_0$ and $\boldsymbol{H}'$ denote large-scale and small-scale fields, respectively. If we substitute (4.57) into the induction equation, (4.15), the following important term appears: that is $\overline{\boldsymbol{v} \times \boldsymbol{H}'}$ where the bar denotes the averaging operation. In order to represent small-scale fields sufficiently well by spherical harmonics, n must be taken as large as possible.

Because of limited computer capability, however, GUBBINS (1973a) first examined the case in which $n = 2$, $\varepsilon = 1/10$. The growth rate described by the real part of $p(= \partial/\partial t)$ was derived for the magnetic Reynolds number $R_m = 10, 20, 30, 40, 50$, and 60, respectively, and it was shown that in this case p has only a real part. As shown in Fig. 4.9, the steady state ($\mathrm{Re}\,p = 0$) is realized at $R_m \approx 53$. It was also shown that $\boldsymbol{S}_1^1$ and $\boldsymbol{T}_1^1$ are dominant among eigenvectors representing various types of magnetic field at the steady state. The convergence was checked up to the 15th degree for 25 divisions of the radial distance. Similarly, the steady state was obtained at $R_m \approx 40$ and 30 for models with $n = 4$, $\varepsilon = 1/30$ and $n = 6$, $\varepsilon = 1/30$, respectively. Magnetic fields expressed by $\boldsymbol{S}_1^1$ and $\boldsymbol{T}_1^1$ are also dominant in these cases.

In this study it is crucial to reduce the number of interactions without significant drawback in approximation. For this purpose, GUBBINS (1973a) considered a two-scale mechanism. Suppose that large-scale fields are characterized by spherical harmonics of degree 1 to s, while small-scale fields are described by harmonics characterizing the velocity field, say n, and those of degrees near n. Here n should be much larger than s. Small-scale fields are likely to be created by interactions between $\boldsymbol{v}$ and $\boldsymbol{H}_0$ (large-scale magnetic field). Then interactions between $\boldsymbol{v}$ and $\boldsymbol{H}'$ (small-scale magnetic field) can be neglected, approximately. This simplification was verified for a model of $n = 6$, $\varepsilon = 1/30$, in which $s = 1$ and three degrees, 5, 6, and 7, were assigned to small-scale fields.

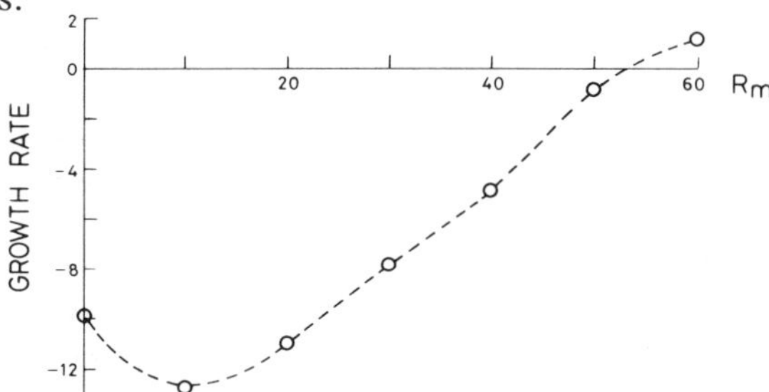

FIG. 4.9. Growth rate for various values of R_m in the case $n = 2$, $\varepsilon = 1/10$ (GUBBINS, 1973a).

4.4.6 *Turbulent dynamo*

As one of the kinematic dynamo models, a dynamo action due to turbulent fluid motion has been investigated by Krause, Rädler, Steenbeck, and others. A series of their papers (written in German) were introduced in an English translation edition by ROBERTS and STIX (1971) and recently a textbook on the turbulent dynamo was provided by KRAUSE and RÄDLER (1980). Although much stress is put on mean-field electrodynamics, the problem is still complicated. In this section, only a brief summary is described and for more details readers should go through a series of their original papers or the recent textbook by KRAUSE and RÄDLER (1980). An article by ROBERTS (1971) and a textbook by MOFFATT (1978) on dynamo theories also include some discussion concerning the turbulent dynamo.

STEENBECK *et al.* (1966) investigated the mean electromotive force associated with turbulent motion with no mean velocity field, on the assumption that the mean magnetic field is weak and consequently the Lorentz force is negligibly small. The mean electric current $\bar{\boldsymbol{i}}$ is written as

$$\bar{\boldsymbol{i}} = \sigma(\bar{\boldsymbol{E}} + \overline{\boldsymbol{v} \times \boldsymbol{H}}). \tag{4.58}$$

Now we can express $\boldsymbol{H}$ and $\boldsymbol{v}$ as

$$\begin{aligned} \boldsymbol{H} &= \bar{\boldsymbol{H}} + \boldsymbol{H}' \\ \boldsymbol{v} &= \bar{\boldsymbol{v}} + \boldsymbol{v}' \end{aligned} \tag{4.59}$$

where the term $\bar{\boldsymbol{v}}$ is assumed to disappear. In general, $\boldsymbol{v} \times \boldsymbol{H} \neq \bar{\boldsymbol{v}} \times \bar{\boldsymbol{H}}$ and in the present case,

$$\overline{\boldsymbol{v} \times \boldsymbol{H}} = \overline{\boldsymbol{v}' \times (\bar{\boldsymbol{H}} + \boldsymbol{H}')} = \overline{\boldsymbol{v}' \times \boldsymbol{H}'}. \tag{4.60}$$

Thus the mean electromotive force $(\overline{\boldsymbol{v} \times \boldsymbol{H}})$ is controlled by the term $\overline{\boldsymbol{v}' \times \boldsymbol{H}'}$. $\boldsymbol{H}'$ can be considered as a perturbation generated by $\boldsymbol{v}'$ in the presence of mean field $\bar{\boldsymbol{H}}$. Thus we have

$$\frac{\partial \boldsymbol{H}'}{\partial t} - \nu_m \nabla^2 \boldsymbol{H}' = \operatorname{curl}\,(\boldsymbol{v}' \times \bar{\boldsymbol{H}}). \tag{4.61}$$

The solution of (4.61) is expressed as

$$\begin{aligned} H_i'(\boldsymbol{x}, t) = (\delta_{il}\delta_{jm} - \delta_{im}\delta_{jl}) \int_0^t \int \frac{\partial}{\partial x_j'} [v_l(\boldsymbol{x}', t')\bar{H}_m(\boldsymbol{x}', t')) \\ \times \Gamma(\boldsymbol{x} - \boldsymbol{x}', t - t')\mathrm{d}\boldsymbol{x}'\mathrm{d}t' \end{aligned} \tag{4.62}$$

where the Einstein summation convention is used and the integration is taken over the whole space. δ_{il} and the like denote the Kronecker tensor. $\Gamma(\boldsymbol{x}, t)$ is given by

$$\Gamma(\boldsymbol{x}, t) = (4\pi\nu_m t)^{-3/2} \exp\,(-x^2/4\nu_m t). \tag{4.63}$$

We can also write as

$$(\overline{\boldsymbol{v}' \times \boldsymbol{H}'})_i = \varepsilon_{ijk}\overline{v_j'(\boldsymbol{x}, t)H_k'(\boldsymbol{x}, t)} \tag{4.64}$$

where

$$\varepsilon_{ijk} = \begin{cases} 1 & \text{(even permutation)} \\ -1 & \text{(odd permutation)} \end{cases} \tag{4.65}$$

$(\varepsilon_{123} = 1)$.

Substituting (4.62) into (4.64),

$$\overline{(\boldsymbol{v}' \times \boldsymbol{H}')}_i = (\varepsilon_{ijl}\delta_{nm} - \varepsilon_{ijm}\delta_{nl})$$

$$\times \int_0^t \int \frac{\partial}{\partial x'_n} R_{jl}(\boldsymbol{x}, \boldsymbol{x}', t, t')\bar{H}_m(\boldsymbol{x}', t')$$

$$\times \Gamma(\boldsymbol{x} - \boldsymbol{x}', t - t')\mathrm{d}\boldsymbol{x}'\mathrm{d}t' \tag{4.66}$$

where

$$R_{jl}(\boldsymbol{x}, \boldsymbol{x}', t, t') = \overline{v'_j(\boldsymbol{x}, t)v'_l(\boldsymbol{x}', t')}\,. \tag{4.67}$$

The term R_{jl} is called the correlation tensor. It is this term that controls the mean electromotive force.

On the assumption that the gradient in turbulent intensity is non-vanishing and turbulent motion is also affected by the Coriolis force, STEENBECK *et al.* (1966) solved the Navier-Stokes equation and derived the expression for $\boldsymbol{v}$. Then it was shown, through complicated calculation, that $\overline{\boldsymbol{v}' \times \boldsymbol{H}'}$ has a component parallel to $\bar{\boldsymbol{H}}$. The generation of this component has been called the "α-effect" (ROBERTS, 1971).

This result is very significant because the α-effect provides a mechanism for a self-exciting electromagnetic field. The mechanism is shown schematically in Fig. 4.10. Suppose that an electric current i flows in ring I, resulting in generation of the magnetic field denoted by H_0. Then the electric current due to the α-effect flows in ring II, creating the magnetic field H_1. Again the α-effect acts so that the current i_2 flows in ring I. Thus it is possible that i and H reinforce each other against decay. The energy required for generation of electromotive force is certainly provided by turbulent motion. However, the growth of the magnetic field will eventually be suppressed by the Lorentz force acting on fluid motion. It was also experimentally demonstrated that the α-effect is actually operative (STEENBECK *et al.*, 1967).

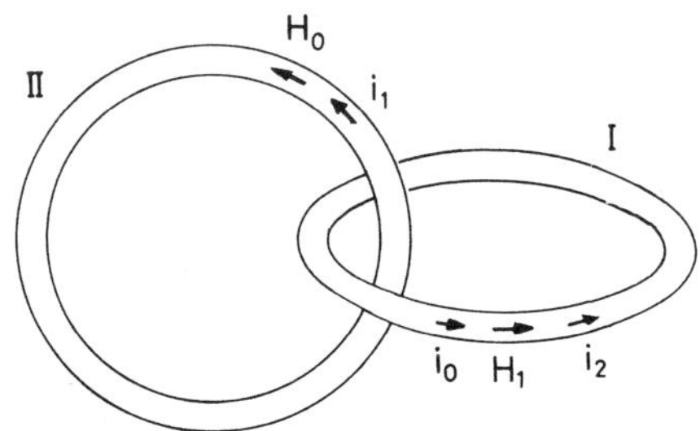

FIG. 4.10. Schematic representation of an α^2-dynamo mechanism (STEENBECK *et al.*, 1966).

The α-effect appearing in the term $\overline{\boldsymbol{v}' \times \boldsymbol{H}'}$ was examined in more detail by RÄDLER (1968). Substituting $\boldsymbol{H} = \bar{\boldsymbol{H}} + \boldsymbol{H}'$ and $\boldsymbol{v} = \bar{\boldsymbol{v}} + \boldsymbol{v}'$ into the induction equation, (4.15), we obtain

$$\frac{\partial}{\partial t}(\bar{\boldsymbol{H}} + \boldsymbol{H}') = \nu_m \nabla^2 (\bar{\boldsymbol{H}} + \boldsymbol{H}') + \mathrm{curl}(\bar{\boldsymbol{v}} \times \bar{\boldsymbol{H}} + \bar{\boldsymbol{v}} \times \boldsymbol{H}' + \boldsymbol{v}' \times \bar{\boldsymbol{H}} + \boldsymbol{v}' \times \boldsymbol{H}'). \tag{4.68}$$

The above equation determines $\boldsymbol{H}'$, if $\bar{\boldsymbol{H}}$, $\bar{\boldsymbol{v}}$ and $\boldsymbol{v}'$ are known. Therefore $\overline{\boldsymbol{v}' \times \boldsymbol{H}'}$ would be expressed in terms of $\bar{\boldsymbol{H}}$, $\bar{\boldsymbol{v}}$, and the average properties of $\boldsymbol{v}'$. It is reasonably assumed that the correlation of turbulent quantities at one point with those at another is vanishingly small unless two points are located sufficiently close to each other. We further assume that local dependence of $\overline{\boldsymbol{v}' \times \boldsymbol{H}'}$ on $\bar{\boldsymbol{H}}$ is represented by the components of $\bar{\boldsymbol{H}}$ and their first spatial derivatives. Turbulent nature also implies that the correlation of turbulent quantities is non-vanishing only over a short time interval. Then we can write

$$(\overline{\boldsymbol{v}' \times \boldsymbol{H}'})_i = a_{ij} \bar{H}_j + b_{ijk} \frac{\partial \bar{H}_j}{\partial x_k} \tag{4.69}$$

where a_{ij} and b_{ijk} are dependent on $\bar{\boldsymbol{v}}$ and average properties of $\boldsymbol{v}'$.

Now we examine a_{ij} and b_{ijk} for a simple case. Suppose that $\bar{\boldsymbol{v}} = 0$ and $\boldsymbol{v}'$ is homogeneous and isotropic, then averages of quantities concerning $\boldsymbol{v}'$ are constant in space and also invariant under rotation about any axes. Therefore, we can write

$$a_{ij} = \alpha \delta_{ij}, \quad b_{ijk} = \beta \varepsilon_{ijk} \tag{4.70}$$

where α and β are scalar quantities determined by $\boldsymbol{v}'$. From (4.69) and (4.70) we obtain

$$\overline{\boldsymbol{v}' \times \boldsymbol{H}'} = \alpha \bar{\boldsymbol{H}} - \beta \,\mathrm{curl}\, \bar{\boldsymbol{H}}. \tag{4.71}$$

Substituting (4.71) into (4.58), we have

$$\boldsymbol{i} = \sigma_T (\boldsymbol{E} + \alpha \bar{\boldsymbol{H}}) \tag{4.72}$$

where

$$\sigma_T = \sigma / (1 + \sigma \beta). \tag{4.73}$$

σ_T is called the "turbulent conductivity." In the case of mirror-symmetry, or reflexional symmetry, it is shown that $\alpha = 0$, and the generation process due to the α-effect cannot be expected. Symmetric properties of tensor and general invariant arguments lead to more general expressions of a_{ij} and b_{ijk} (see ROBERTS, 1971; KRAUSE and RÄDLER, 1980).

If turbulent motion is strong and its energy is efficiently transformed into magnetic field energy, a dynamo action may be generated by the α-effect only, as demonstrated schematically in Fig. 4.10. Such a dynamo has been called the α^2-dynamo, which is characterized by the creation of a toroidal magnetic field from a poloidal magnetic field. On the other hand, in the dynamo action considered by Elsasser, Parker, Bullard, and others, differential rotation is required for the creation of a toroidal magnetic field from a poloidal magnetic field. However, the process of poloidal field generation from a toroidal field can be regarded as the α-effect. In this sense, a dynamo including differential rotation has been called the "$\alpha\omega$-dynamo."

The relation between the α^2- and $\alpha\omega$-dynamos can be examined by using cylindrical coordinates (r, ϕ, z) as follows (MOFFATT, 1978). We can express the induction equation in terms of the mean fields (here bars are abbreviated) as

$$\frac{\partial \boldsymbol{H}}{\partial t} = \text{curl}\ (\boldsymbol{v} \times \boldsymbol{H}) + \text{curl}\ \boldsymbol{\varepsilon} + \nu_m \nabla^2 \boldsymbol{H}$$
$$\boldsymbol{\varepsilon} = \alpha \boldsymbol{H} - \beta\ \text{curl}\ \boldsymbol{H}. \tag{4.74}$$

If $\boldsymbol{v}, \boldsymbol{H}$ and $\boldsymbol{\varepsilon}$ are axisymmetric, we can write

$$\boldsymbol{v} = r\omega(r,z)\boldsymbol{k}_\phi + \boldsymbol{v}_P$$
$$\boldsymbol{H} = H(r,z)\boldsymbol{k}_\phi + \boldsymbol{H}_P \tag{4.75}$$
$$\boldsymbol{\varepsilon} = \varepsilon_\phi \boldsymbol{k}_\phi + \boldsymbol{\varepsilon}_P$$

where $\boldsymbol{k}_\phi$ is the unit vector in the ϕ direction and $\boldsymbol{H}_P$ can be expressed as

$$\boldsymbol{H}_P = \text{curl}\ (A(r,z)\boldsymbol{k}_\phi). \tag{4.76}$$

Substituting (4.75) into (4.74), we obtain

$$\frac{\partial H}{\partial t} + r(\boldsymbol{v}_P \cdot \text{grad})\left(\frac{H}{r}\right)$$
$$= r(\boldsymbol{H}_P \cdot \text{grad})\omega + \text{curl}(\alpha \boldsymbol{H}_P) + (\nu_m + \beta)\left(\nabla^2 - \frac{1}{r^2}\right)H \tag{4.77}$$

$$\frac{\partial A}{\partial t} + \frac{1}{r}(\boldsymbol{v}_P \cdot \text{grad})(rA) = \alpha H + (\nu_m + \beta)\left(\nabla^2 - \frac{1}{r^2}\right)A. \tag{4.78}$$

(4.77) is the equation expressing the creation of the toroidal magnetic field H from the poloidal magnetic field $\boldsymbol{H}_P$ through the terms $r(\boldsymbol{H}_P \cdot \text{grad})\omega$ and curl $(\alpha \boldsymbol{H}_P)$. The former term represents the effect of differential rotation (the ω-effect), while the latter represents the α-effect. The ratio of the ω- to α-effects is given as $O(L^2\omega_0'/\alpha_0)$, where α_0 and ω_0' are typical values of α and $|\text{grad}\ \omega|$, respectively. L indicates the characteristic length scale. If $L^2\omega_0' \gg \alpha_0$, the ω-effect dominates, while the α-effect dominates if $\alpha_0 \gg L^2\omega_0'$. In any case, the creation of poloidal magnetic field is due to the α-effect as clearly shown by (4.78). In the case of the earth, the ω-effect has been supposed to be dominant in the process of generation of a toroidal magnetic field from a poloidal magnetic field and consequently the $\alpha\omega$-dynamo is likely to work in the earth's core.

4.4.7 Herzenberg and Backus dynamo models

Mathematically rigorous proof of dynamo action was pursued by HERZENBERG (1958) and also by BACKUS (1958). However, their discussions are based on rather unrealistic models for the sake of mathematical treatment. Nevertheless, their work is significant at least in the sense that the possibility of a dynamo mechanism within the earth's core is suggested.

In Herzenberg's model, two rotating rigid spheres are embedded in a stationary sphere. The working principle of this system lies in the amplification of magnetic energy by the rotation. In one sphere a toroidal field is generated by the ω-effect and then it diffuses through the conducting medium surrounding the sphere. The magnetic field from this origin would be weak at another sphere but it would possibly be amplified by the ω-effect in the second sphere. If the speed of rotation is sufficiently fast for both spheres, the magnetic field can be maintained against dissipation or decay.

The theory of the Herzenberg dynamo was further developed by GIBSON (1968a, 1968b, 1969). The Herzenberg model was also examined experimentally by LOWS and WILKINSON (1963, 1968) using two rotaing cylinders instead of spheres.

A dynamo model for which BACKUS (1958) attempted mathematically rigorous treatment is similar to the Bullard-Gellman model. In his case, however, the fluid motion is stopped from time to time so that the decay of the magnetic field can be estimated exactly. It is then proved that magnetic fields of short wave-length decay more rapidly than those of long wave-length. In this way, the truncation of spherical harmonics of higher degrees is justified, in contrast to the case of the B-G model.

4.5 Hydromagnetic Dynamo Models

In kinematic dynamo models, the effect of the Lorentz force on fluid motion has been assumed to be negligibly small. However, as the magnetic field grows, the Lorentz force becomes stronger and stronger, and eventually it should begin to suppress the magnetic field amplification due to dynamo action. In this sense, it must be essential to treat simultaneously the induction equation and the Navier-Stokes equation. Such a treatment is particularly necessary for investigation of time-dependent behavior of a dynamo action. In this case, the appearance of non-linear terms makes mathematical treatment much more difficult. However, some dynamo models in which the equation of motion is also taken into consideration have been proposed, and we may call them "hydromagnetic dynamo models."

As an extension of his work on a kinematic dynamo, GUBBINS (1975) attempted to include the effect of the Lorentz force. He assumed a fluid motion characterized by a velocity field expressed by (4.55), and then examined evolution of the magnetic field. According to the result, the magnetic field seems to approach the steady state in an oscillatory manner, and in the extreme case the oscillation becomes marked; such a behavior may suggest a phenomenon such as geomagnetic reversal.

WATANABE (1977) estimated the influence of the Lorentz force on the α-effect. In this case, the coefficient α depends on $\boldsymbol{H}$, and consequently the induction equation for turbulent motion, (4.74), becomes a non-linear equation in $\boldsymbol{H}$. Watanabe divided the term α into terms corresponding to generating and degenerating effects, respectively. The latter represents the suppression of amplification of the magnetic field imposed by the Lorentz force. The requirement that the dynamo action must be maintained leads to the condition that the generating term must be stronger than the degenerating one. This condition sets bounds on the magnitude of the geomagnetic field in a hydromagnetic $\alpha\omega$-dynamo model; the toroidal field must be weaker than 30 gauss, while the poloidal field in the core must not be stronger than 10 gauss.

WATANABE (1977) solved the induction equation for turbulent motion numerically on the basis of his estimation of α-value derived as mentioned above. In the calculation, the Bullard-Gellman formalism was used, although some new coupling terms must be added. Instead of treating the case of the steady state ($\partial/\partial t = 0$), Watanabe integrated the differential equation numerically and examined the evolution of the magnetic field. The existence of a steady state was suggested by the equilibrium value which the estimated magnetic field approaches.

Meanwhile, BUSSE (1975, 1976, 1977) pursued quite a new approach to a hydromagnetic dynamo by taking into account a new pattern of convection as deduced from the Navier-Stokes equation for a rotating spherical body. Busse's procedure of solving the induction and Navier-Stokes equations consists of the following four steps. (I) Assuming that the magnetic field and the convection velocity field are both weak, the Lorentz force is neglected in (4.13) and then $\boldsymbol{v}$ is derived as a solution of (4.13) without the Lorentz force term. (II) $\boldsymbol{v}$ thus obtained is substituted into the induction equation (4.15) and the equation is solved. From those solutions which do not lead to any

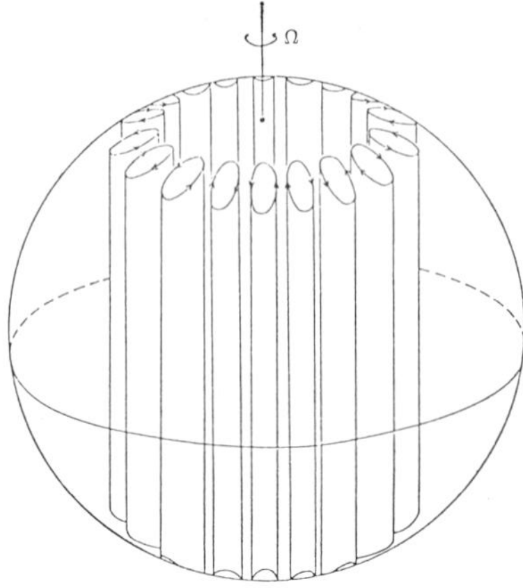

FIG. 4.11. Pattern of convection in a fluid sphere (BUSSE, 1970).

eventual decay, the $\boldsymbol{H}$ derived from the smallest convection velocity is taken as the magnetic field to be realized. (III) The Lorentz force estimated from $\boldsymbol{H}$, derived in step II, is adopted in (4.13) and the equation is solved. (IV) (4.15) is again solved for the $\boldsymbol{v}$ given as a solution in step III and the $\boldsymbol{H}$ thus obtained is considered to be a final solution. Since the effect of the Lorentz force is manifested only as a reduction in the magnitude of fluid velocity, steps I and II are sufficient for a dynamo problem.

As a possible solution of step I, BUSSE (1970) obtained a convection pattern as shown in Fig. 4.11. In his mathematical analysis, a cylindrical annulus as shown in Fig. 4.12 was used. It is assumed for mathematical convenience that

$$r_1 - r_0 \ll r_1. \tag{4.79}$$

Then the Cartesian coordinates (x, y, z) can be used as a local coordinate system. In the model, the top and bottom surfaces are approximated, respectively, by

$$z = \pm\frac{1}{2}e^{-2\eta x} \tag{4.80}$$

where length is made non-dimensional using a characteristic length scale, that is $2z_0$ in the present case. η can be regarded as a perturbation parameter and we assume $\eta \ll 1$, $\eta \ll 1/D$, where D is the ratio of width to height of the annulus. The boundary conditions for velocity $\boldsymbol{v}$ are given as

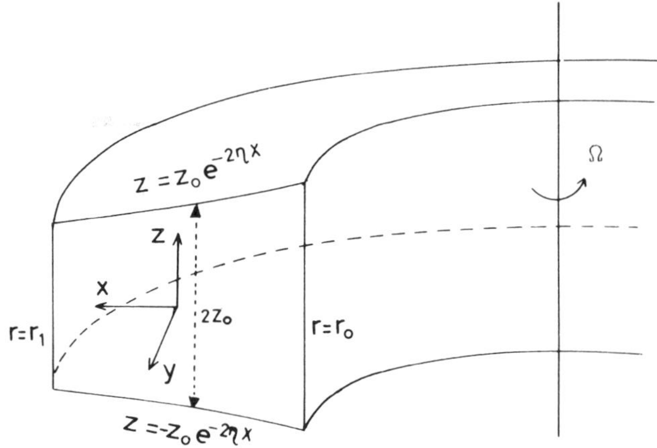

FIG. 4.12. A cylindrical annulus model for mathematical analysis (BUSSE, 1975).

$$
\begin{aligned}
&v_x = 0 && \text{at } x = \pm\frac{D}{2} \\
&\pm v_z + \eta e^{-2\eta x} v_x = 0 && \text{at } z = \pm\frac{1}{2}e^{-2\eta x}.
\end{aligned}
\qquad (4.81)
$$

Now we express $\boldsymbol{v}$ satisfying (4.13) without the Lorentz force term and (4.81) as

$$\boldsymbol{v} = \text{curl}\,(\boldsymbol{k}\psi) + \text{curl curl}\,(\boldsymbol{k}\phi) \qquad (4.82)$$

where $\boldsymbol{k}$ denotes the unit vector in the z direction. Then ψ and ϕ are expressed by the expansions in power of η as

$$
\begin{aligned}
\psi &= (\psi_0 + \eta\psi_1 + \eta^2\psi_2 + \ldots\ldots) \exp\{i(\omega t + \alpha y)\} \\
\phi &= (\eta\phi_1 + \eta^2\phi_2 + \ldots\ldots) \exp\{i(\omega t + \alpha y)\}
\end{aligned}
\qquad (4.83)
$$

where homogeneity in the y direction is taken into consideration and time-dependence is assumed to be exp $(i\omega t)$ where $\omega \ll 1$ in non-dimensional form. Here Ω^{-1} is taken as the characteristic time scale.

In the next step (step II), assuming $D \gg 1$, the magnetic field is expressed as

$$\boldsymbol{H} = H_A[-\boldsymbol{j}G(z,t) + \boldsymbol{i}F(z,t) + \text{curl curl}\,(\boldsymbol{k}h) + \text{curl}\,(\boldsymbol{k}g)] \qquad (4.84)$$

where $\boldsymbol{i}$ and $\boldsymbol{j}$ denote the unit vectors in the x and y directions, respectively. In (4.84), the method as used in the turbulent dynamo models is taken into account; G and F are large-scale magnetic fields, while h and g represent small-scale magnetic fields and consequently $\bar{h} = \bar{g} = 0$, where the bar indicates the averaging procedure over the x-y cross-section. In this model the width of the annulus is assumed to be much larger than its height, so that G and F are functions of z only. Substituting (4.82), which satisfies (4.13) for some orders of η, and (4.84) into (4.15), analytical expressions for G, F, h, and g are derived (BUSSE, 1975).

The Busse dynamo mechanism is shown schematically for the northern hemisphere in Fig. 4.13. z_1 and z_2 correspond to lower and higher latitudes, respectively. The meridional magnetic field at high-latitudes is distorted by the horizontal components of convection and then the downward fluid flow pulls down the distorted magnetic lines of force, resulting in the creation of an azimuthal magnetic field at low-latitudes from the magnetic field which was originally meridional at high-latitudes. A similar process occurs for the azimuthal magnetic field at low-latitudes; in this case, the magnetic lines of force distorted by the horizontal components of convection are pulled up by the upward convective motion and a meridional magnetic field is created at high-latitudes. Thus the meridional magnetic field at high-latitudes and the azimuthal magnetic field at low-latitudes reinforce each other. Such a mechanism would belong to the α^2-dynamo action. It is expected, therefore, that in Busse's model toroidal and poloidal magnetic fields are of the same order of magnitude.

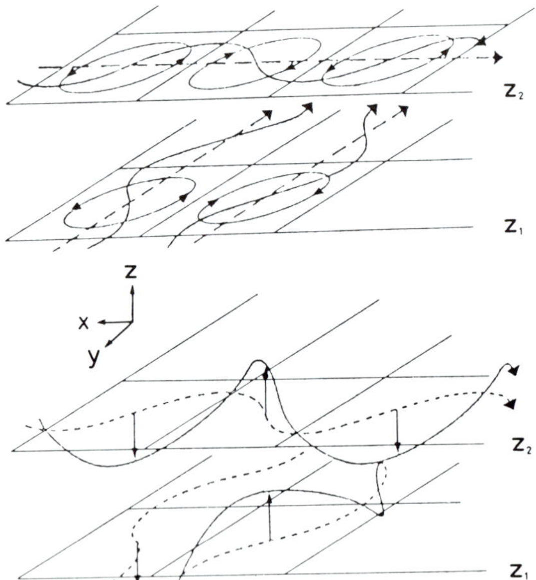

FIG. 4.13. Schematic representation of the Busse dynamo mechanism for the northern hemisphere (BUSSE, 1975).

4.6 Non-Steady Behavior of Some Dynamo Models and Mechanism of Magnetic Field Reversal

Even in a simple disk dynamo, the time-dependent behavior is not simple because of the coupling between the magnetic field and the mechanical force. If a disk system can be considered to correspond to a small convective or a small-scale turbulent motion in the earth's core, studies of the coupling between some disk systems would provide important information on time-dependent nature of the geomagnetic field. In view of this, RIKITAKE (1958) examined a non-steady state for a coupled disk dynamo model as shown in Fig. 4.14.

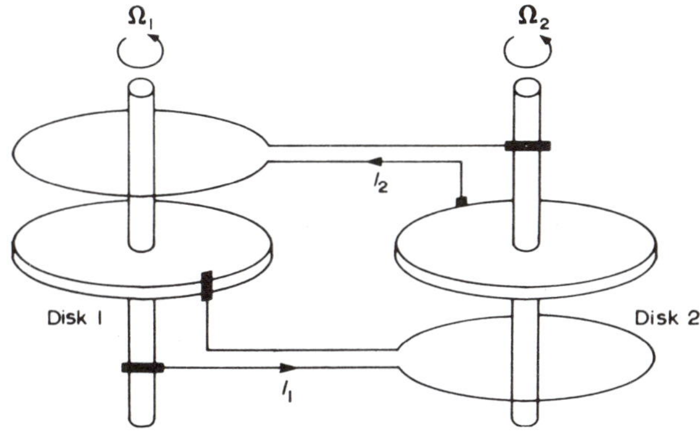

FIG. 4.14. The Rikitake coupled disk dynamo model.

In this system, equations similar to (4.3) and (4.6) are derived as

$$\begin{aligned} L_1 \mathrm{d}I_1/\mathrm{d}t + R_1 I_1 &= \omega_1 M I_2 \\ L_2 \mathrm{d}I_2/\mathrm{d}t + R_2 I_2 &= \omega_2 N I_1 \\ C_1 \mathrm{d}\omega_1/\mathrm{d}t &= G_1 - M I_1 I_2 \\ C_2 \mathrm{d}\omega_2/\mathrm{d}t &= G_2 - N I_1 I_2 \end{aligned} \tag{4.85}$$

where subscripts 1 and 2 denote quantities belonging to the disks 1 and 2, respectively, and $2\pi M$ and $2\pi N$ are the mutual-inductance. Suppose that two disk systems are equivalent, then

$$\begin{aligned} L_1 &= L_2 = L \\ R_1 &= R_2 = R \\ N &= M \\ C_1 &= C_2 = C \\ G_1 &= G_2 = G. \end{aligned} \tag{4.86}$$

Now quantities appearing in (4.85) are transformed to

$$I_1 = (G/M)^{1/2}x_1, \quad I_2 = (G/M)^{1/2}x_2$$
$$\omega_1 = (GL/CM)^{1/2}y_1, \quad \omega_2 = (GL/CM)^{1/2}y_2 \tag{4.87}$$
$$t = (CL/GM)^{1/2}\tau, \quad \mu = (R/L)(LC/MG)^{1/2}.$$

and we have a set of differential equations in dimensionless form:

$$\mathrm{d}x_1/\mathrm{d}\tau + \mu x_1 = y_1 x_2$$
$$\mathrm{d}x_2/\mathrm{d}\tau + \mu x_2 = y_2 x_1 \tag{4.88}$$
$$\mathrm{d}y_1/\mathrm{d}\tau = \mathrm{d}y_2/\mathrm{d}\tau = 1 - x_1 x_2.$$

RIKITAKE (1958) solved (4.88) numerically and examined the evolution of the magnetic field. Numerical results clearly demonstrated reversal of polarity of the magnetic field.

This coupled disk dynamo has been called the Rikitake model and further numerical integration of (4.88) was carried out by ALLAN (1962). In this model, the magnetic field is characterized by oscillatory fluctuations with gradually increasing amplitude and polarity reversal occurring from time to time. COOK and ROBERTS (1970) also examined the Rikitake model and obtained a result similar to Allan's result. Figure 4.15 shows evolution of the

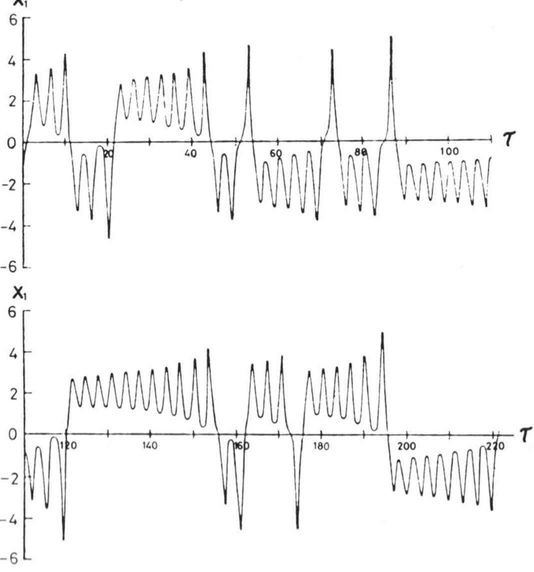

FIG. 4.15. Evolution of the electric current for one loop, represented by x_1, in the Rikitake model (COOK and ROBERTS, 1970).

electric current for one loop (COOK and ROBERTS, 1970). Reversal of the magnetic field (equivalent to the electric current) takes place rather at random. The condition for polarity reversal seems to lie in the amplitude of oscillation. A state with a very weak magnetic field is temporarily realized as a result of oscillation of large amplitude and then the magnetic field grows in the opposite direction. This must be an important implication for the mechanism of reversal of the earth's magnetic field.

The Rikitake model illustrates well the α^2-dynamo mechanism as pointed out by INGLIS (1981). The interaction between the magnetic field, say H_0, and ω_1 creates the electric current I_1 and consequently H_1 is generated. Then the interaction between H_1 and ω_2 creates I_2 from which the magnetic field initially characterized by H_0 is generated. If this process is compared with that described for the α-effect in Subsection 4.4.6, a similarity between them is easily recognized. In this sense, the Rikitake model suggests the possibility of polarity reversal for a more general α^2-dynamo model which includes the effect of Lorentz force.

Recently the Rikitake model attracted some attention from a mathematical point of view. HOLMS (1977) pointed out that the Rikitake model is one of the typical *strange* phenomena in dynamical systems; motions are caused by a deterministic system yet their time-dependence is complicated and can even be regarded as random. ITO (1980) investigated statistical properties of polarity reversal in the Rikitake model and also revealed a feature characterized by chaotic dynamics. These results imply that the apparently complicated fluctuation of the geomagnetic field may be a manifestation of such a chaotic nature and the dynamic system responsible for the dynamo action in the earth's core may not be a random one but rather a deterministic one.

Time-dependent behavior of the magnetic field has also been examined for the B-G dynamo model (RIKITAKE and HAGIWARA, 1968a, b) and for the B-G-L dynamo model (UNO, 1972). Their studies are based on numerical integration of (4.32) for the velocity field (4.44) in the B-G model and for (4.52) in the B-G-L model. The results obtained for the B-G-L model are shown in Fig. 4.16 for the poloidal (S_1) field at the core's surface ($r = 1$) and for the toroidal (T_2) field at $r = 0.85$ (UNO, 1972). In this case, the radial distance is divided into 20 equal parts and spherical harmonics representing various types of magnetic fields are truncated for $n \geqq 3$. The initial condition is given by $A(r, 0) = r^2/2$, where $A(r, t)$ denotes the radial function of the S_1 field. In the case of $V = 25.5$, for which a steady state is realized (LILLEY, 1970), the S_1 and T_2 fields approach the steady state about 6,000 years after the initial state. For other values of V, all the fields examined by UNO (1972) tend to decay or increase monotonically.

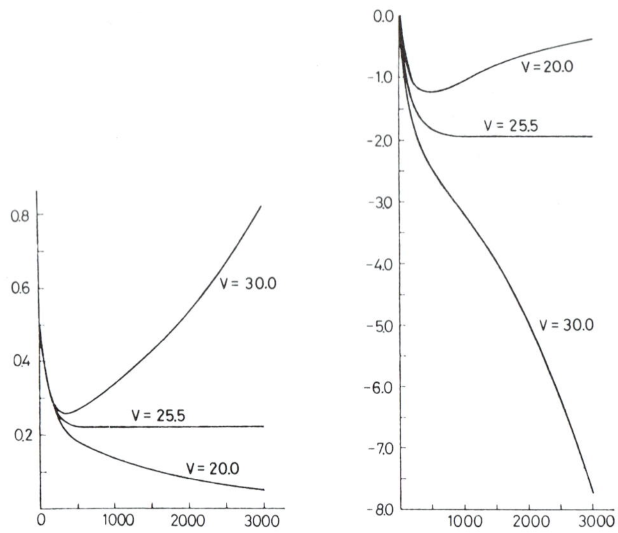

FIG. 4.16. Left: evolution of the S_1 magnetic field at the core's surface ($r = 1$). Right: evolution of the T_2 magnetic field at $r = 0.85$. The time scale is shown by a dimensionless parameter. The case $V = 25.5$ corresponds to the steady state in the B-G-L model (UNO, 1972).

The studies of non-steady state for various dynamo models have provided some information on the theoretically expected time-dependent nature of the geomagnetic field. These results provided RIKITAKE (1972, 1973a) with a possible insight into the mechanism of geomagnetic field reversal as will be described in the following. The stochastic nature of polarity reversal, as disclosed by paleomagnetic studies, does not necessarily indicate a mechanism of random nature, as implied by the chaotic behavior in the Rikitake model. The following reversal mechanism may be supposed to be likely; a reduction in dynamo action efficiency due to a slight change in fluid motion (for instance, a change into a more axisymmetric flow) would cause decay of a poloidal field and then, with the recovery of fluid motion providing efficient dynamo action, a new poloidal field begins to grow in either direction with equal probability. If the above mechanism occurs, however, the time required for each reversal process should be much longer than the free-decay time for the earth's core ($\sim 10^4$ years) but this is inconsistent with paleomagnetic results, as described in Section 3.5. Moreover, a poloidal magnetic field is unlikely to become stronger than the initial field if fluid motion is more or less the same as in the steady state, as implied by time-dependent behavior in the B-G and B-G-L models. In view of these points, RIKITAKE (1972, 1973a) pointed out that energy conversion from magnetic fields to fluid motions and vice versa must be involved in the mechanism of magnetic field reversal.

For a simplified hydromagnetic $\alpha\omega$-dynamo, WATANABE (1981) examined a non-steady state; for the same model the existence of steady state had already been shown (WATANABE, 1977). As shown in Fig. 4.17, the S_1 field (characterized by the Gauss coefficient g_1^0) undergoes a variation with occasional polarity reversal. A polarity reversal seems to be completed in some thousand years, which is in good agreement with paleomagnetic results. Watanabe also pointed out that a polarity reversal is closely related with a change in the pattern of non-uniform rotation in his model.

On the other hand, YOSHIMURA (1980) showed that polarity reversals could be represented well in evolution of the magnetic field derived from a numerical solution of simplified non-linear equations. One of the characteristics of his treatment is a time-delayed feedback process; that is, the velocity field at a certain time is considered to be a function of magnetic fields at an earlier time. Although a detailed mechanism for the time-delayed feedback process is unknown, magnetic field reversals are clearly demonstrated if parameters describing the feedback process are properly given.

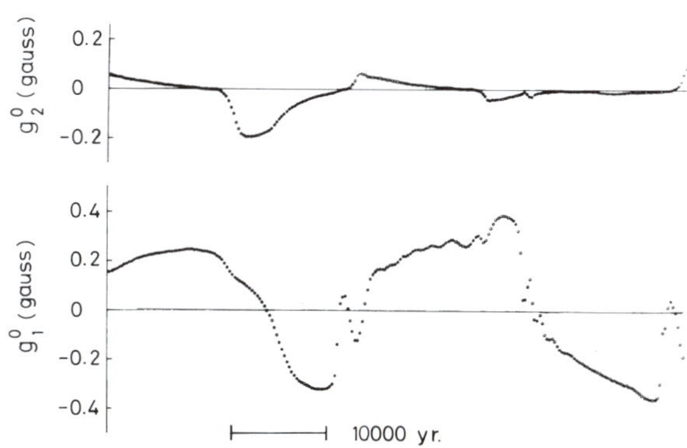

FIG. 4.17. Evolution of the S_1 (dipole) and S_2 (quadrupole) fields represented by Gauss coefficients g_1^0 and g_2^0 (WATANABE, 1981).

4.7 *Energy for the Geomagnetic Dynamo*

Most geomagnetic dynamo models are based on the assumption that convective or turbulent motion is taking place in the earth's liquid core. In the meantime, HIGGINS and KENNEDY (1971) pointed out the possibility that the outer liquid core might be stably stratified and consequently no convective circulation of fluid would be expected. KENNEDY and HIGGINS (1973) later noted that in the lower one-third of the liquid core the temperature distribution may be along the adiabatic gradient, so that fluid circulation is possible. However, there are a variety of discussions on the melting and adiabatic gradients of core material, and the above claim is not firmly accepted as really reflecting the state of the core.

It has been widely presumed that convective motion would be driven thermally by a heat source of radioactive origin, although no rigorous evidence supporting such an idea has been put forward. On the basis of the present knowledge of heat flux at the earth's surface and some physical properties of core material, GUBBINS (1976, 1977) examined this hypothesis and imposed bounds on heat flux coming from the earth's core. It was then concluded that at least 10^{11} watt and most probably 10^{13} watt must come from the core, if convective motion prevails in the core. However, such a large amount of heat flux, if we accept a probable estimate of 10^{13} watt, seems to be unlikely in view of the total surface heat flux estimated from surface heat flux data, at present of the order of 10^{13} watt, and the high concentration of radioactive material in the crust.

Then GUBBINS (1976, 1977) considered another kind of energy: more efficient energy released as a result of chemical differentiation related to the cooling of the core. LOPER (1978) also put forward a hypothesis of gravitationally powered dynamo action. Since dense material in the liquid core would be accreted by the inner solid core, rearrangement of liquid material is likely to occur, resulting in the generation of the fluid motion necessary for dynamo action. GUBBINS (1977) pointed out that this kind of energy can be converted to magnetic field energy more effectively than the thermal energy. If the estimated gravitational energy is adopted, a toroidal magnetic field as strong as some hundred gauss is not at all impossible, and LOPER (1978) favors an $\alpha\omega$-dynamo action rather than an α^2-dynamo action for the mechanism of geomagnetic field generation.

As another energy source for a dynamo action, precessional power was proposed by MALKUS (1963, 1968) and also by STACEY (1973). However, ROCHESTER *et al.* (1975) showed that precessional energy available in the core would be less than 10^8 watt, while the energy necessary for a dynamo action is estimated as 10^9–10^{12} watt, depending on types of dynamo models. Therefore, ROCHESTER *et al.* (1975) argued against the precessionally driven dynamo model.

4.8 Non-Dipole Field and Its Westward Drift

4.8.1 Electromagnetic core-mantle coupling

If the core is tightly coupled to the mantle, an observer fixed to the mantle will see no drift of the non-dipole field which is supposed to originate in the near-surface part of the core. If the coupling is viscous, the tidal deceleration of the mantle will cause an eastward drift, contrary to the observed westward one of the non-dipole field (see Section 2.4). No theory that can account for such a westward drift had been proposed until BULLARD *et al.* (1950) put forward a theory based on electromagnetic coupling between the core and the mantle.

As has been the case for an $\alpha\omega$-dynamo action, convective fluid motion gives rise to a differential rotation in the core, resulting in the generation of a toroidal magnetic field in the core. Another toroidal field is also created at the core-mantle boundary because of a differential rotation at the boundary. ROCHESTER (1960) treated a simplified model as shown in Fig. 4.18, in which ω_2 and ω_m are the angular velocities of the outer core and the mantle, respectively. The electrical conductivities in the core and the mantle are denoted by σ and σ_1. The upper mantle for $c < r < d$ is assumed to be non-conducting. The inner core for $r < a$ rotates with an angular velocity larger than ω_2. ROCHESTER (1960) called the interface, $r = a$, the Bullard discontinuity.

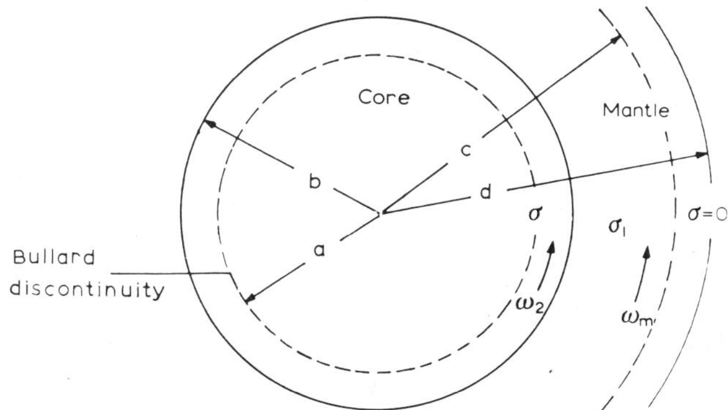

FIG. 4.18. A model of differential rotation at the core-mantle boundary (ROCHESTER, 1960).

In the presence of a dipole field represented by S_1, which is perpendicular to the equatorial section in Fig. 4.18, a T_2 field is generated at the Bullard discontinuity because of the differential rotation. The steady value of the ϕ component of that field is calculated as

$$H^{(a)}_{T_2,\phi} = -\frac{2}{3}M\left(\frac{b}{r}\right)^3 \frac{1 - r^5/c^5}{1 - b^5/c^5}\frac{\mathrm{d}P_2}{\mathrm{d}\theta} \tag{4.89}$$

where M represents the maximum value of $H^{(a)}_{2,\phi}$ at the core-mantle interface ($r = b$). The r and θ components are not generated by the differential rotation.

The slip at the core-mantle interface produces another T_2 field. Its steady ϕ component has been calculated in the case of $\sigma_1/\sigma \ll 1$ (see RIKITAKE, 1966a, pp. 95–102) as

$$H^{(b)}_{T_2,\phi} = \frac{4}{3}\pi\sigma_1(\omega_m - \omega_2)b^2 g_1^0 \left(\frac{d}{r}\right)^3 \frac{1 - r^5/c^5}{1 + (3/2)(b^5/c^5)} \frac{\mathrm{d}P_2}{\mathrm{d}\theta} \tag{4.90}$$

where g_1^0 is the Gauss coefficient.

The mechanical couple exerted on the mantle is generally given by

$$\Gamma = (1/4\pi) \int_b^c \int_0^\pi \int_0^{2\pi} r^3 \sin^2\theta (\mathrm{curl}\, \boldsymbol{H} \times \boldsymbol{H})_\phi \mathrm{d}r\mathrm{d}\theta\mathrm{d}\phi. \tag{4.91}$$

In the steady state, $\Gamma = 0$ and we obtain

$$H_{T_2,\phi} = H^{(a)}_{T_2,\phi} + H^{(b)}_{T_2,\phi} = 0 \tag{4.92}$$

and consequently

$$\omega_m - \omega_2 = \frac{M}{2\pi\sigma_1 b^2 g_1^0}\left(\frac{b}{d}\right)^3 \frac{1 + (3/2)b^5/c^5}{1 - b^5/c^5}. \tag{4.93}$$

As the righthand-side of (4.93) is positive, $\omega_m > \omega_2$, which implies that the core moves westwards relative to an observer fixed to the mantle. Substituting $\omega_m - \omega_2 = 0.2°/yr$ (approximate westward drift velocity for the non-dipole field) and $\sigma_1 = 10^{-9}$ emu (a value widely believed for the lower mantle; see Section 10.4) into (4.93), we obtain $M \approx 0.1$ gauss. Since the maximum value of the T_2 field in the core is supposed to be the order of 10^2 gauss, the above value of M would not be unreasonable.

Thus the theory of electromagnetic coupling between the core and the mantle seems to account for the westward drift of the non-dipole field which is supposed to originate near the core surface. Theories including the non-steady state have further been developed by ROCHESTER (1960, 1962, 1968), ROCHESTER and SMYLIE (1965), RODEN (1963), KAKUTA (1961, 1965), RIKITAKE (1962b), YUKUTAKE (1972), ROBERTS (1972), WATANABE and YUKUTAKE (1975), and others.

Suppose that the toroidal field due to the differential rotation at the Bullard discontinuity changes from its steady state at $t = 0$ as

$$M = M^{(0)} + M^{(1)}(t). \tag{4.94}$$

Similarly the angular velocities can be written as

$$\begin{aligned} \omega_m &= \omega_m^{(0)} + \omega_m^{(1)}(t) \\ \omega_2 &= \omega_2^{(0)} + \omega_2^{(1)}(t). \end{aligned} \tag{4.95}$$

Then the equation of motion for the mantle rotation is given by

$$C_m \mathrm{d}\omega_m^{(1)}/\mathrm{d}t = \Gamma^{(1)}(t) \tag{4.96}$$

where C_m is the moment of inertia of the mantle. Assuming that the diffusion of the magnetic field is completed, we obtain from (4.91)

$$\Gamma^{(1)}(t) = \frac{8}{15} d^3 g_1^0 M^{(1)} - \frac{16}{15}\pi\sigma_1 (g_1^0)^2 b^5 \left(\frac{d}{b}\right)^6 (\omega_m^{(1)} - \omega_2^{(1)}) \frac{1 - b^5/c^5}{1 + (3/2)b^5/c^5}. \tag{4.97}$$

On the other hand, the law of conservation of angular momentum leads to

$$C_m\omega_m^{(1)} + C_c\omega_2^{(1)} = 0 \tag{4.98}$$

where C_c is the moment of inertia of the core. From (4.96), (4.97) and (4.98), we obtain

$$C_m \frac{\mathrm{d}\omega_m^{(1)}}{\mathrm{d}t} + \frac{16 C\pi\sigma_1 b^5}{15 C_c (g_1^0)^{-2}} \left(\frac{d}{b}\right)^6 \frac{1 - b^5/c^5}{1 + (3/2)b^5/c^5}\omega_m^{(1)} = \frac{8}{15} d^3 g_1^0 M^{(1)} \tag{4.99}$$

where C is defined by $C = C_c + C_m$.

If $M^{(1)}$ is a step function, that is $M^{(1)} = 0$ for $t < 0$ and $M^{(1)} = 1$ for $t \geqq 0$, $\omega_m^{(1)}$ is proportional to $e^{-t/\tau}$, where the time constant τ is given by

$$\tau = \frac{15}{16} \frac{C_c C_m}{C} \frac{(g_1^0)^{-2}}{\pi\sigma_1 b^5} \left(\frac{b}{d}\right)^6 \frac{1 + (3/2)b^5/c^5}{1 - b^5/c^5}. \tag{4.100}$$

τ is estimated to be 40 years for likely values of the various parameters involved. However, ROCHESTER (1960) obtained $\tau \sim 25$ years, taking into account the effect of the non-dipole field on the core-mantle coupling.

Suppose that the toroidal field in the core suddenly changes, for instance by 10%, then the length of the day changes by 0.6–0.3 ms within ten years. It is possible, therefore, to assume that irregularity of fluid motion in the core gives rise to a change in the toroidal field and consequently an irregular change in the earth's rotation speed. A change in the toroidal field is also caused by a change in the dipole field, since the distortion of magnetic lines of force representing the dipole field due to the differential rotation is the mechanism of toroidal field generation in the core.

In view of the good correlation between the observed change in the dipole field and the change in the length of the year, YUKUTAKE (1972) estimated theoretically the electromagnetic core-mantle coupling for a sinusoidally oscillating dipole field. The result shows that the rotational velocity of the mantle increases as the dipole moment decreases. However, the dependence of the coupling on the period of oscillatory variation as derived from the model is not in harmony with the observed results (see Section 2.7).

WATANABE and YUKUTAKE (1975) examined the coupling for a core consisting of several shells. In their model, electromagnetic interactions between a dipole field change and shear motions between shells in the core are also taken into account. In this case, the calculated results account for the observed data in both magnitudes and phases of mantle oscillation for periods of 8000, 400 and 65 years, respectively, provided that the dipole field changes only in the surface layer, extending to a depth of only a few hundred kilometers, and the lower mantle conductivity amounts to 10^{-9}–10^{-8} emu. WATANABE and YUKUTAKE (1975) also pointed out that surface motion of the core deduced from their model calculation was in good harmony with changes in the westward drift velocity of geomagnetic secular variation.

It should be mentioned, however, that HIDE (1966, 1970) is of the opinion that the westward drift of the geomagnetic field is caused by M.H.D. (magneto-hydrodynamic) waves in the core. It has been speculated that such a wave is most likely excited by bumps or topographical irregularities at the core-mantle interface. HIDE and MALIN (1970) pointed out the correlation between the global gravitational and geomagnetic potentials that might be a manifestation of M.H.D. interactions between core motions and topographical features of the core-mantle interface, although no outstanding evidence showing the existence of such a bump has been brought to light by seismology.

4.8.2 *Oscillation of quadrupole field*

As the apparent northward shift of the dipole field is closely related to an increase of $|g_2^0|$, NAGATA and RIKITAKE (1963) examined the possibility of M.H.D. oscillation of the S_2 field (quadrupole field), represented by g_2^0, as a physical mechanism of the northward shift. The absolute value of Gauss coefficient g_2^0 underwent a marked increase (with minus sign) over the past 100 years.

Assuming that a steady homogeneous dynamo is in operation in the earth's core, NAGATA and RIKITAKE (1963) discussed the time-dependent behavior of an infinitesimally small magnetic field $\boldsymbol{h}$ and velocity $\boldsymbol{v}$ which are superimposed on the steady state. If the second- and higher-order small quantities are ignored, the equations of magnetic field and fluid motion are given by

$$[\partial/\partial t - (4\pi\sigma)^{-1}\nabla^2]\boldsymbol{h} = \text{curl}\,(\boldsymbol{V}_0 \times \boldsymbol{h}) + \text{curl}\,(\boldsymbol{v} \times \boldsymbol{H}_0) \tag{4.101}$$

$$\rho\partial\boldsymbol{v}/\partial t + 2\rho(\boldsymbol{\omega}_0 \times \boldsymbol{v}) = (4\pi)^{-1}\{\text{curl}\,(\boldsymbol{h} \times \boldsymbol{H}_0) + \text{curl}\,(\boldsymbol{H}_0 \times \boldsymbol{h})\} - \text{grad}\,p \tag{4.102}$$

where ρ, ω_0 and p denote the density, the angular velocity of earth's rotation and the pressure, respectively. $\boldsymbol{V}_0$ and $\boldsymbol{H}_0$ are the velocity and the magnetic field in the steady state. In the above equations, fluid viscosity is neglected. (4.101) and (4.102) are fundamental equations for a study of small-amplitude M.H.D. oscillation in the core, although only approximate solutions have so far been obtained (e.g. RIKITAKE, 1966a, pp. 47–58).

On the basis of the steady state of the Bullard-Gellman dynamo model, NAGATA and RIKITAKE (1963) examined the S_2 field arising from interactions between magnetic fields and fluid motions of various types. An oscillatory solution was found and the period for an infinitely conducting core was estimated as 77 years. It is significant that the possibility of an oscillation of the S_2 field was suggested, although the solution was derived only from a simple model.

4.8.3 *Near-surface motion in the core*

If the core is treated as a perfect conductor, the non-dipole field $\boldsymbol{H}$, its secular variation $\partial \boldsymbol{H}/\partial t$, and the fluid velocity $\boldsymbol{v}$ are controlled by

$$\partial \boldsymbol{H}/\partial t = \mathrm{curl}\ (\boldsymbol{v} \times \boldsymbol{H}). \tag{4.103}$$

Then $\boldsymbol{v}$ can be calculated from values of $\boldsymbol{H}$ and $\partial \boldsymbol{H}/\partial t$ at the core's surface. These values would be estimated by extrapolation from the non-dipole field and its secular variation observed at the earth's surface. VESTINE (1965), VESTINE and KAHLE (1966), and KAHLE *et al.* (1967) applied this method to the data of non-dipole fields and inferred near-surface motion in the core. In these studies, however, no attention was paid to the boundary conditions. Similar research was undertaken by MUTH and BENTON (1981) on the basis of Ohm's law instead of the induction equation.

NAGATA and RIKITAKE (1961) studied the interaction between the T_2 field arising from the differential rotation and S_n^{mc} and S_n^{ms} motions representing convective flows. It was shown that such an interaction gives rise to the S_{n-1}^{mc}, S_{m+1}^{mc}, S_{n-1}^{ms}, and S_{n+1}^{ms} magnetic fields. RIKITAKE (1967a) used this method to estimate the surface motion in the core from the Gauss coefficients. A marked down-flow with a velocity of the order of 0.001 cm/sec has thus been found beneath the Pacific area, while an up-flow was found beneath Africa. HONKURA and RIKITAKE (1972) applied this method to the drifting and standing parts of the non-dipole field. As shown in Fig. 4.19a, surface motion derived for the drifting part is characterized by a large-scale convection of S_1^1 type. Similarly Fig.4.19b shows surface motion for the standing part. Down-flow is evident beneath the Pacific and also beneath the Indian Ocean, while the fluid seems to be upwelling beneath Central or South America. As a whole, the surface motion is more complicated for the standing part than for

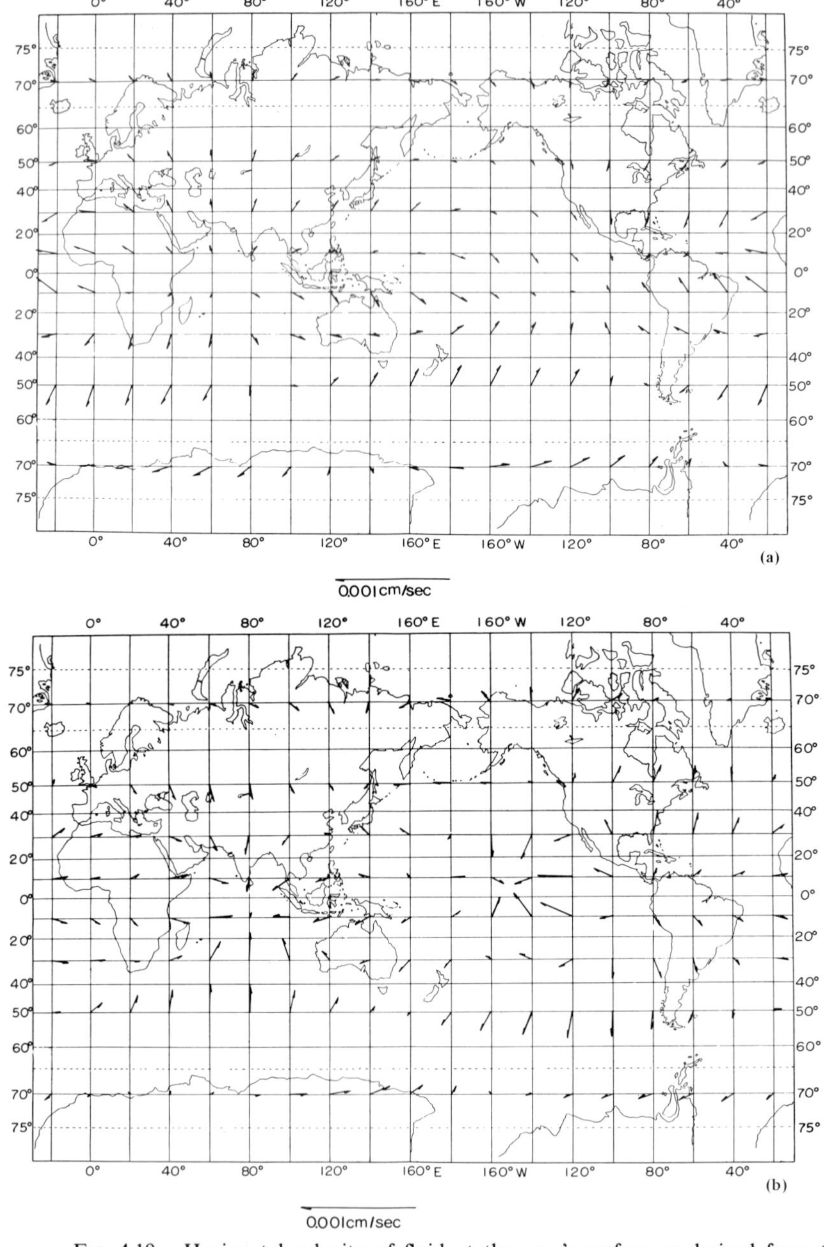

FIG. 4.19. Horizontal velocity of fluid at the core's surface as derived from the drifting (a) and standing (b) non-dipole fields for epoch 1945.

the drifting part.

In an attempt to interpret the finding that the drift velocity is nearly independent of latitude, YUKUTAKE (1981) presumed that the core would have a stratified surface layer and in this layer, toroidal motion would be predominant. Such a spherical shell is supposed to rotate westwards relative to the mantle, causing a westward drifting part. YUKUTAKE (1981) also speculated that in the deeper part of the core convection of Busse type might be predominant and standing magnetic fields would be generated. However, it is not known whether or not such a combination of completely different fluid motions is dynamically possible in a rotating sphere.

CHAPTER 5

CHAPTER 5

MAGNETIC FIELDS OF THE MOON AND PLANETS

The development of space probe technology has enabled us to measure the magnetic field of the moon and major planets. As the internal structure of the moon and planets can be surmised, to some extent, from the results of other measurements, information about the magnetic field of those celestial bodies provides some clue to understanding the origin of the geomagnetic field which is attributed to dynamo action taking place within the earth's fluid core.

5.1 History of Measurement of the Lunar Magnetic Field

When the techniques of rocket and artificial satellite design were so developed, toward the late 1950's, that it was not beyond man's dream to send a sounder to the moon, measurement of the moon's magnetic field was chosen as one of the major studies to be conducted near and on the moon. According to the structure then supposed, on the basis of its size, mass and so on, to be present within the moon, the moon's central core, if any, must be very small. In that case, no dynamo action could effectively occur there, so that most geomagneticians guessed that no large-scale magnetic field would be associated with the moon.

Under the circumstances, it was extremely important and interesting to test whether or not the moon has a magnetic field by direct measurements in its proximity. Should the moon possess a significant magnetic field, it is certain that the dynamo theory of the geomagnetic field would lose its uniqueness.

A Soviet moon rocket named the Lunik 1 was sent to the moon on Jan. 2, 1959. Although a magnetometer was on board the rocket, no significant information about the lunar magnetic field was obtained as the rocket passed by the moon at a distance of several hundred miles.

The Lunik 2, which was subsequently sent to the moon by the Soviet Union on Sept. 12, 1959, measured the magnetic field until it reached an altitude of 55 km above the lunar surface immediately before its collision with the moon. It was reported, however, that no magnetic field exceeding

50–100 γ was observed. It was then concluded that the magnetic moment of the moon is smaller than a 10,000 th of that of the earth. In April, 1966, the Luna 10 rocket from the Soviet Union orbited the moon. It detected a magnetic field ranging from 24 to 40 γ at an altitude of 350 km above the lunar surface. Explorer 35, which was sent by the U.S.A. on July 22, 1967, became a lunar orbiter and approached to an altitude of 800 km above the lunar surface. The probe was equipped with two magnetometers which could measure magnetic fields with an accuracy of $\pm 0.2\,\gamma$. The magnetic field measured by the magnetometers fluctuated over a certain range because of the influence of the field associated with the solar wind and the earth's magnetosphere. It turned out, however, that the magnetic field proper to the moon is so small that it was concluded that the magnetic moment of the moon is smaller than 5×10^{-6} times that of the earth.

Extensive measurement of lunar magnetic field was conducted by the command modules of Apollo 15 and 16 which encircled the moon many times. According to the results obtained by the Apollo 16 command module, the lunar magnetic moment is smaller than the earth's magnetic moment by a factor 1×10^{-8} or thereabout. It was further disclosed that a magnetic field smaller than 0.5 γ is often observed at an altitude of 100 km above the lunar surface. However, fairly intense local fields reaching 2.5 γ were also observed at some places.

Two astronauts on board the Apollo 11 lunar module landed on the moon on July 21, 1969. Many rock samples were brought back to the earth by this and following missions. A brief account of the magnetic properties of lunar rocks will be given later in this chapter.

The Apollo 12 mission, which was launched on Nov. 15, 1969, set a three-component flux-gate magnetometer on the lunar surface. Magnetic surveys on the lunar surface, by means of a portable magnetometer, were even carried out on later missions, and a local field amounting to as much as 320 γ was detected.

The magnetometers set on the moon's surface by the Apollo 12, 14, 15 and 16 missions recorded fluctuations of magnetic field mostly related to the field carried by the solar wind. Analysing these fluctuations and comparing them to the field fluctuations observed by Explorer 35, which was some distance from the moon, it became possible to consider electromagnetic induction within the moon, and so an account of the electrical conductivity in the moon's interior was put forward.

5.2 *Magnetic Moment of the Moon*

On the basis of a number of measurements as described in the last section, the magnetic moment of the moon has been estimated as shown in Table 5.1, on the assumption that the observed field arises from a centered dipole.

As can be seen in the table, the magnetic moment of the moon is negligibly small compared to that of the earth. It can then be concluded that no large-scale planetary magnetic field exists on the moon. This point is very important for inferring the internal structure of the moon.

TABLE 5.1. Magnetic moment of the moon.

Magnetic moment (e.m.u.c.g.s.)	Magnetic moment in units of the earth's moment	Probe	Literature
$<6 \times 10^{21}$	$<1 \times 10^{-4}$	Lunik 2	DOLGINOV *et al.* (1961)
$<(2-4) \times 10^{20}$	$<(2.5-5) \times 10^{-6}$	Explorer 35	SONETT *et al.* (1967), BEHANNON (1968)
$<10^{18}$	$<1 \times 10^{-8}$	Apollo 16 subsatellite	COLEMAN *et al.* (1972)
$<1.1 \times 10^{19}$	$<1 \times 10^{-7}$	Apollo 15, 16 subsatellites	COLEMAN and RUSSELL (1977)

(1 e.m.u.c.g.s. = 10^{-3} Am^2)

5.3 *Magnetic Field Measurement of the Lunar Surface*

DYAL *et al.* (1970a, b, 1974) and DYAL and PARKIN (1971) reported on the magnetic measurements carried out on the moon's surface. The magnetometer set on the lunar surface is one of the elements of a measuring complex, including a seismometer and the like, called the 'Apollo lunar surface experiments package' (ALSEP) which was developed by the Ames Research Center, NASA.

As can be seen in Fig. 5.1 (DYAL and PARKIN, 1971), a flux-gate sensor is

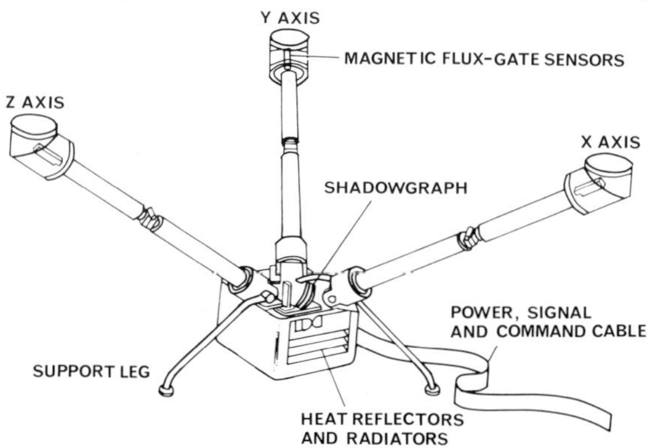

FIG. 5.1. Lunar surface magnetometer (LSM) (DYAL and PARKIN, 1971).

attached at each tip of the three arms which are perpendicular with one another. The sensors are set at a height of about 75 cm above the lunar surface. The package is so constructed that the system of sensors can be rotated by a command from the control center on the earth. The motor and the data processing unit were put in a container protected against temperature change over a wide range by a special thermal insulation device. The astronauts could set the magnetometer in a proper orientation with the aid of a bubble level and an azimuth indicator attached to the upper surface of the container. The power, commands and digital signals were controlled by the ALSEP receiver-transmitter system to which the magnetometer was connected by a ribbon cable.

In the case of the Apollo 14 and 16 missions, the astronauts conducted magnetic measurements by means of a portable magnetometer at considerable distances from the lunar module. Figure 5.2 (Dyal and Parkin, 1971) shows the construction of the portable lunar magnetometer. Three flux-gate sensors are put in a square box which is mounted on a tripod. The sensors are connected to a controlling unit containing the oscillators, amplifiers and so on, by a ribbon cable of about 15 m in length. The measurements were made by the astronauts who transmitted the readings of each component-magnetometer orally, by radio to the control center on the earth. It is important to appreciate that the magnetic fields measured by the portable device are free from the influence of the lunar module.

Dyal *et al.* (1974) presented a summary of magnetic measurements on the lunar surface, including the survey results using the portable magneto-

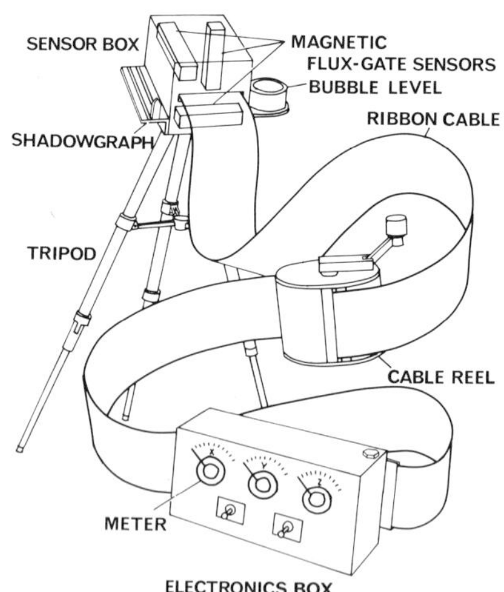

Fig. 5.2. Lunar portable magnetometer (Dyal and Parkin, 1971).

TABLE 5.2. Summary of lunar surface remanent magnetic field measurements (DYAL *et al.*, 1974)

Site	Field magnitude, γ	Magnetic field components, γ		
		Up	East	North
Apollo 16; 8.9°S, 15.5°E				
ALSEP site	235 ± 4	−186 ± 4	−48 ± 3	+135 ± 3
Site 2	189 ± 5	−189 ± 5	+3 ± 6	+10 ± 3
Site 5	112 ± 5	+104 ± 5	−5 ± 4	−40 ± 3
Site 13	327 ± 7	−159 ± 6	−190 ± 8	−214 ± 6
LRV* final site	113 ± 4	−66 ± 4	−76 ± 4	+52 ± 2
Apollo 15; 26.1°N, 3.7°E				
ALSEP site	3.4 ± 2.9	+3.3 ± 1.5	+0.9 ± 2.0	−0.2 ± 1.5
Apollo 14; 3.7°S, 17.5°W				
Site A	103 ± 5	−93 ± 4	+38 ± 5	−24 ± 5
Site C′	43 ± 6	−15 ± 4	−36 ± 5	−19 ± 8
Apollo 12; 3.2°S, 23.4°W				
ALSEP site	38 ± 2	−25.8 ± 1.0	+11.9 ± 0.9	−25.8 ± 0.4

*Lunar Rover.

meter, as reproduced in Table 5.2. It is striking that a magnetic field well over 300 γ is observed at one point. Although it appears that the moon has no associated large-scale planetary field, it now becomes evident that anomalous magnetic fields, probably of local origin, seem to exist on the lunar surface. This point is particularly interesting when one takes into account the fact that some of the rock samples brought back to the earth from the moon possess natural remanent magnetism, as will be mentioned in a later section.

5.4 Subsatellite Measurement of the Lunar Magnetic Field

Explorer 35, which orbited the moon, definitely observed that there are localities on the lunar surface where the magnetization must be unusually intense. Such magnetic anomalies were shown more clearly by the Apollo 15 and 16 command modules. These subsatellites were equipped with two perpendicularly-set flux-gate sensors. It is possible to obtain the magnetic field components from the amplitude and sign of the magnetic field in the spin axis, along with the amplitude and phase in the spin plane.

The magnetic field near the lunar surface is always disturbed because of the magnetic field trapped by the solar wind, except when the moon is in the

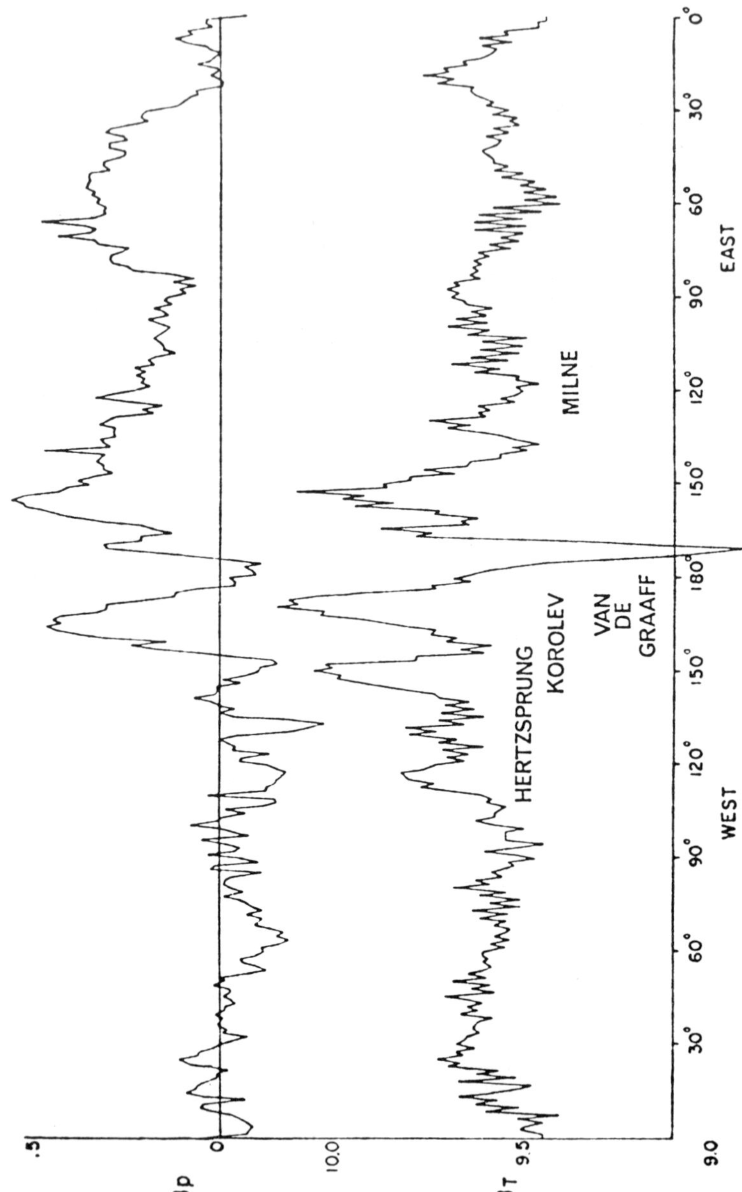

FIG. 5.3. The magnetic anomaly over the Van de Graaff crater on the far-side of the moon (SHARP *et al.*, 1973).

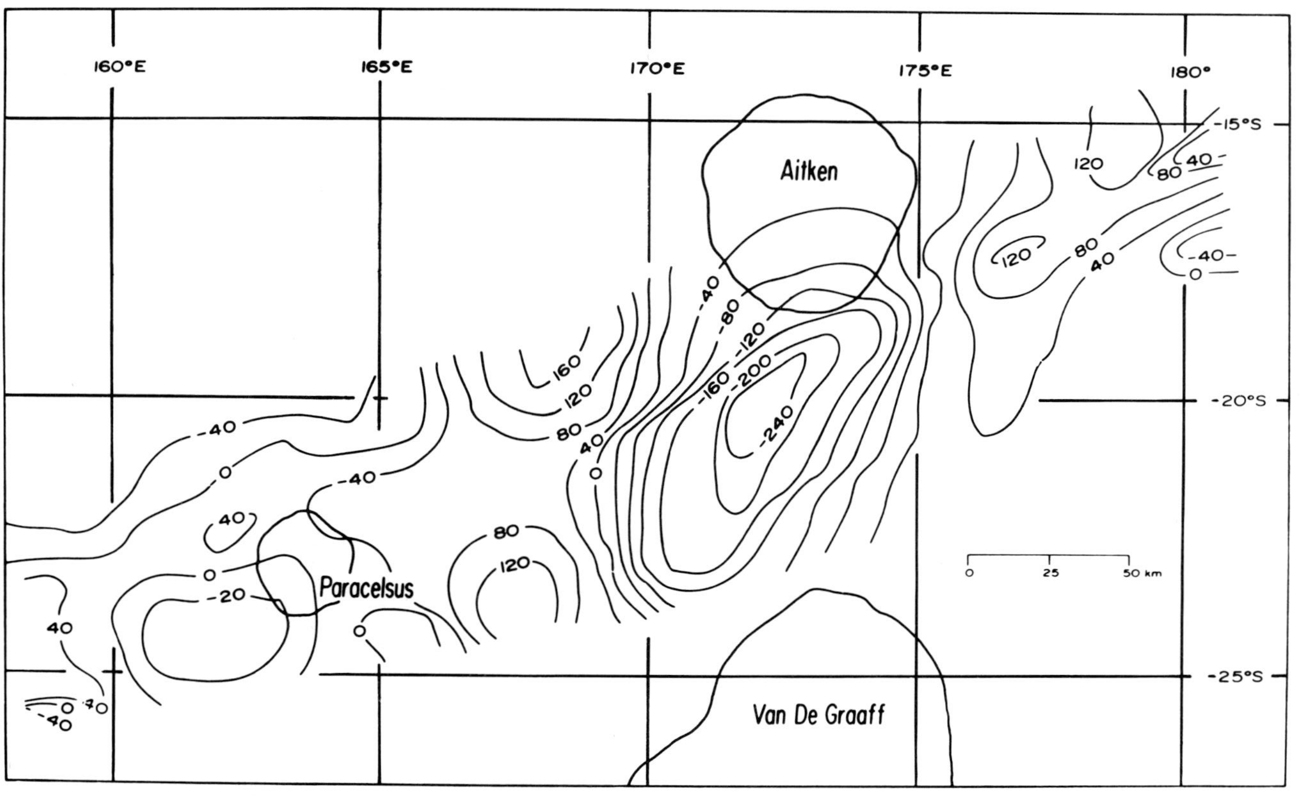

FIG. 5.4. Van de Graaff-Aitken anomaly at an altitude of 67 km (SHARP *et al.*, 1973).

geomagnetic tail. Even in such a case, the disturbing field is generally larger than that at a height of 100 km above the lunar surface. It is possible, however, to observe a large-scale anomaly in the lunar magnetic field when the fluctuations due to the solar wind are small. It is also possible to detect rather small-scale anomalies by taking an average of many observations made over repeated orbits.

Figure 5.3 (SHARP *et al.*, 1973) is the anomaly over the Van de Graaff crater on the far-side of the moon, which can never be seen from the earth. The spatial distribution of the anomaly is also shown in Fig. 5.4 (SHARP *et al.*, 1973) which is obtained on the basis of the 10 consecutive flights at an altitude of 67 km. The field shown in the figure is almost normal to the lunar surface in the vicinity of 180° in longitude. Should the anomaly be interpreted to be caused by a magnetized disk, the radius, thickness and intensity of magnetization would amount to 80 km, 10 km and 6×10^{-5} e.m.u./cm^3 (6×10^{-2} A/m), respectively. Such an interpretation seems to be compatible with the fact that many rock samples brought back from the moon have a natural remanent magnetization of the order of 10^{-5} e.m.u./cm^3.

Figure 5.5 shows the magnetic anomaly in the radial component over the east half of the moon, facing the earth, at a height of 100 km above the lunar surface. It may be said that the magnetic anomaly is stronger over the far side of the moon than over the near side facing the earth. This fact might be correlated with the observation that the topography on the far side is more rugged than that on the near side.

5.5 *Magnetic Permeability of the Moon as a Whole*

It is interesting to compare the field values observed on the moon, measured by the lunar surface magnetometer, to those simultaneously observed by Explorer 35, which is located at a very large distance from the moon. Such a comparison makes it possible to infer the magnetic permeability of the moon as a whole. DYAL *et al.* (1974) and Parkin *et al.* (1974), who used such a data set, amounting to 2,703 in number, concluded that the moon as a whole is paramagnetic or weakly ferromagnetic, the average permeability obtained being

$$\mu = 1.012 \pm 0.006$$

in electromagnetic units. PARKIN *et al.* (1974), who assumed certain temperature and composition distributions within the moon, concluded from the above result that the weight percentages of metallic and ionic iron are 2.5 ± 2.0 and $9.0 \pm 4.7\%$, respectively.

Further studies related to the magnetic permeability of the moon as a

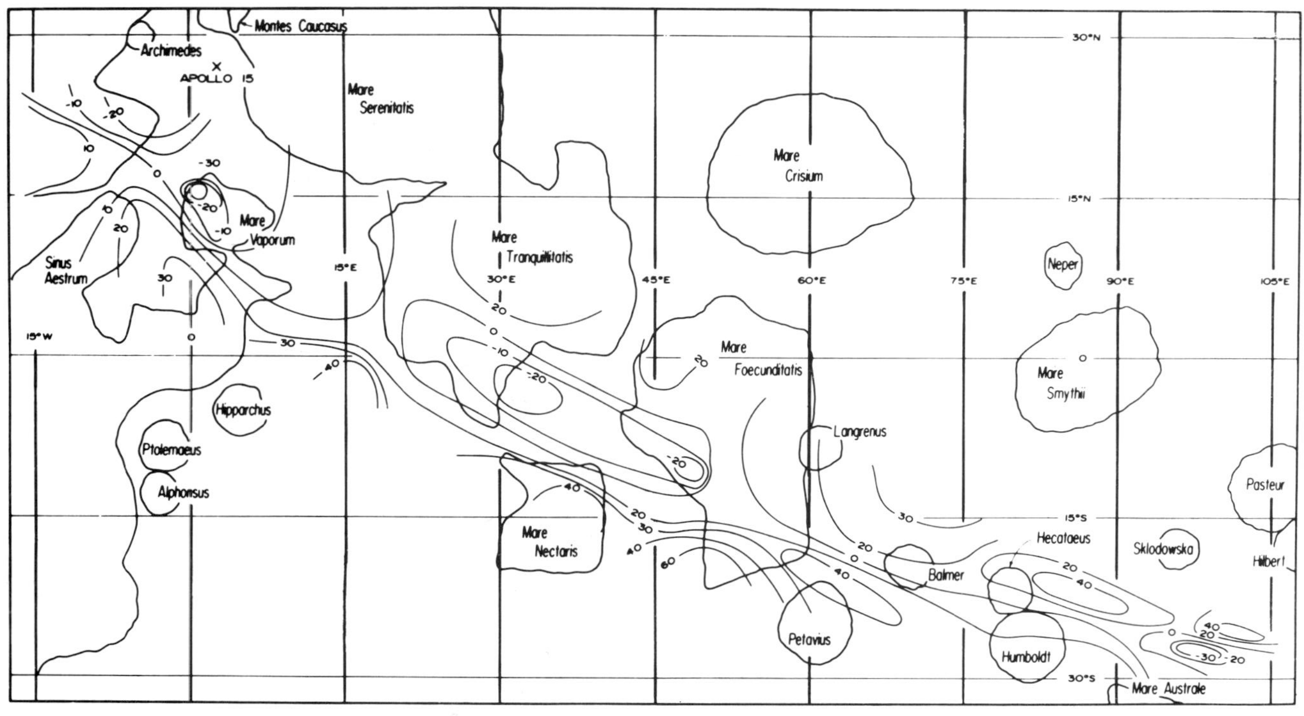

FIG. 5.5. Magnetic anomaly in the radial component over the east-half of the moon, facing the earth, at an altitude of 100 km (SHARP *et al.*, 1973).

whole are made by RUSSELL *et al.* (1974), GOLDSTEIN *et al.* (1976) and others, although no detailed account is given here.

5.6 General Magnetic Properties of the Apollo Rock Samples

Magnetic properties of rock samples brought back to the earth by the Apollo missions have been extensively studied (DOELL *et al.*, 1970; HELSLEY, 1970; NAGATA *et al.*, 1970; RUNCORN *et al.*, 1970; STRANGWAY *et al.*, 1970). Many of the results have also been reviewed by FULLER (1974), STRANGWAY *et al.* (1971, 1974) and NAGATA *et al.* (1972a). All through these studies, it has become apparent that many lunar rock samples have stable natural remanent magnetization (NRM) of the order of 10^{-5} e.m.u./g (10^{-5} Am^2/kg).

One of the most important points in relation to lunar rock magnetism is certainly to see which minerals are responsible for the magnetization measured. Fig. 5.6 (NAGATA *et al.*, 1972a) shows the changes in the saturation magnetization of an Apollo 14 igneous rock sample (NASA No. 14053–48) with temperature. The Curie point, where the ferromagnetism vanishes, is estimated from the graph as 765 ± 5°C, so that it is apparent that the major constituent of the ferromagnetic materials which carry the magnetization, must be metallic iron.

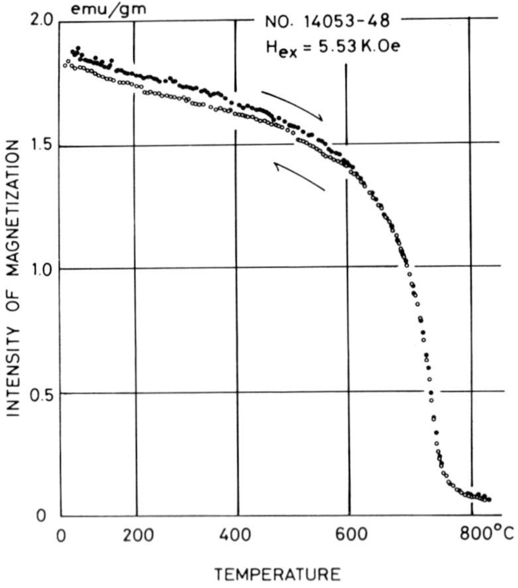

FIG. 5.6. Changes in the saturation magnetization of an Apollo 14 igneous rock sample with temperature (NAGATA *et al.*, 1972a).

For some samples (e.g. No. 14053–35), the saturation magnetization versus temperature curves differ considerably from one another on heating and on cooling. Such a discrepancy is thought to be caused by the α-γ transition of Fe–Ni alloys. A typical example of this is the Apollo 11 breccia sample (STRANGWAY *et al.*, 1970).

NAGATA *et al.* (1974) presented a histogram of Curie points for lunar rocks from which we see that the materials bearing the magnetization of lunar rocks are mostly metallic iron and Fe–Ni alloys. It is possible to estimate the percentage of metallic iron contained in a lunar rock sample by comparing the saturation magnetization actually measured to that of iron. According to NAGATA *et al.* (1972b), the weight percentage amounts to 0.05–1.0% for fines and breccias. For igneous rock samples, it ranges from 0.01 to 0.1%, except for No. 14053. This value seems a little smaller than that obtained from the magnetic permeability as a whole in the last section.

5.7 *Natural Remanent Magnetism of the Apollo Rock Samples*

5.7.1 *Results of measurement and test*

As stated in the foregoing sections, it has become clear that lunar rocks possess NRM. Table 5.3 is a summary of the NRM and its stability in lunar

TABLE 5.3. Natural remanent magnetization and its stability in lunar materials (NAGATA *et al.*, 1972a).

Sample No.	NRM (e.m.u./g)	$\bar{H}_0$ (Oe)	$\bar{H}$ (Oe)	h (Oe)
(Igneous rocks)				
10024-22	7.5×10^{-6}	25	30	14
12053-47	2.3×10^{-6}	8	7	9
12038-29	8.1×10^{-6}	55	40	17
12038-32	6.3×10^{-6}	12	5	15
14053-31	2200.0×10^{-6}	19	80	50
14063-47	14.1×10^{-6}	145	150	47
14310-159	2.4×10^{-6}	>25	>25	12
(Breccias or clastic rocks)				
10021-32	15.0×10^{-6}	35	40	—
10048-55	56.0×10^{-6}	~400	>100	—
10085-16	153.0×10^{-6}	~1400	>500	—
14047-47	8.2×10^{-6}	45	30	7.5
14301-65	41.2×10^{-6}	48	40	23
14303-35	131.2×10^{-6}	18	40	29
14311-23	8.1×10^{-6}	8	10	14

(1 e.m.u./g = 1 Am^2/kg)

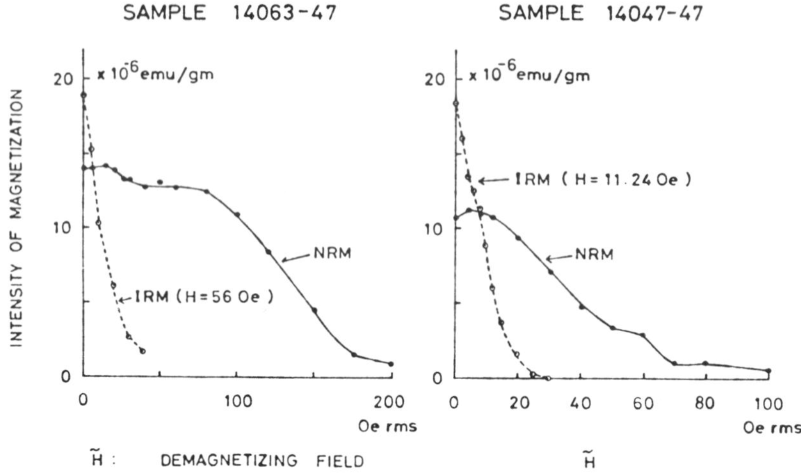

FIG. 5.7. a.c. demagnetization of lunar rock samples (NAGATA *et al.*, 1972a).

rock samples as summarized by NAGATA *et al.* (1972a). When a.c. demagnetization is applied to a lunar rock sample, the amplitude of the a.c. magnetic field at which the NRM falls by a factor $1/e$ is denoted by $\bar{H}_0$. $\bar{H}$ denotes the amplitude of the a.c. magnetic field when the direction of NRM is almost unchanged. Meanwhile, h denotes the magnetic field which is required to produce an isothermal remanent magnetization (IRM) with a magnitude equal to the NRM.

In Table 5.3, we see that there are samples for which $\bar{H}_0$ and $\bar{H}$ are smaller than 60 Oe. In some extreme cases, they are smaller than 20 Oe. Such magnetizations, with a very small coercivity, indicate that the origin of NRM of the sample may not be a thermal-remanent magnetization (TRM). As can be seen in Fig. 5.7 (NAGATA *et al.*, 1972a), the NRM is much more stable than the IRM. However, it is hard to interpret the NRM as an IRM or artificial contaminating magnetization acquired by the sample after it was brought back to the earth.

Figure 5.8 (NAGATA *et al.*, 1972a) is an example of thermal demagnetization of a lunar rock sample. Each point representing the magnetization in the graph is obtained by heating the sample up to the relevant temperature and cooling it down to room temperature in a zero magnetic field. Judging from the curve in the figure, it seems likely that the magnetization consists of two components; the one which vanishes by demagnetization at relatively low temperatures and the other which is fairly stable up to rather higher temperatures. The latter component may be accounted for by a TRM. It is surmised that the highly stable NRM for Nos.

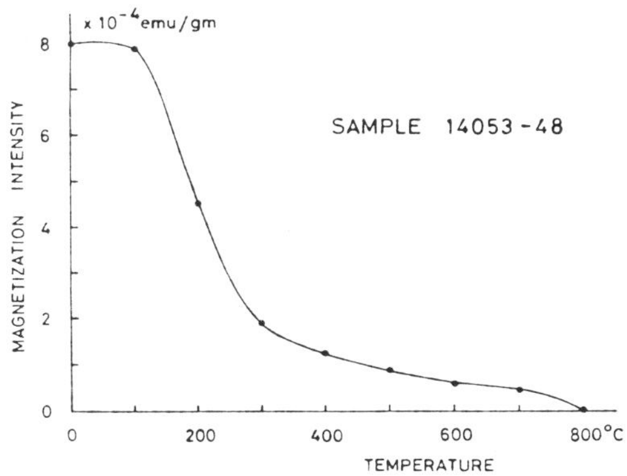

FIG. 5.8. Thermal demagnetization of a lunar rock sample (NAGATA *et al.*, 1972a).

10085–16 and 10048–55 may have been acquired as TRM.

In order to see the extent of possible magnetic contamination, a test for a demagnetized lunar rock sample was made during the Apollo 16 mission (STRANGWAY *et al.*, 1974). The sample proved to have acquired an IRM, which was likely to be produced in a magnetic field of 10–20 Oe, during its round trip to the moon. It may be said, therefore, that a NRM, that can be demagnetized by an a.c. magnetic field amounting to 10 Oe or thereabout, can hardly be regarded as an intrinsic NRM in a lunar rock sample.

5.7.2 Origin of magnetization of lunar rocks

On the basis of the studies as summarized in the last subsection, possible origins of lunar rock magnetization can be attributed to the following three mechanisms:

(1) Magnetic contamination during transportation, careless handling and so on.

(2) Secondary magnetization on the lunar surface.

(3) Primary magnetization when the rock was formed.

Much of mechanism (1) has already been presented in the last subsection. Samples that are likely to have been contaminated should be rejected.

As for mechanism (2), a number of possible causes of secondary magnetization have been proposed. Acquisition of magnetization by irradiation by cosmic rays as proposed by BUTLER and COX (1971) and others is one theory. According to this hypothesis, lattice defects are produced in lunar rocks under cosmic-ray exposure, so stresses arise in the rocks. Such stresses may convert IRM into a magnetization of high coercivity. It is reported that the coercive force of a lunar rock sample exposed to a neutron beam, of intensity about the same as that on the lunar surface, is increased by a factor of 16%.

FULLER (1974) pointed out, however, that it seems difficult to account for the NRM of lunar rocks only on the basis of this mechanism.

According to FULLER (1972), the daily temperature variation on the lunar surface ranges from 400 to 250 K. Accordingly, it may be possible that lunar surface rocks acquire a partial thermo-remanent magnetization (PTRM) (NAGATA, 1961). It is also concluded by FULLER (1972) that such PTRM is not enough to account for most NRM of lunar rocks.

A number of researchers (e.g. NAGATA *et al.*, 1970) suggested that NRM of lunar rocks must have been produced by a shock such as caused by meteorite collision against the lunar surface in a weak magnetic field. Such magnetization may be called the shock remanent magnetization (SRM). CISOWSKI *et al.* (1973) conducted an experiment to give a shock of 10–100 kbar to the lunar soil. They found that a fairly stable SRM is produced by the solidification due to collision. It is reported that a magnetization of the order of $10^{-4} - 10^{-5}$ e.m.u./g is acquired by shock experiments in a magnetic field of the order of $10^2 - 10^3$ γ. Tests of the stability of the SRM thus obtained are by no means complete, but it appears that SRM may be one possible candidate for NRM in lunar rocks.

There are ample instances of NRM of igneous and breccia lunar rocks that make us believe that they are doubtless produced as TRM on the lunar surface when the lunar rocks were subjected to cooling from a temperature higher than their Curie points. It may therefore be said that TRM is highly important for primary magnetization of lunar rocks. But magnetizations of secondary origin superposing on the primary magnetization play an important role for some samples. SRM, especially, seems particularly important.

5.7.3 Paleointensity of the lunar magnetic field

The technique for inferring the intensity of the geomagnetic field in ancient times, as used extensively in paleomagnetic studies of terrestrial rocks (see Section 3.2), can also be applied to lunar rocks. In spite of considerable difficulties in conducting actual experiments due to the effects of oxidation and other causes, a few results have so far been reported. STRANGWAY *et al.* (1974) reported that the NRM of No. 15498–36 must have been acquired in a magnetic field of 2,100 $\pm$ 80 γ. Meanwhile, a value amounting to 1.20 Oe was reported for No. 62235–53 by COLLINSON *et al.* (1973).

It is sometimes possible to find the approximate intensity of the lunar magnetic field by comparing the NRM of a lunar rock sample to the saturation IRM (SIRM). The intensity of the lunar magnetic field when the NRM was formed may thus be guessed to some extent. Although the intensities obtained scatter over a wide range, many rocks seem to have been magnetized in a magnetic field of the order of 10^3 γ.

5.8 *Origin of the Lunar Magnetic Field*

We are now in a position to make some conjecture on origin of the lunar magnetic field, based on what we have learned in the foregoing sections in this chapter. The most important points for such a conjecture can be summarized as follows;

(1) The moon has no planetary field of large scale.

(2) There are local anomalous magnetic fields on the lunar surface amounting to 300 γ or thereabout at maximum, according to present knowledge.

(3) Some of the local anomalies cover an area of several tens of kilometers in their spatial extent, as revealed by the subsatellite observations.

(4) Many of the lunar rock samples possess a NRM which seems likely to have been acquired in a magnetic field of the order of 10^3 γ.

5.8.1 *Magnetization by an external magnetic field*

(1) Solar wind hypothesis

Magnetic fields near the sun's surface may sometimes be frozen in highly-conducting plasma streams or the solar wind, and carried to the moon. Such magnetic fields might have magnetized lunar rocks some time in the past when they happened to be intense. As it is highly likely that the direction of such fields fluctuates over a wide range, no magnetization in a steady magnetic field can be predicted by the solar wind hypothesis.

(2) Geomagnetic field hypothesis

It is assumed in this hypothesis that the moon was magnetized either when the geomagnetic field was extremely intense or when the moon's orbit was quite close to the earth. Since the ages of lunar rocks, which have an appreciable NRM, range over several hundred million years, the authors feel that this hypothesis is hard to believe.

5.8.2 *Magnetization by an internal magnetic field*

(1) Dynamo theory

STRANGWAY *et al.* (1971), RUNCORN *et al.* (1971), RUNCORN (1978) and others proposed that a sort of dynamo action must have prevailed in the moon's core in order to account for the NRM of lunar rock samples, which are likely to have been magnetized in a magnetic field of the order of 10^3 γ. As the moon has at present no planetary magnetic field, it is clear that no dynamo action is currently operating. It is surmised, however, that dynamo action was going on in the moon's central core, when its physical state was different from that at present, over 10^9 years or so ago.

It is hard to evaluate the validity of such a dynamo mechanism in ancient times because nothing is known about the internal structure of the moon at that time. LEVY (1972) pointed out, however, that a comparison between the moon and the earth leads us to a conclusion that, in order to have a workable dynamo, the rotation velocity of the moon would be so large that it would result in a stress which exceeds the mechanical strength. In spite of this, no mechanism that adequately explains the NRM of lunar rock samples and related effects can be found.

RUNCORN *et al.* (1977), who support the lunar dynamo theory, supposed that the moon had a systematic magnetic field for a period from the epoch a few hundred million years after the birth of the moon to that about 320 million years before present. In order to have a heat source sufficient to drive a dynamo in a core of rather small size, it is speculated that an element with atomic number 116–124 must have existed in the lunar core. Such a speculation is interesting although there is no way to prove it.

When a uniform shell is magnetized by a magnetic field originating inside the shell, the magnetization does not produce a magnetic field outside the shell. If the moon was once magnetized by such an internal magnetic field, probably due to the dynamo action, which later ceased to operate at an early stage, the lunar crust may have some magnetization even now, although no field is observed outside the moon. RUNCORN (1975) speculated that this is the reason why the moon has no planetary field in spite of the fact that lunar rocks possess NRM.

If one admits such a magnetization of the lunar crust, we may well expect some local magnetic fields over portions of the moon's surface where the topography deviates from a sphere. Such anomalies may therefore be observed over lunar craters where the crust is broken by meteorite collision.

(2) Multiple shallow dynamo and thin-shell dynamo

MURTHY and BANERJEE (1973) proposed a hypothesis that a kind of dynamo action was at one time effective in a fluid mass of Fe–FeS alloy. There were a number of such masses near the lunar surface, the orders of spatial extent and time constant of those fluid bodies being supposed to be 100 km and 230 years.

SMOLUCHOWSKI (1973) studied a dynamo action in a thin layer of molten basalt. The dynamo is supposed to be driven by the precessional motion of the moon and the tidal force due to the earth although it is not entirely clear that such mechanisms are really workable.

5.8.3 Fossil magnetization

RUNCORN and UREY (1973) assumed that the moon was magnetized by a solar magnetic field of the order of 20 Oe in the early period following the moon's birth. It is supposed that the moon was surrounded by a cloud in which contraction could have given magnetic field enhancement. The method of acquisition of magnetization is something like that for IRM and DRM (detrital remanent magnetization, see Subsection 3.1.3) in the case of terrestrial rocks. It is further assumed that the internal temperature of the moon has not exceeded the Curie point of rocks since then, so that it is possible to assume that the NRM of lunar surface rocks is formed in such a magnetic field.

Probably, such a fossil magnetization hypothesis was best worked out by STRANGWAY *et al.* (1974), who also assumed a solar magnetic field of the order of 20 Oe. They pointed out that it is possible to have had a magnetic field of the order of 1,000 γ at some time on the lunar surface. As the temperature within the moon went up as time went on, this magnetic field became gradually smaller and it finally vanished, but the NRM of lunar surface rocks remained.

5.8.4 Transient local field

(1) Unipolar generator

NAGATA *et al.* (1972b) pointed out a possibility that TRM of lunar surface rocks may be produced in a magnetic field arising from the so-called unipolar generator (SONETT and COLBURN, 1968). When the moon is electrically conducting, the induced electric field generated by the interaction between the solar wind plasma and the magnetic field around the moon is short-circuited forming a unipolar generator. In this case a toroidal magnetic field will be created. Such an effect might be important for a molten, highly-conducting magma reservoir, if any, near the lunar surface. It is difficult, however, to think of generating a steady magnetic field on the basis of this mechanism only.

(2) Plasma generation by collision

CAP (1972) suggested that a flow of volcanic ash may give rise to ionization, and that the motion of charged particles may generate a magnetic field of the order of 1,000 γ. That such a process may be workable to some extent seems to be proved by the fact that the lunar surface magnetometer recorded some magnetic disturbances when the lunar module took off from the lunar surface. It is not known, however, whether all the NRM of lunar rocks can be accounted for by this mechanism. The possibility of ionization by collisions of meteorites with the lunar surface has also been suggested. If this occurs, production of a magnetic field might be effected, although nothing concrete can be said about the detailed process.

(3) Thermoelectric current

DYAL *et al.* (1971) put forward a hypothesis that thermoelectric currents might have been driven by an electromotive force due to temperature differences when the lunar surface was covered by molten magmas. Such currents are capable of producing a magnetic field in which the NRM of lunar rocks must have been acquired. The authors feel, however, that there is no way to check such a hypothesis.

5.8.5 Zero field magnetization

A remanent magnetization of the order of 10^{-5} e.m.u./g was obtained by an experiment in which the soil taken by the Apollo 11 mission was deposited in zero field conditions (FULLER, 1972). Such a phenomenon is well known in paleomagnetic study as a random moment. It is doubtful, however, whether such a mechanism can account for the NRM of lunar rocks.

It is also hypothesized that SRM may be produced even in the case of a zero magnetic field although it is hardly expected that such a mechanism prevailed widely on the lunar surface.

It is not known which theories mentioned in the last five subsections can be applied to account for the lunar magnetism. When we exclude hypotheses which involve too many unknown factors, it is natural to reach a conclusion that there may have been a sort of dynamo in the moon some time in the past.

5.9 Magnetic Field of Planets

5.9.1 Summary of space probe observations

In view of the fact that the moon has no planetary magnetic field, though there are some local fields on the lunar surface, and many of lunar rock samples possess NRM, whether or not other planets have a magnetic field became one of the most interesting topics of planetology in recent years.

At present, some knowledge about the magnetic fields of Mercury, Mars, Venus, Jupiter and Saturn have been obtained through intensive space probing programs in the U.S.A. and the U.S.S.R., although it is beyond the scope of this book to present a detailed account of space probe techniques. RUSSELL (1979) presented a comprehensive review of planetary magnetism. What follows in this section largely relies on this review, as well as on other results of later observations.

The most up-to-date estimates of planetary magnetic moment are summarized in Table 5.4. On the basis of the Mariner 10 observations, RUSSELL (1979) concluded that the magnetic moment of Mercury may amount to 6 ± 10^{22} e.m.u.c.g.s. (6 ± 10^{19} Am^2). Some other references (e.g. NESS *et al.*, 1975) for the same problem may be found in RUSSELL (1979).

RUSSELL (1976), who reanalysed the Venera 4 data, estimated the magnetic moment of Venus as 6.5×10^{22} e.m.u.c.g.s., which more or less coincides with the estimates by others, as can also be seen in RUSSELL (1979). In 1978, observations by the Pioneer Venus orbiter led to the conclusion that the

TABLE 5.4. Magnetic moments of planets.

Planet	Magnetic moment (e.m.u.c.g.s.)	Magnetic moment in units of the earth's moment	Probe	Year	Literature
Moon	$<1.1 \times 10^{19}$	$<1 \times 10^{-7}$	Apollo 15 & 16 subsatellite	1972	COLEMAN and RUSSEL (1977)
Mercury	$6 \pm 2 \times 10^{22}$	$7 \pm 2 \times 10^{-4}$	Mariner 10	1974	RUSSELL (1979) and others
Mars	2.5×10^{22}	3.2×10^{-4}	Mars 3 & 5	1972, 1974	DOLGINOV *et al.* (1976), RUSSELL (1979)
Earth	7.9×10^{25}	1	MAGSAT	1979	LANGEL *et al.* (1980b)
Venus	6.5×10^{22}	8.2×10^{-4}	Venera 4	1967	RUSSELL (1976, 1979)
Jupiter	1.55×10^{30}	1.96×10^{4}	Pioneer 11	1974	RUSSELL (1979) and others
Saturn	$4.3 \pm 0.2 \times 10^{28}$	$5.40 \pm 0.30 \times 10^{2}$	Pioneer 11	1979	ACUÑA and NESS (1980)

(1 e.m.u.c.g.s. = 10^{-3} Am^2)

TABLE 5.5. Radius, density and rotation speed of planets.

Planet	Equatorial radius (km)	Mean density (g/cm^3)	Rotation speed (day)
Moon	1,738	3.35	27.32
Mercury	2,440	5.41	59
Venus	6,060	5.24	245
Earth	6,378	5.52	0.9973
Mars	3,397	3.94	1.0260
Jupiter	71,460	1.33	0.4101
Saturn	60,400	0.68	0.4264

magnetic moment of Venus is smaller than 10^{22} e.m.u.c.g.s. (Russell *et al.*, 1979, 1980), probably something around 5.5×10^{21} e.m.u.c.g.s.

The only U.S. measurements of the magnetic field of Mars were made on the occasion of the Mariner 4 flyby in 1965. Although it was concluded by Smith *et al.* (1965) that the magnetic moment of Mars is less than 3×10^{-4} in units of the earth's moment, no accurate estimate could be made because the probe passed by the planet at too great a distance. It appears that the estimate by Dolginov *et al.* (1976), who analysed the data taken by Mars 3 and 5, is more reliable. The magnetic moment of Mars is then estimated as 2.5×10^{22} e.m.u.c.g.s. (Dolginov *et al.*, 1976; Russell, 1979).

It has long been known, from observations of the polarization of radio waves from Jupiter, that the planet has a strong magnetic field. Such a field was observed directly by Pioneer 11 in 1974. Acuña and Ness (1976) estimated the magnetic moment of Jupiter as 1.5×10^{30} e.m.u.c.g.s. Russell (1979) adopted 1.55×10^{30} e.m.u.c.g.s. as the value agreed by two groups. The magnetic dipole of Jupiter is located highly eccentrically relative to the center of the planet.

Acuña and Ness (1980), who analysed the Pioneer 11 data taken in September, 1979, reported that the magnetic moment of Saturn amounts to $4.3 \pm 0.2 \times 10^{28}$ e.m.u.c.g.s. which is smaller by a factor 5 than the value hitherto estimated from radio wave observation. The tilt of the dipole axis is surprisingly small and is estimated as $2° \pm 1°$ against the rotation axis. No data taken by the Voyagers 1 and 2 missions were available when this manuscript was being prepared.

5.9.2 Physical significance of planetary magnetic field in relation to the internal structure of planets

Jupiter and Saturn are entirely different from the terrestrial planets, i.e. Moon, Mercury, Venus, Earth and Mars, in their size and constitution. For these gigantic planets, the main constituents are H_2 and He and so a core that consists of metallic hydrogen may exist. Turbulence also prevails there, so that dynamo action may well be effective; it is no surprise that these planets have strong magnetic fields.

As can be seen in Table 5.5, in which various constants related to the planets are summarized, among the terrestrial planets, only Venus has a size comparable to that of the earth. It should be seriously taken into account, however, that the rotation speed of Venus is extremely small. Even if Venus has a fluid core, therefore, it may not be possible to have workable dynamo action because the Coriolis force is not large enough. This might be the reason why Venus has no appreciable planetary field.

As for the Moon and Mars, the core, if any, must be fairly small, judging from their size. In such a case, the dissipation of a magnetic field becomes large, resulting in it being difficult to sustain dynamo action.

Although much remains to be studied about the magnetic field of Mecury, a magnetic field of a few hundred gammas was observed and a shock-front related to its magnetosphere was also detected. It is surmised that there was a dynamo in Mercury when its rotation speed was high, some time in the past. Accordingly, a magnetized crust has been postulated by SHARPE and STRANGWAY (1976) and STEPHENSON (1976).

In summary, it has become clear that the earth is the only one among the terrestrial planets which has a large-scale planetary magnetic field. This may be the result of a combination of favorable conditions: the size and electrical conductivity of the core and the earth's rotation speed. It may thus be said that the geomagnetic field is maintained by a dynamo under rather special conditions.

5.9.3 Magnetic Bode's law

By analogy with Bode's law, which indicates the regularity of mean distance of planets from the sun, the so-called magnetic Bode's law has been proposed (e.g. RUSSELL, 1978, 1979; ACUÑA and NESS, 1980). Both the laws are only empirical, no physical explanation having been postulated.

In the case of the magnetic Bode's law, the logarithmic magnetic moment of each planet measured in units of that of the earth is plotted against the logarithmic angular momentum, also measured in units of that of the earth. Figure 5.9 is such a graph drawn on the basis of the data given in Table 5.4, along with the known angular momenta of the planets. It appears that the plots for Mars and Mercury deviate considerably from the linear relation which otherwise seems, very roughly, to hold.

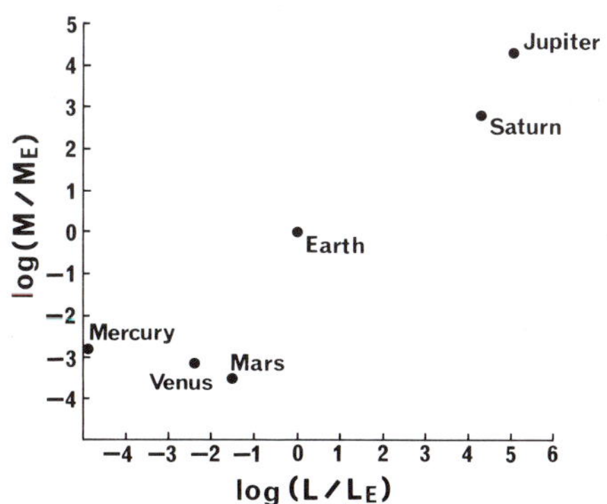

FIG. 5.9. Magnetic Bode's law from the most up-to-date data.

CHAPTER 6

CHAPTER 6

LOCAL GEOMAGNETIC ANOMALIES

The distribution of the geomagnetic field over the earth is actually so complicated because of anomalies of local nature that its representation in terms of spherical harmonics is by no means adequate. In view of the limited spatial extent of local anomalies, their detailed features can be investigated separately from the global geomagnetic field by carrying out some kind of intensive geomagnetic survey.

6.1 Magnetic Anomalies on Land

6.1.1 Aeromagnetic survey and magnetic anomalies

In studies of magnetic anomalies on land, data obtained from magnetic measurements on the earth's surface are generally inappropriate because of the irregular surface topography. Therefore, measurements at a fixed altitude by a proton precession magnetometer installed in an aircraft are now usual. This means of measurement of local geomagnetic field has been called the aeromagnetic survey. Extensive aeromagnetic surveys have been carried out in Canada, the U.S.S.R., the U.S.A., Japan, and many other countries, and regional aeromagnetic maps have been derived from the observed data. There is a variety of purposes for aeromagnetic surveys; investigation of crystalline basement, mafic intrusives, fault and fracture zones, in addition to mineral exploration and geothermal studies.

Many reports have been issued on mapping of the crystalline basement, which is usually covered by non-magnetic or weakly magnetized sediments and sedimentary rocks (e.g. HINZE, 1979). In such cases, the aeromagnetic method is often combined with the gravity method. Magnetic studies of core samples from drill holes sometimes provide additional information. Marked anomalies of high amplitude and small spatial extent are often found in association with intrusive igneous rock bodies generally having strong magnetization.

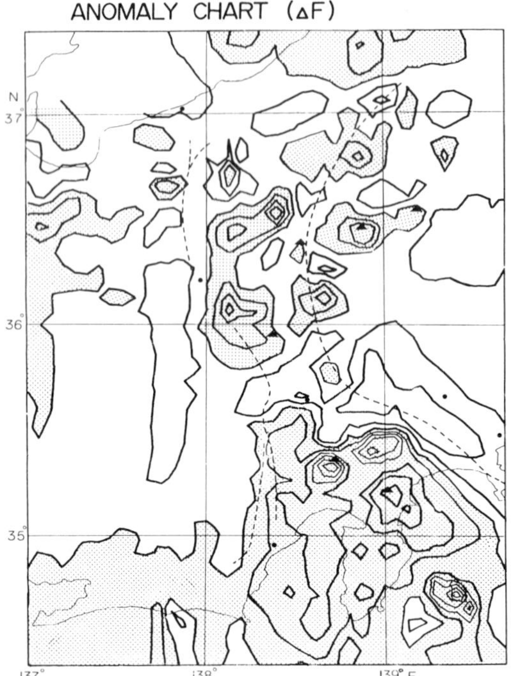

FIG. 6.1. Total intensity anomaly map in the central part of Japan. Stippled portions indicate positive anomalies. The contour interval is 100 gammas (ISHIKAWA, 1979).

In the central part of Japan, where a tectonic zone called the Fossa Magna extends in the north-south direction, the Geographical Survey Institute conducted an aeromagnetic survey at an altitude of 3 km. The data from this survey were analyzed by ISHIKAWA (1979) and a map of the total intensity anomaly was derived as shown in Fig. 6.1. HAGIWARA (1980) pointed out a good correlation between the pattern of magnetic anomalies and that of Bouguer anomalies of short wave-length and compared the magnetic anomaly map with the pseudomagnetic anomaly map which was derived from calculation by using the gravity data. On the basis of the method proposed by LOURENCO and MORRISON (1973), ISHIKAWA (1979) also derived anomaly maps for the three components from the total intensity data. A detailed interpretation of anomalies in terms of geological or tectonic features, however, has not yet been undertaken.

A correlation between a local magnetic anomaly and local seismicity has sometimes been recognized (e.g. HINZE, 1979). It is not unlikely that intrusives are associated, more or less, with present tectonic activity or an ancient tectonic structure which may be reactivating.

Linear patterns have in some cases been recognized in magnetic anomaly maps. Such magnetic lineaments are often associated with faults, metamorphic belts, fracture zones, and other linear tectonic features. A metamorphic belt extending over 1,800 km in a nearly north-south direction was disclosed partly by means of magnetic lineaments in the North American Central Plains (LIDIAK, 1971; COLES *et al.*, 1976). This metamorphic belt is also supposed to be the origin of an electrical conductivity anomaly called the North American Central Plains conductivity anomaly (ALABI *et al.*, 1975; see Subsection 12.4.2). Another notable example of magnetic lineament is the one found in the Appalachian basin. This lineament extends about 1,600 km in the north-south direction from Alabama to New York and coincides well with the gravity anomaly found in this region. This anomaly has been interpreted as indicating a boundary of the Grenville basement province (KING and ZIETZ, 1978).

Magnetic minerals such as titanomagnetite contained in crustal rocks are known to lose their magnetism at a temperature higher than the Curie point. It is expected, therefore, that magnetic anomaly data provide information on the Curie point isotherm in the crust. In areas of shallow Curie point isotherm, heat flow would be high and geothermal studies there might be promising. In this respect, magnetic surveys aimed at investigation of the Curie point isotherm have drawn much attention.

In the western United States, maps of Curie point depth were derived from spectral analyses of magnetic anomaly data (BHATTACHARYYA and LEU, 1975; BYERLY and STOLTZ, 1977; SHUEY *et al.*, 1977; SMITH *et al.*, 1977). In the transition zone between the Basin and Range and the Colorado Plateau, a contrast in the isotherm depth between these areas was found; the isotherm is shallower in the Basin and Range area (SHUEY *et al.*, 1973). This result is in good agreement with an electrical conductivity anomaly in this area (SCHMUCKER, 1964, 1970); that is, a highly conducting zone exists at a shallower depth in the Basin and Range area, implying higher temperature there (see Subsection 12.4.2). Heat flow is also higher in the Basin and Range area (ROY *et al.*, 1972). BHATTACHARYYA and LEU (1975) and SMITH *et al.* (1977) pointed out that magnetic anomaly data at the Yellowstone National Park could be interpreted as indicating a shallow location, of the order of 10 km, of the Curie point isotherm there.

In shield areas, the Curie point isotherm is deeper than the Moho discontinuity and in this case magnetic anomalies are related to the thickness of the crust. In fact, HALL (1974) showed that there is a good correlation between the observed magnetic anomaly and crustal thickness in the Canadian shield area. A similar correlation was also pointed out for the Ukranian shield area (KRUTIKHOVSKAYA and PASHKEVICH, 1977). In the southern part of the Siberian platform, the depth to the Curie point isotherm was determined from magnetic anomalies as 32.5 km, while it is located closer to the earth's surface, 18.5 km depth, beneath the Baikal rift zone (NOVOSELOVA, 1978). Although this rift zone may be a structure related to a kind of new crust creation, magnetic lineations such as the ones found in the oceanic crust (see Subsection 6.2.1) have not been observed there.

6.1.2 Magnetic anomalies over volcanoes

The rocks comprising a volcano often have strong magnetization; 0.01–0.05 e.m.u./cm^3 (10–50 A/m) for basalt and about one-tenth of these values for andesite (e.g. NAGATA, 1961). When a lava flow solidifies in the geomagnetic field, magnetization is acquired in the direction of the geomagnetic field. Therefore, relatively new volcanoes, mostly of Quaternary age, should be magnetized as a whole in the direction of the present geomagnetic field, if the intensity of magnetization can be considered to be uniform within each volcano.

,At Volcano Mihara, a basaltic volcano forming an island called Oshima Island about 100 km south of Tokyo, magnetic surveys have repeatedly been carried out since the 1930's. Figure 6.2 shows the distribution of magnetic declination on the volcano as derived by land survey data (RIKITAKE *et al.*, 1951). Perturbation in declination can be interpreted in terms of a magnetic

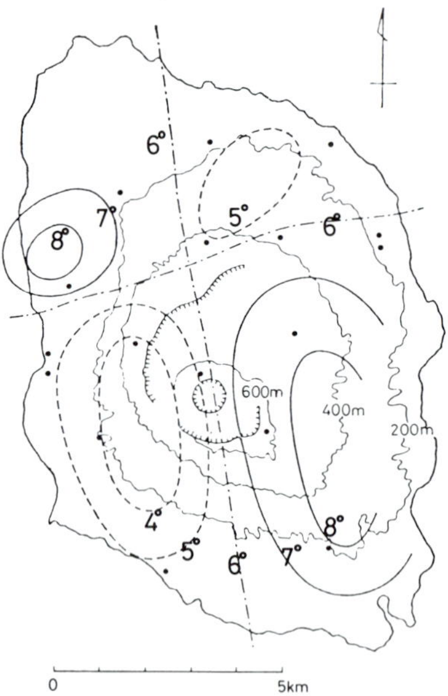

FIG. 6.2. Distribution of declination on Oshima Island (RIKITAKE *et al.*, 1951).

dipole approximating the magnetization of the volcanic body. The magnetic dip distribution is also strongly anomalous on the island. It increases from about 47° near the coastline towards the summit and takes a value of 56° at the top of the volcano. The total intensity distribution was derived from aeromagnetic survey data at altitudes of 4,000 feet and 6,000 feet, respectively, over the island. Figure 6.3 shows the distribution at these altitudes (Y. Kato, personal communication, 1968).

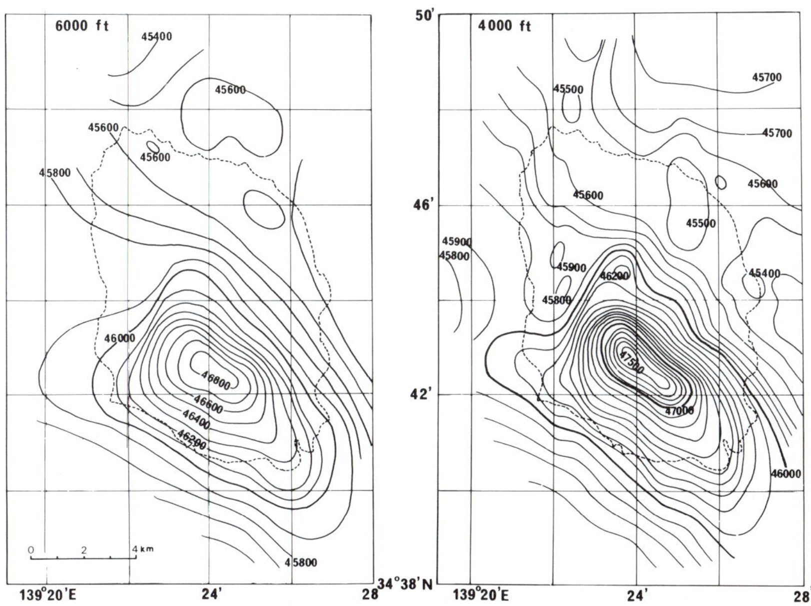

FIG. 6.3. Total intensity distribution at altitudes of 4,000 and 6,000 feet, respectively, over Oshima Island. Contours are given in gammas (from KODAMA and UYEDA, 1979).

Assuming that all the rocks constituting the volcanic body are uniformly magnetized, KODAMA and UYEDA (1979) estimated the intensity and direction of the magnetization vector by applying the Talwani method (see Subsection 6.3.3) to Volcano Mihara. The intensity of magnetization was estimated as about 0.013 emu/cm^3. This value seems to be reasonable for the basaltic rocks constituting the volcano. However, the estimated declination amounts to 20° and 30° east for aeromagnetic data obtained at altitudes of 4,000 feet and 6,000 feet, respectively. Such an easterly declination value is not consistent with the current declination in this region (slightly west). On the other hand, the inclination was estimated as about 50°. This value agrees well with the present inclination in this region (about 47°).

Aeromagnetic surveys were also carried out over nearby volcanic islands: Miyake-jima, Nii-jima, and Kozu-shima Islands (Japanese National Committee for Upper Mantle Project, 1972). Similar estimations of magnetization vectors disclosed that the inclination ranges from 40° to 53° and the declination points slightly westwards. The magnetization vector thus obtained can be interpreted in terms of magnetization acquired in geologically recent years.

The declination of 20° ~ 30° east in the case of Oshima Island is regarded as anomalous on the grounds that Volcano Mihara is supposed to have been formed some tens of thousand years ago and the declination at nearby islands points slightly westwards. This anomalous feature could reasonably be accounted for by KODAMA and UYEDA (1979) by assuming a non-uniform distribution of magnetization intensity for Volcano Mihara. If a part of the eastern side of the volcano is assumed to be weakly magnetized compared with the other areas, the anomaly in declination disappears and the magnetization vector is now in good agreement with those for nearby islands.

A similar discrepancy in direction of the magnetization vector was also reported for Volcano Aso in Kyushu, Japan. In this case, also, an inhomogeneous distribution of magetization intensity had to be postulated (VACQIUER and UYEDA, 1967). Aeromagnetic survey data over Mount Shasta, one of the Quaternary volcanoes in the High Cascade province, U.S.A., were analyzed by BLAKELY and CHRISTIANSEN (1978), who found that the location of a virtual geomagnetic pole (VGP) is determined as 46°N in latitude and 32°W in longitude, if a uniformly magnetized model is adopted. This result is unreasonable for a Quaternary volcano. BLAKELY and CHRISTIANSEN (1978) then proposed two reasonable interpretations; stronger magnetization in the western part of the volcano or partial demagnetization due to temperature higher than the Curie point beneath the summit area.

6.2 Marine Magnetic Anomalies

6.2.1 Marine magnetic survey and magnetic lineation

Measurements of the geomagnetic total intensity in ocean areas by a ship-towed proton precession magnetometer can be considered to be equivalent to an aeromagnetic survey on land. Because of a simple measuring procedure, marine magnetic surveys have been extensively carried out and a vast amount of marine magnetic data has been accumulated, not only for major oceans but also for some marginal seas. The most remarkable feature of marine magnetic anomalies is a strip-like pattern; that is, lineations of magnetic high and low appearing by turns and extending, in some cases, over a distance of a thousand kilometers.

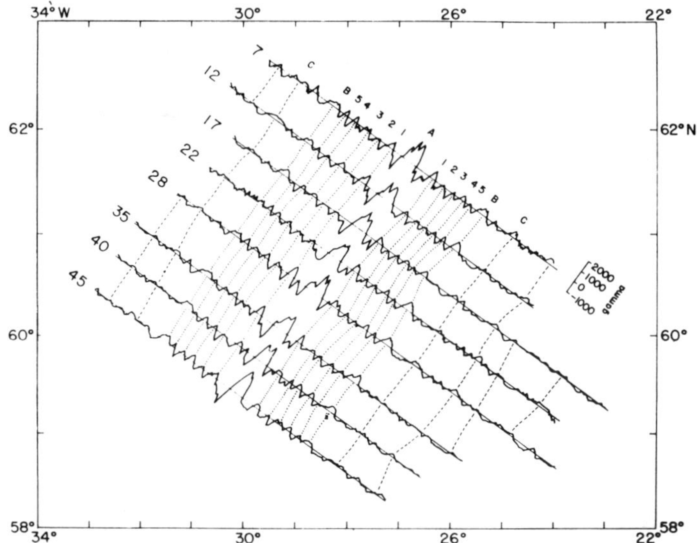

FIG. 6.4. Total intensity profiles across the Reykjanes ridge in the north Atlantic (HEIRTZLER *et al.*, 1966).

A typical example of magnetic lineations is shown in Fig. 6.4 (HEIRTZLER *et al.*, 1966). The figure shows total intensity profiles across the Reykjanes ridge south of Iceland. A remarkably symmetric feature is easily recognized with the center of symmetry coinciding with the ridge axis. Similar lineations have been discovered in most of the major oceans.

The lineations, which are symmetric with respect to the ridge axis, were interpreted by VINE and MATTHEWS (1963) as indicating a process of magnetization acquisition by oceanic crust at the ridge axis. Basaltic rocks are supposed to be formed from hot mantle material coming up at the ridge. These rocks will acquire magnetization (TRM), while they are cooling, in the direction of the ambient geomagnetic field, the polarity of which may be normal or reverse. If this process occurs successively over a geological time scale, the ocean crust already created must move away, most likely in both directions, from the axis, leaving room for new crust creation at the ridge axis. During this process the geomagnetic field undergoes polarity reversals and the direction of crustal magnetization changes accordingly. Such a direction change is manifested by magnetic highs and lows at the sea surface that correspond to normally and reversely magnetized oceanic crusts, respectively. In this way the symmetric, strip-like pattern is created.

Studies of magnetic lineations have made it possible to determine the ages of ocean floors, because the time-scale for geomagnetic reversal has been known (see Chapter 3). Although the global nature of magnetic anomalies in the major oceans is now fairly clear, some problems still remain to be solved through detailed studies of magnetic anomalies, as will be described below.

In the Vine and Matthews hypothesis a block model is used in which a magnetized layer is divided into many blocks and within each block the magnetization vector is assumed to be uniform. As for the thickness of this magnetic layer, TALWANI *et al.* (1971) and ATWATER and MUDIE (1973) claimed that the layer should be fairly thin, 0.5 km say, and it should be located at the basement immediately below the sedimentary layer. However, recent paleomagnetic results in the Deep Sea Drilling Project (DSDP) cast doubt on the assumption of uniform magnetization within each block, because some cores from drill holes showed divergent magnetic properties and even different polarities within one sequence of core (JOHNSON and MERRILL, 1978).

DSDP cores also revealed that the intensity of magnetization is weaker than had been expected (HARRISON, 1976) and hence suggested reconsideration of the thickness of the magnetic layer. In fact, a two-layer model was proposed by BLAKELY (1976). In this model, the upper layer is supposed to be strongly magnetized initially, but it is subject to subsequent intensity reduction due to low-temperature oxidation of basalts. On the other hand, magnetization of the second layer is supposed to be kept constantly weak.

Measurements of the geomagnetic total intensity have so far been carried out at the sea surface, thus usually some kilometers above magnetic sources. In this situation, high resolution of anomalies of short wave-length cannot be expected. In the meantime, measurements by a deep-tow magnetometer have become possible (e.g. BLAKELY *et al.*, 1975), and extensive measurements at a height less than 200 meters from the seafloor have been carried out, particularly in ridge areas (e.g. JOHNSON, 1979). Magnetic data thus obtained provide information on the fine magnetic structure such as the width of the transition zone for a geomagnetic reversal and the lateral variation of magnetization intensity. It would be expected that measurements by a deep-tow magnetometer combined with paleomagnetic work for DSDP cores would play an important role in understanding the spreading process at ridge axes.

Recently much attention has been paid to measurements of the total intensity in marginal seas with particular reference to regional tectonic history there (see Section 6.4). In some marginal basins, magnetic lineations have been found, which suggests a spreading process in such basins (e.g. KOBAYASHI and NAKADA, 1978).

6.2.2 Magnetic anomalies over sea-mounts

Local magnetic anomalies as derived from marine magnetic surveys are often associated with sea-mounts which are considered as extinct volcanoes rising from the seafloor. Therefore, a magnetic survey over a sea-mount resembles an aeromagnetic survey over a volcano as described in Subsection 6.1.2. Figure 6.5 shows the topography and the total intensity distribution over a sea-mount called Z-4-2 in the western Pacific (VACQUIER and UYEDA, 1967). The magnetic anomaly pattern in this figure indicates that the sea-mount is magnetized approximately in a north-south direction.

Paleomagnetic studies are possible for a sea-mount of known age on the assumption that it was magnetized in the direction of the geomagnetic field when it was formed. For example, paleomagnetic results from a group of sea-mounts in the western Pacific indicate that these sea-mounts have moved northward by 37° in latitude since late Cretaceous. However, some attention must be paid to the possible effect of non-uniform magnetization when this kind of discussion is made, as noted in the case of magnetic anomalies of volcanoes (Subsection 6.1.2).

An island made of coral reef is also considered to be a kind of sea-mount,

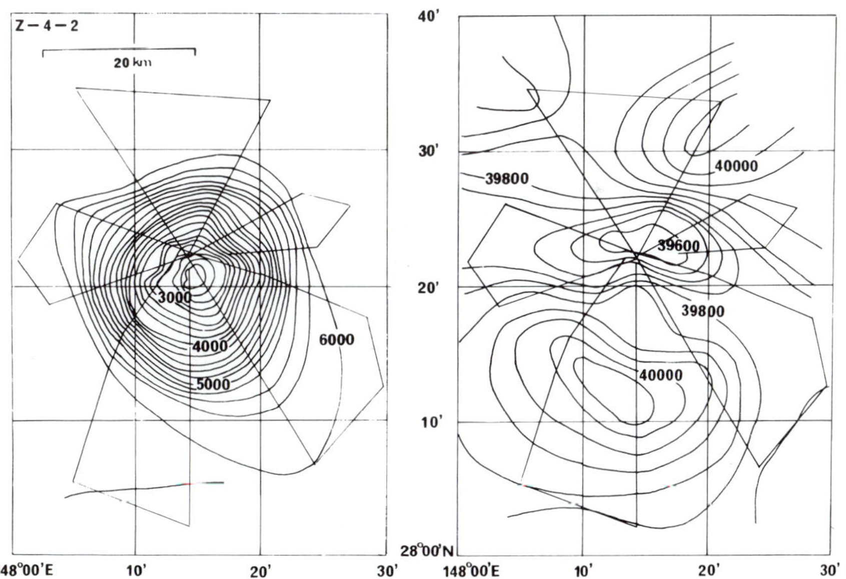

FIG. 6.5. Left: topography of the sea-mount Z-4-2 as indicated by contours of sea depth in meters. Right: total intensity distribution at sea level over the sea-mount. Contours are given in gammas (VACQUIER and UYEDA, 1967).

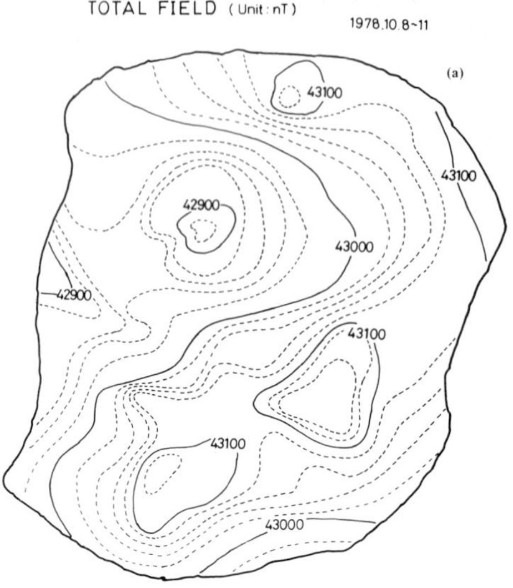
TOTAL FIELD (Unit: nT)
1978.10.8~11
(a)
43100
43100
42900
43000
42900
43100
43100
43000

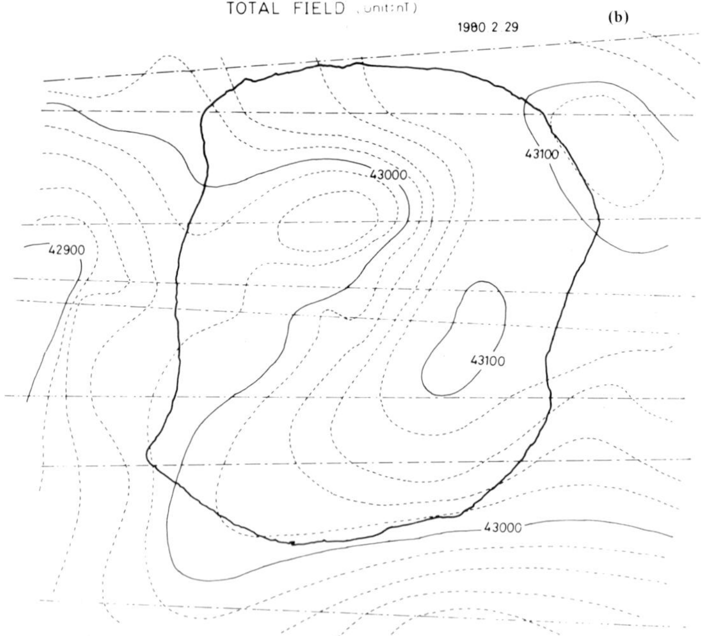
TOTAL FIELD (Unit: nT)
1980 2 29
(b)
43100
43000
42900
43100
43000

although the depth to the basement is unknown in most cases. Magnetic surveys over coral-reef islands called the Minami-daito and Kita-daito Islands were carried out in an attempt to determine the depth to the basement, in addition to paleomagnetic work (K. Yaskawa, personal communication, 1980). In view of the narrow coverage of a land survey, an aeromagnetic survey was also conducted. The results of both surveys are shown in Fig. 6.6. Preliminary results of data analysis give an inclination shallower than the present inclination in this region, implying that these islands may have moved northward after their formation.

6.3 *Magnetized Bodies and Their Magnetic Fields*

6.3.1 *Magnetic fields produced by a magnetic body*

The magnetic potential W at a point (x, y, z) due to a magnetized body, where the magnetization of volume element $\mathrm{d}v$ is specified by $\boldsymbol{J}\mathrm{d}v$, is given by

$$W = \int_V \frac{(\boldsymbol{J} \cdot \boldsymbol{r})}{r^3} \mathrm{d}v \tag{6.1}$$

where the integration is performed over the whole volume of the magnetic mass and $\boldsymbol{r}$ denotes the vector connecting the volume element (ξ, η, ζ) to the point (x, y, z). Obviously we have the relation

$$r^2 = (x - \xi)^2 + (y - \eta)^2 + (z - \zeta)^2. \tag{6.2}$$

If the magnetization can be regarded as uniform, W can be obtained from the so-called "Poisson relation." Using the gravitational potential W_g of the mass V as given by

$$W_g = k \int_V \frac{\rho}{r} \mathrm{d}v \tag{6.3}$$

where ρ is the density and k is the gravitational constant, W is obtained as

$$W = \frac{J}{k\rho} (\alpha\partial/\partial x + \beta\partial/\partial y + \gamma\partial/\partial z) W_g \tag{6.4}$$

where α, β and γ are the direction cosines of the uniform magnetization vector $\boldsymbol{J}$ with its intensity denoted by J.

FIG. 6.6. (a) Total intensity contours in gammas on Minami-daito Island. (b) Total intensity contours in gammas at 1500 feet altitude over the island. Chained lines indicate flight paths (K. Yaskawa, personal communication, 1980).

The magnetic field components in the x, y and z directions are then calculated as

$$X = -\frac{J}{k\rho}(\alpha\partial^2 W_g/\partial x^2 + \beta\partial^2 W_g/\partial x\partial y + \gamma\partial^2 W_g/\partial x\partial z)$$
$$Y = -\frac{J}{k\rho}(\alpha\partial^2 W_g/\partial x\partial y + \beta\partial^2 W_g/\partial y^2 + \gamma\partial^2 W_g/\partial y\partial z) \quad (6.5)$$
$$Z = -\frac{J}{k\rho}(\alpha\partial^2 W_g/\partial x\partial z + \beta\partial^2 W_g/\partial y\partial z + \gamma\partial^2 W_g/\partial z^2).$$

It had been no easy matter, however, to calculate the magnetic field arising from a magnetic body of arbitrary shape until computers came into general use. We can nowadays calculate the magnetic field due to a magnetized body having a fairly complicated shape. As a typical example, the magnetic field produced by a magnetized circular cone will be given in the next section.

6.3.2 *Magnetic field over a circular cone*

When considering the local magnetic anomaly due to a volcano, it is of interest to examine the magnetic field due to a uniformly magnetized circular cone which, in many cases, will approximate the shape of a volcano. RIKITAKE and HAGIWARA (1965) calculated the magnetic field over a circular cone as will be outlined in the following.

The gravitational potential W_g of a circular cone having radius a_0 at its bottom and height h can be obtained, by integrating over a series of circular disks, as

$$W_g = 2\pi k\rho\int_z^h a\mathrm{d}z_1\int_0^\infty \alpha^{-1}\mathrm{e}^{-\alpha(z_1-z)}J_0(\alpha r)J_1(\alpha a)\mathrm{d}\alpha$$
$$+2\pi k\rho\int_0^z a\mathrm{d}z_1\int_0^\infty \alpha^{-1}\mathrm{e}^{-\alpha(z-z_1)}J_0(\alpha r)J_1(\alpha a)\mathrm{d}\alpha \quad (6.6)$$

where ρ and r denote the density and radial distance from the central axis, which is taken as the z axis, of the cone. J_0 and J_1 are Bessel functions. a is the radius of the cone at $z = z_1$. Assuming that the cone is uniformly magnetized in a direction specified by an inclination θ in the $x - z$ plane, the x, y and z components of the magnetic field are calculated as

$$\Delta X = 2\pi J\{\cos\theta(F/r\cdot\cos 2\phi - G\cos^2\phi) - H\sin\theta\cos\phi\}$$
$$\Delta Y = 2\pi J\{\cos\theta\sin 2\phi(F/r - G/2) - H\sin\theta\sin\phi\} \quad (6.7)$$
$$\Delta Z = 2\pi J(G\sin\theta - H\cos\theta\cos\phi)$$

where J is the intensity of magnetization and ϕ is the azimuthal angle of a cylindrical coordinate with its origin taken at the center of the bottom plane. The x axis lies in a plane specified by $\phi = 0$.

In (6.7), F, G and H are functions involving infinite integrals of Bessel functions. For the convenience of computer work, these functions are transformed to

$$\begin{aligned} 2\pi F &= 2a_0 \rho_1^{-1/2} \int_0^{h/a_0} \alpha^{1/2} \left(\frac{2-\chi^2}{\chi} K - \frac{2}{\chi} E \right) d\zeta_1 \\ 2\pi G &= \rho_1^{-1/2} \int_0^{h/a_0} \chi \alpha^{-1/2} \left\{ K + \frac{\chi^2 \alpha/\rho_1 - (2-\chi^2)}{2(1-\chi^2)} E \right\} d\zeta_1 \\ 2\pi H &= \rho_1^{-3/2} \int_0^{h/a_0} \chi \alpha^{-1/2} (\zeta - \zeta_1) \left\{ -K + \frac{2-\chi^2}{2(1-\chi^2)} E \right\} d\zeta_1 \end{aligned} \tag{6.8}$$

in which

$$\begin{aligned} \chi^2 &= 4\alpha\rho_1 / \{(\alpha + \rho_1)^2 + (\zeta - \zeta_1)^2\} \\ a/a_0 &= \alpha, \quad r/a_0 = \rho_1, \quad z_1/a_0 = \zeta_1, \quad z/a_0 = \zeta \end{aligned} \tag{6.9}$$

and K and E are the complete elliptic integrals.

When the direction of magnetization (x-direction in the present case) is deflected from the north by ψ, the northward and westward components of the magnetic field become

$$\left. \begin{aligned} \Delta X' &= \Delta X \cos\psi - \Delta Y \sin\psi \\ \Delta Y' &= \Delta X \sin\psi + \Delta Y \cos\psi. \end{aligned} \right\} \tag{6.10}$$

Usually the anomalous field is much weaker than the normal geomagnetic field, and hence the anomaly in the total intensity is given by

$$\Delta F = (\Delta X' \cos D + \Delta Y' \sin D) \cos I + \Delta Z \sin I \tag{6.11}$$

where D is the declination and I the inclination of the geomagnetic field.

RIKITAKE and HAGIWARA (1965) gave many examples of the magnetic anomaly over a magnetized circular cone on the basis of the above theory. Figure 6.7 shows the distribution of ΔY and ΔF, respectively, due to a cone as indicated by the shaded circle located at a height of 0.3 in units of the base radius. In the calculation, the direction of magnetization is assumed to lie in the magnetic meridian plane and its inclination is taken as 48°. The slope angle of the cone is assumed to be 10°. The model approximates well the topography of Oshima Island for which magnetic anomalies were discussed in Subsection 6.1.2.

It is interesting to compare Fig. 6.7 with Figs. 6.2 and 6.3 showing the

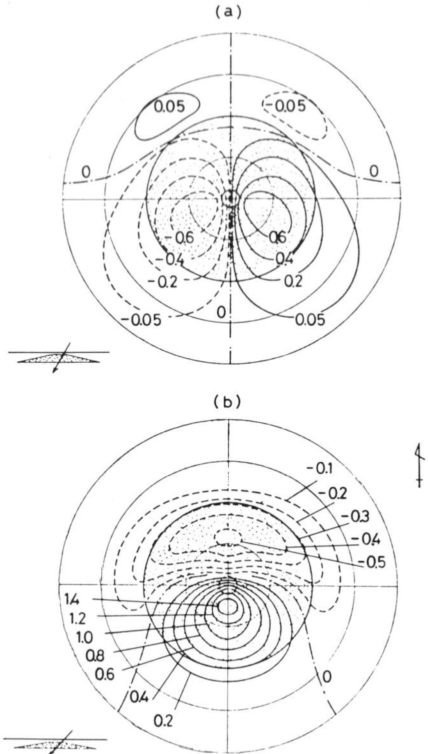

FIG. 6.7. Distribution of the westward component (a) and the total intensity (b) in units of magnetization intensity over a circular cone denoted by the shaded circle (RIKITAKE and HAGIWARA, 1965).

observed anomalies in the declination and total intensity. We see a good agreement, as a whole, between the calculated and observed anomalies, although a more detailed study was undertaken by KODAMA and UYEDA (1979) on the basis of the Talwani method which will be introduced in the next Subsection.

6.3.3 *Magnetic field produced by a magnetic body of arbitrary shape*

There are many volcanoes and sea-mounts that are not reasonably approximated by circular cones. In such cases, it is necessary to solve (6.1) for V representing a body of arbitrary shape. Using the relation

$$\Delta \boldsymbol{H} = - \operatorname{grad} W \tag{6.12}$$

where $\Delta \boldsymbol{H}$ denotes the magnetic field produced by the magnetic body and is often referred to as a magnetic anomaly, the three components of anomalous magnetic field are given as

$$\begin{aligned}\Delta X &= J_x W_1 + J_y W_2 + J_z W_3 \\ \Delta Y &= J_x W_2 + J_y W_4 + J_z W_5 \\ \Delta Z &= J_x W_3 + J_y W_5 + J_z W_6\end{aligned} \quad (6.13)$$

where the Cartesian coordinate system is used and magnetization is assumed to be uniform within V. $W_1, \ldots\ldots, W_6$ are expressed as

$$\begin{aligned}W_1 &= \iiint \frac{3x^2 - r^2}{r^5} \mathrm{d}x\,\mathrm{d}y\,\mathrm{d}z \\ W_2 &= \iiint \frac{3xy}{r^5} \mathrm{d}x\,\mathrm{d}y\,\mathrm{d}z \\ W_3 &= \iiint \frac{3xz}{r^5} \mathrm{d}x\,\mathrm{d}y\,\mathrm{d}z \\ W_4 &= \iiint \frac{3y^2 - r^2}{r^5} \mathrm{d}x\,\mathrm{d}y\,\mathrm{d}z \\ W_5 &= \iiint \frac{3yz}{r^5} \mathrm{d}x\,\mathrm{d}y\,\mathrm{d}z \\ W_6 &= \iiint \frac{3z^2 - r^2}{r^5} \mathrm{d}x\,\mathrm{d}y\,\mathrm{d}z.\end{aligned} \quad (6.14)$$

Now the problem is reduced to the evaluation of $W_1, \ldots\ldots, W_6$ for a given shape.

One of the methods of evaluating $W_1, \ldots\ldots, W_6$ is to divide V into a number of elementary rectangular prisms and perform analytical integration with respect to a certain axis and numerical integration for the other axes (e.g. RICHARDS *et al.*, 1967). This method was extended by EMILIA and MASSEY (1974) to the case in which magnetization intensity is variable from one prism to another, while the direction of magnetization is common for all the prisms.

In another method, usually referred to as the Talwani method (TALWANI, 1965), the magnetic body V is represented topographically by contours and then each contour is approximated by a polygonal lamina. In this case, double integration can be performed analytically with respect to the x and y axes and the remaining integration along the z axis is performed numerically. The analytical expressions were given by TALWANI (1965) in terms of parameters that can be determined from a set of coordinates of corners of polygons.

BOTT (1967) and EMILIA and BODVARSSON (1969) applied the Talwani method to the calculation of the magnetic field arising from a magnetic layer of the oceanic crust. In this case, the direction of magnetization is specified. ISEZAKI (1973) extended the method to the case in which the direction of magnetization is treated as a parameter to be determined from the observed data.

6.3.4 Reduction of aeromagnetic data to those at different altitude

Data obtained from aeromagnetic surveys include anomalies of short as well as long wave-length. In order to investigate anomalies of deep-seated origin, it is necessary to deal with anomalies of long wave-length. The most straightforward way of achieving this is to make an aeromagnetic survey at higher altitude. It is unnecessary to carry out magnetic surveys at various altitudes; on the basis of potential theory, it is possible to obtain the magnetic field distribution at a certain level from data at another level (e.g. HENDERSON, 1960; HAGIWARA, 1965; KONTIS, 1971; TSAY, 1978).

The method proposed by HAGIWARA (1965) is based on the Fourier transform technique. Let $W(x, y, z)$ and $\tilde{W}(\omega_1, \omega_2, z)$ be a magnetic potential function at altitude z above sea level and its Fourier transform, respectively. The relations between the two functions are given by

$$\tilde{W}(\omega_1, \omega_2, z) = \int_{-\infty}^{\infty} \int_{-\infty}^{\infty} W(x, y, z) e^{-i(\omega_1 x + \omega_2 y)} dx dy \tag{6.15}$$

and

$$W(x, y, z) = \frac{1}{4\pi^2} \int_{-\infty}^{\infty} \int_{-\infty}^{\infty} \tilde{W}(\omega_1, \omega_2, z) e^{i(\omega_1 x + \omega_2 y)} d\omega_1 d\omega_2. \tag{6.16}$$

According to potential theory, the magnetic potential at $z = z_2$ can be obtained from that at $z = z_1$ $(z_2 > z_1)$ by making use of the relation

$$W(x, y, z) = \int_{-\infty}^{\infty} \int_{-\infty}^{\infty} w(x - x', y - y', z_2 - z_1) W(x', y', z_1) dx' dy' \tag{6.17}$$

where the weighting function w is given by

$$w(x, y, z) = \frac{1}{2\pi} |z| (x^2 + y^2 + z^2)^{-3/2}. \tag{6.18}$$

The Fourier transform of (6.17) becomes

$$\tilde{W}(\omega_1, \omega_2, z_2) = \tilde{w}(\omega_1, \omega_2, z_2 - z_1) \tilde{W}(\omega_1, \omega_2, z_1) \tag{6.19}$$

where $\tilde{w}(\omega_1, \omega_2, z)$ is the Fourier transform of $w(x, y, z)$. $\tilde{w}$ is derived as

$$\tilde{w}(\omega_1,\omega_2,z) = \frac{1}{2\pi}\int_{-\infty}^{\infty}\int_{-\infty}^{\infty} |z|\,(x^2+y^2+z^2)^{-3/2} \times \exp[-i(\omega_1 x+\omega_2 y)]\mathrm{d}x\mathrm{d}y = \exp[-|z|\,(\omega_1^2+\omega_2^2)^{1/2}]. \tag{6.20}$$

Hence, (6.19) can be written as

$$\tilde{W}(\omega_1,\omega_2,z_2) = \tilde{W}(\omega_1,\omega_2,z_1)\exp[-|z_2-z_1|\,(\omega_1^2+\omega_2^2)^{1/2}]. \tag{6.21}$$

A magnetic anomaly $\boldsymbol{F}(x,y,z)$ is defined by

$$\boldsymbol{F}(x,y,z) = -\,\mathrm{grad}\,W(x,y,z). \tag{6.22}$$

Suppose that the anomalous field is much weaker than the averaged or normal field $\boldsymbol{F}_0$, then the anomalous total intensity derived from observation can be given approximately by

$$F = |\boldsymbol{F}+\boldsymbol{F}_0| - |\boldsymbol{F}_0| \approx (\boldsymbol{F}_0\cdot\boldsymbol{F})/|\boldsymbol{F}_0| \tag{6.23}$$

so that, taking (6.21) and (6.22) into consideration, the Fourier transform of F can be obtained approximately as

$$\tilde{F}(\omega_1,\omega_2,z_2) \approx \tilde{F}(\omega_1,\omega_2,z_1)\exp[-|z_2-z_1|(\omega_1^2+\omega_2^2)^{1/2}] \tag{6.24}$$

The inverse transform of (6.24) is obtained as

$$F(m,n,z_2) = \sum_{\mu=-N}^{N}\sum_{\nu=-N}^{N} \Phi(\pi(m-\mu),\pi(n-\nu), -\pi(z_2-z_1)/s)F(\mu,\nu,z_1) \tag{6.25}$$

where m and n are integers indicating x and y coordinates with respect to the grid spacing s. The weighting function Φ is defined by

$$\Phi(\xi,\eta,\zeta) = \int_0^1\int_0^1 \exp[\zeta(m^2+n^2)^{1/2}]\cos m\xi\cos n\eta\,\mathrm{d}m\,\mathrm{d}n. \tag{6.26}$$

Hagiwara (1965) demonstrated a number of examples of upward and downward continuation by making use of aeromagnetic survey data over Japanese volcanoes. A typical example is shown in Fig. 6.8. This figure demonstrates that a deep-seated magnetized body can easily be disclosed by upward continuation of the total intensity anomaly. On the other hand, downward continuation is greatly affected by small errors involved in the original data.

A more general reduction method was proposed by Bhattacharyya and Chan (1977), with application to effective removal of terrain effects. If

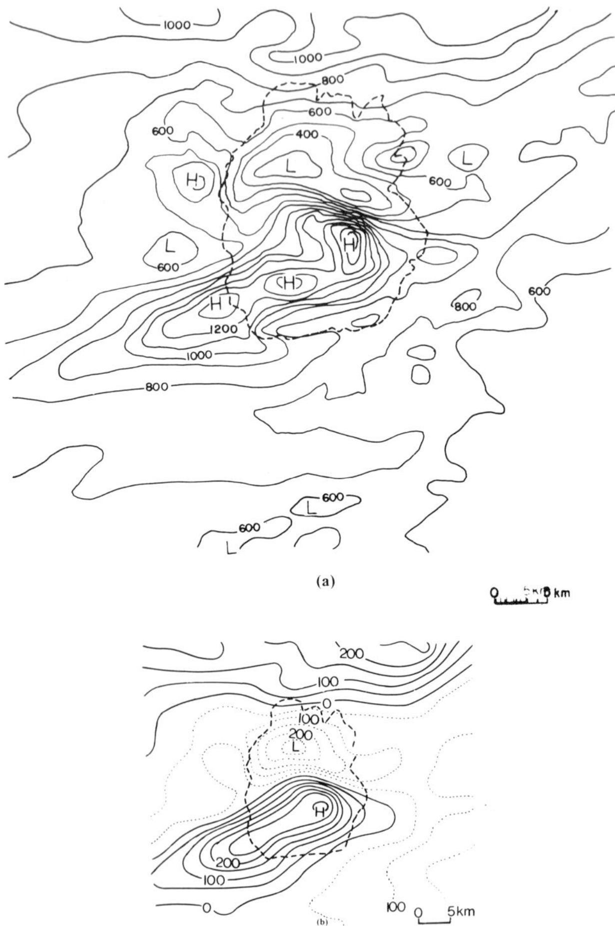

FIG. 6.8. (a) Total intensity distribution at 2,100 meters altitude over Volcano Aso. Contours of relative values are shown in gammas. (b) Total intensity distribution reduced to an altitude of 4,100 meters as derived by upward continuation of aeromagnetic data shown in (a) (HAGIWARA, 1965).

magnetic surveys are made on the earth's surface where there is high topographic relief, the terrain effect is expected to become appreciable, depending on the intensity of magnetization of the terrain. In their method, which is based on the expression of the surface field in terms of magnetic dipoles normal to the surface everywhere, continuation of the magnetic field at one general surface to that at another surface is possible. Reduction from one horizontal plane to another is a special case of this general problem. Based on a model calculation of terrain effect, BATTACHARYYA and CHAN (1977) pointed out that the terrain effect can be removed effectively if upward reduction is made on a curved surface resembling the topography of the earth's surface.

6.4 Magnetic Anomalies and Plate Tectonics

The interpretation of magnetic lineations in oceans by VINE and MATTHEWS (1963) has led to the hypothesis of ocean floor spreading. As described in Subsection 6.2.1, the age of the ocean floor has been determined for the major oceans from the lineation pattern and geomagnetic reversal sequence. The age increases in both directions away from the ridge axis and consequently ocean floor spreading has been confirmed for most of the major oceans. Recent results of age determination for DSDP cores generally verify the ages of seafloor determined from magnetic lineation patterns.

Moreover, a large-scale displacement in the magnetic lineation pattern, as first found off the west coast of North America, provoked WILSON (1965) to the new concept of the transform fault. Today, in the framework of plate tectonics, transform faults, as well as ridges and trenches, are believed to be boundaries of rigid plates with which the entire earth's surface is covered (e.g. COX, 1973; LE PICHON *et al.* 1973; BIRD, 1980).

Magnetic lineation patterns have also provided important information on spreading rates for oceanic plates. Thus, it was found that the spreading rate varies systematically along a spreading ridge. Combining various kinds of information such as spatial dependence of spreading rate, strikes of transform faults, and seismic focal mechanisms, MCKENZIE and PARKER (1967), MORGAN (1968), and LE PICHON (1968) claimed that movements of rigid plates relative to each other exist and such relative movement can be expressed by the rotation of a plate with respect to a certain axis.

Extensive marine magnetic surveys revealed that the lineation pattern extending in an approximately north-south direction off the west coast of North America bends sharply, by nearly 90 degrees, in the Gulf of Alaska. This is called the great magnetic bight of the north-eastern Pacific Ocean. The lineation pattern tends to be older towards the southwest. However, a ridge system which must have produced these lineations has not been found in the Pacific Ocean. PITMAN and HAYES (1968) put forward a hypothesis that there once existed a triple junction of ridge-ridge-ridge type, but it has already subducted beneath the North American Continent, leaving a portion of a plate created by this triple junction system.

Subduction of a ridge system has also been inferred in the eastern Pacific area in interpreting the fact that at present there is no spreading ridge corresponding to magnetic lineations off the California coast. The above subduction hypothesis stimulated discussions on the tectonic history of the western part of North America (McKenzie and Morgan, 1969; Atwater, 1970). In a similar manner, magnetic lineation patterns have contributed to studies of oceanic plate evolution for the Atlantic Ocean (e.g. Pitman and Talwani, 1972), the Indian Ocean (e.g. Johnson *et al.*, 1976), the Pacific Ocean (e.g. Hilde *et al.*, 1976), and so on.

Much attention has recently been paid to magnetic anomalies in more complex areas, particularly in marginal basins (e.g. Blakely and Cande, 1979). There are many marginal basins behind island arc—trench systems in the western Pacific. Such a marginal basin is supposed to have something to do with an extensional feature of the crust (Karig, 1971) and seafloor spreading may be involved in evolution of a basin. In fact, magnetic lineations have been found in marginal basins such as the Shikoku basin (Tomoda *et al.*, 1975), the South Fiji basin (Weissel and Watts, 1975), and the West Philippine basin (Watts *et al.*, 1977). However, in some basins no marked magnetic lineations could be resolved. In these cases, the magnetic anomaly pattern is characterized by rather diffuse features, presumably reflecting complicated tectonics (Lawver and Hawkins, 1978).

6.5 Magnetic Anomaly of Long Wavelength

Aeromagnetic surveys from the Pacific to Atlantic coasts along transcontinental paths over the U.S.A. were carried out during the Upper Mantle Project. As one of the results, it was discovered that regions west of the Rocky Mountains are characterized by magnetic anomalies of long wavelength (Zietz *et al.*, 1969).

In the northern and western parts of Canada, extensive aeromagnetic surveys have been carried out, generally at an altitude of 3.5 km, and some magnetic anomalies of long wavelength were discovered (Hall, 1974; Coles *et al.*, 1976). The upward continuation technique was applied to the aeromagnetic data so as to isolate anomalies of long wavelength. Figure 6.9 shows magnetic anomalies derived from the data reduced to an altitude of 300 km (Coles and Hains, 1979). Some positive anomalies can be recognized in the figure; for instance, an anomaly over the Alpha ridge in the Arctic basin. There also exist some negative anomalies in the southern Canada basin, the western Cordillera, and other regions.

The above aeromagnetic data were reduced further upward to an altitude of 500 km so as to be compared with the Pogo data (see Section 1.6). Correlation between these two data sets turned out to be generally good, in

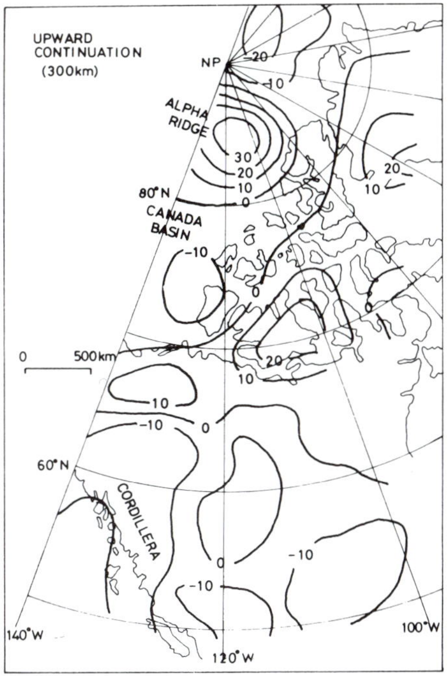

Fig. 6.9. Magnetic anomalies reduced to an altitude of 300 km by upward continuation of aeromagnetic data in northern and western Canada. Contours are given in gammas (Coles and Hains, 1979).

anomaly pattern as well as amplitude. The magnetic high over the Alpha ridge is still distinct at this altitude. Similar anomalies of long wavelength are known to exist over the Lord Howe rise in the southwestern Pacific and over the Broken ridge in the Indian Ocean (Regan *et al.*, 1975): these areas are characterized by a crust of continental composition. Taking these anomalies over the oceans into account, Langel *et al.* (1980a) interpreted the anomaly over the Alpha ridge as being due to high magnetization of the crust in this region. On the other hand, negative anomalies as found, for instance, in the west of the Cordillera were ascribed to a regional upheaval of the Curie point isotherm due to high heat flow.

Satellite magnetic surveys are expected to provide valuable data for studies of magnetic anomalies of long wavelength. In fact, some anomalies of long wavelength have been derived from the Pogo data; one of the most remarkable anomalies is the Bangui anomaly over central Africa (see Section 6.1). Mayhew (1979) applied the technique of equivalent source representation to the Pogo data over the U.S.A. and adjacent areas and obtained the distribution of crustal magnetization.

A large amount of marine magnetic survey data which has been accumulated for the western Pacific was used by NOMURA (1979) for investigating magnetic anomalies of long wavelength. As a general tendency in the western Pacific, negative anomalies are found in marginal basins behind island arcs, while positive anomalies exist on the oceanic side of island arcs.

There are some discussions on the origin of magnetic anomalies of long wavelength. HALL (1974) and KRUTIKHOVSKAYA and PASHKEVICH (1977) presumed, in their interpretation of magnetic anomalies, that the lower crust should be highly magnetized and hence the magnetic boundary should coincide with the Moho discontinuity. If the Curie point isotherm happens to be located within the crust, negative anomalies would be expected because of demagnetization of the crust below the isotherm. This seems to be the case for the Baikal rift zone (NOVOSELOVA, 1978). It should be noted, however, that the Curie point depends strongly on the type of magnetic mineralogy (WASILEWSKI *et al.*, 1979), while the Curie point for magnetite, about 580°C, is usually the one used in mapping the Curie point isotherm.

As a magnetic mineral responsible for high magnetization of the lower crust, HAGGERTY (1978) proposed alloys of Fe–Ni–Co–Cu, which are supposed to be associated with serpentinization of mafic and ultramafic rocks. They seem to be stable under the conditions prevailing in the lower crust and upper mantle. Moreover, their Curie point is fairly high, ranging 620°–1,100°C.

On the other hand, NOMURA (1979) presumed that a deep-seated origin for the magnetization might be possible if the Hopkinson effect is taken into consideration; that is, enhancement of magnetic susceptibility just below the Curie point. In this case, strong induced magnetization would be expected. However, it is not clear how such an idea also accounts for anomalies in areas of high heat flow, particularly negative anomalies in marginal basins, without any contradiction.

CHAPTER 7

CHAPTER 7

TECTONOMAGNETISM

Precise magnetic surveys have become possible since the development of portable magnetometers having high accuracy and stability. Results of such precise surveys, carried out repeatedly at exactly the same locations, have suggested secular changes of the geomagnetic field of crustal origin. Some of the reported changes seemed to be well correlated with tectonic events such as earthquakes and volcanic eruptions. Hence it has been speculated that secular changes of crustal origin may be a manifestation of rock magnetization changes due to changes in the stress state associated with earthquakes and volcanic eruptions. Rock magnetization changes due to stress have also been disclosed experimentally. NAGATA (1969) proposed the term "tectonomagnetism" for a research field dealing with geomagnetic changes of crustal origin.

7.1 Local Geomagnetic Changes Associated with Earthquakes and Volcanic Eruptions

7.1.1 Geomagnetic changes associated with earthquakes

Numerous examples of geomagnetic changes, seemingly associated with earthquake occurrences, have been reported so far, some of them amounting to as much as a hundred gammas or more. However, the amount of change has apparently decreased and, since the development of proton precession magnetometers, changes exceeding 10 gammas have seldom been reported (RIKITAKE, 1968a, 1976). Today, it is generally believed that geomagnetic changes associated with earthquake occurrences would be of the order of 10 gammas at most and usually less than that, as will be described in the following.

A typical example of geomagnetic change was obtained during the Matsushiro earthquake swarm which persisted through 1965 and 1966 in central Japan. On this occasion, continuous measurements of the total intensity were undertaken by installing proton precession magnetometers in

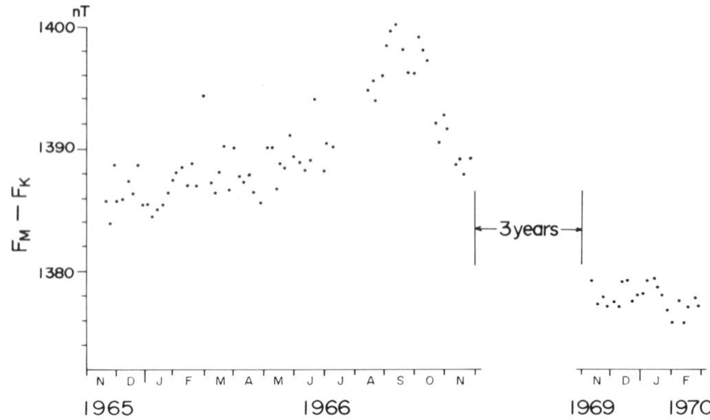

FIG. 7.1. Changes in five-day means of the total intensity difference between the Matsushiro (M) and reference (K) stations (YAMAZAKI and RIKITAKE, 1970).

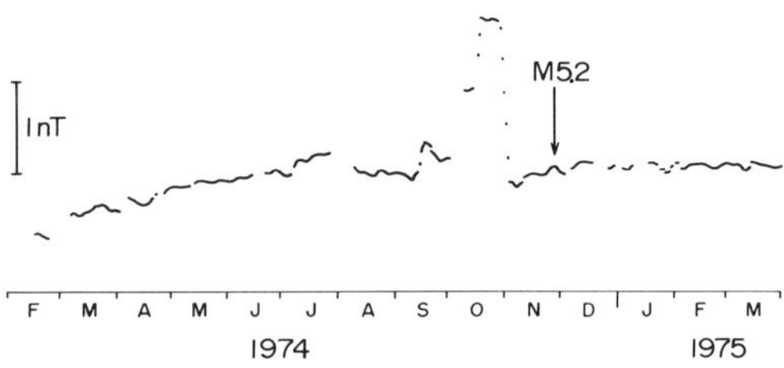

FIG. 7.2. Changes in five-day running means of the total intensity difference between a pair of stations. An arrow and a bar indicate an earthquake of magnitude 5.2 and a scale length for one gamma (1 nT), respectively (SMITH and JOHNSTON, 1976).

the swarm region. Figure 7.1 shows changes in the total intensity at one of the observation sites, Matsushiro, relative to the reference station, in this case, the Kanozan Observatory about 200 km distant from the Matsushiro site (YAMAZAKI and RIKITAKE, 1970). As recognized in this figure, the appearance of anomalous change in the total intensity seems to coincide with the period of high seismic activity.

In the U.S.A. also, a remarkable change in the total intensity was detected at one of the continuous observation sites that had been established along the San Andreas fault (SMITH and JOHNSTON, 1976). The change, amounting to two gammas or so, appeared about two months prior to an earthquake of magnitude 5.2 as shown in Fig. 7.2. The epicentral distance was about 11 km in this case. This change has been considered as a typical precursory change in the geomagnetic field. More detailed analyses based on a sophisticated data reduction technique (see Section 7.5) have disclosed precursory variations in the total intensity, similar to the one above, at two other sites located more remotely than the above site (DAVIS *et al.*, 1980).

An intensive geomagnetic observation system has recently been put into operation in the Izu Peninsula, Japan, and some remarkable changes in the total intensity have been observed in association with earthquakes that have taken place in and around the Izu Peninsula (RIKITAKE *et al.*, 1980). About two months prior to an earthquake of magnitude 7.0, the total intensity difference between two continuous observation sites in the Izu Peninsula underwent a precursory change as shown in Fig. 7.3 (HONKURA, 1978b). The epicentral distance was about 30 km at one of the sites (SGH) and about 40 km at the other (MTZ). It is interesting to note that the electric self-potential at another site (NKZ), which is very close to SGH, underwent a precursory change similar in time-dependent behavior to the change in the total intensity difference.

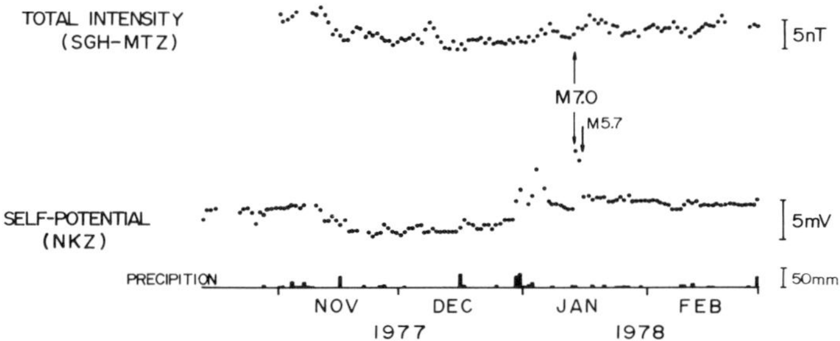

FIG. 7.3. Changes in total intensity difference between two stations (SGH and MTZ) and in self-potential at a station NKZ. Arrows indicate an earthquake of magnitude 7.0 and its largest aftershock.

In association with the above earthquake (*M* 7.0), results of repeated surveys of the total intensity in the Izu Peninsula also revealed remarkable changes. Figure 7.4 shows the spatial distribution of the observed changes in the total intensity during the periods (a) from July, 1977 to January, 1978 and (b) from January, 1978 to July, 1978 (RIKITAKE *et al.*, 1980). The survey in January, 1978, was carried out immediately after the earthquake occurrence. The overall pattern of changes shown in the figure can be accounted for by the seismomagnetic effect, as will be shown later.

Another typical example of precursory and co-seismic changes was also obtained at one of sites for continuous observation of the total intensity in the Izu Peninsula. In this case, an earthquake of magnitude 5.0 took place within 5 km of the site. As shown in Fig. 7.5, the difference in the total intensity between the site (KWZ) and the reference station, the Kanozan Observatory

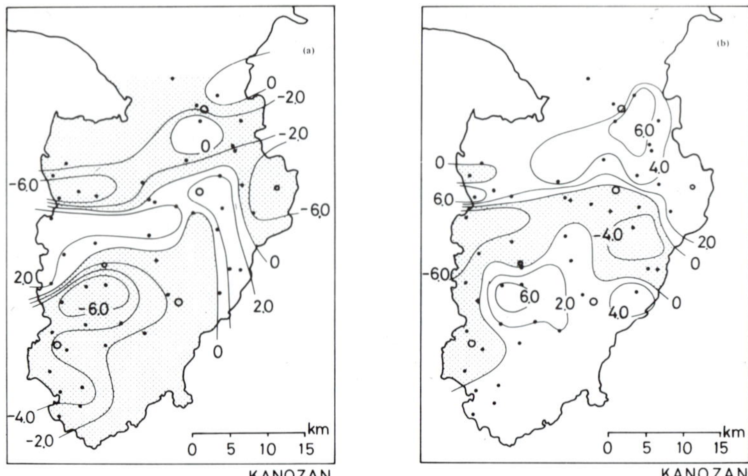

FIG. 7.4. Distribution of secular changes in the total intensity in the Izu Peninsula relative to the reference station (Kanozan Observatory) during the periods from July, 1977 to January, 1978 (a) and from January, 1978 to July, 1978 (b), respectively. Contours are given in gammas (RIKITAKE *et al.*, 1980).

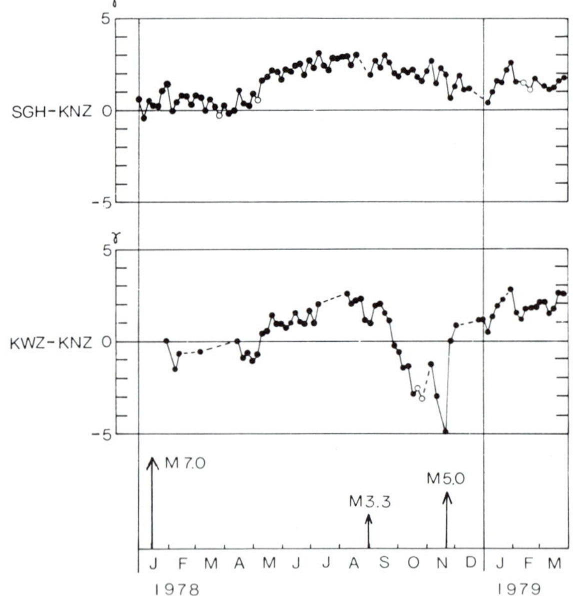

FIG. 7.5. Changes in total intensity difference between SGH and the reference station (KNZ) and also between KWZ and KNZ. Arrows indicate earthquakes of various magnitudes (SASAI and ISHIKAWA, 1980).

(KNZ), about 100 km distant from the Izu Peninsula, underwent a precursory change about two months prior to the earthquake occurrence (SASAI and ISHIKAWA, 1980). In contrast to the other cases shown above, a co-seismic change also appeared at the time of earthquake occurrence.

There are many other reports of changes in the geomagnetic field that appeared in association with earthquakes in various regions over the world, as reviewed by HONKURA (1981). It should be pointed out, also, that particularly in China, some of the changes observed before earthquake occurrences actually contributed, at least partly, to practical prediction of some of the earthquakes.

7.1.2 Geomagnetic changes associated with volcanic eruptions

A fairly large change in the geomagnetic field was observed at the time of eruption of Volcano Mihara, Japan. On Oshima Island where Volcano Mihara rises, surveys of magnetic dip were carried out twice in July and September, 1950. During this period a big eruption took place and a considerable amount of lava flowed out from the crater. Figure 7.6 shows changes in magnetic dip, in minutes of arc, during this period (RIKITAKE, 1951b; RIKITAKE and YOKOYAMA, 1955a). The spatial distribution of changes

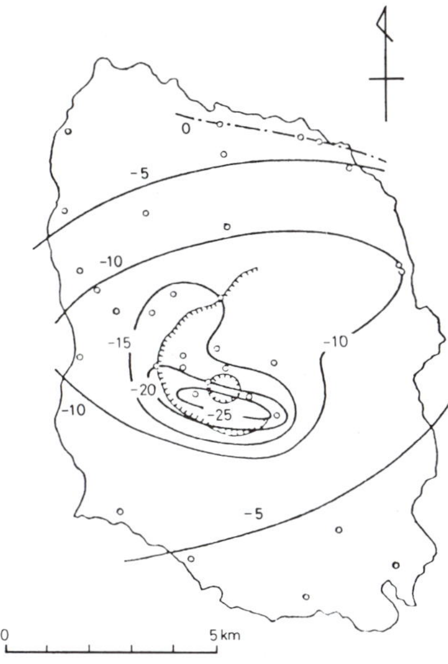

FIG. 7.6. Changes in magnetic dip in minutes of arc on Oshima Island during the period from July to September, 1950.

in magnetic dip associated with the eruption was found to be in good agreement with that to be expected from thermal demagnetization of basaltic rocks beneath the crater. In a theoretical interpretation, the region in which the temperature exceeds the Curie point of basaltic rock was approximated by a sphere of 2 km radius located at a depth of 5 km beneath the volcano (RIKITAKE, 1951b).

Continuous measurements of magnetic declination have also been carried out at a station located on the western flank of the volcano. It was then found that changes appeared in association with volcanic activity and in some cases the declination tended to deflect slightly westward before eruptions (YOKOYAMA, 1969). Such a pattern of change is in good agreement with the interpretation in terms of thermal demagnetization of basaltic rocks at the time of eruption, although no quantitative interpretation has been attempted.

JOHNSTON and STACEY (1969) reported on changes in the total intensity in association with activity of Mt. Ruapehu, an active volcano in New Zealand. In this case, however, the observed changes were ascribed to a piezomagnetic effect rather than a thermal effect; that is, rock magnetization changes due to stress field changes associated with volcanic activity.

A volcanomagnetic effect, that is a geomagnetic change at the time of volcanic eruption, was also examined by DAVIS *et al.* (1973) for activity of Kilauea Volcano, Hawaii, although no significant change was detected in association with its eruption. Meanwhile, a more effective data reduction technique was applied to the total intensity data obtained by DAVIS *et al.* (1973) and a slight change, one gamma or so, could be detected in association with the eruption (DAVIS *et al.*, 1979, 1981). It was also discovered that the change is in good agreement with an anomaly in ground tilt.

Three magnetometers had been in operation on Mt. St. Helens, the U.S.A., when it underwent a violent eruption in 1980 (M.J.S. Johnston, personal communication, 1981). Although two of them were unfortunately lost, the last one, which had been installed about 5 km west of the crater, provided invaluable data. The comparison of data thus obtained with those at the Victoria Observatory, Canada, disclosed transient changes in the total intensity exceeding 10 gammas in association with eruptions as well as a rather permanent change of about 9 gammas. Johnston and his co-workers interpreted the latter change in terms of stress released at the time of eruptions.

7.2 Rock Magnetization Change due to Stress

As one of the physical properties of rocks constituting the earth's crust, their magnetization has been investigated in fair detail (e.g. NAGATA, 1961; NAGATA and OZIMA, 1967). The aspect pertaining to tectonomagnetism is the effect of mechanical stress on rock magnetization, as is implied in the preceding section, as well as thermal demagnetization at a temperature higher than the Curie point.

NAGATA (1970a, b) reviewed studies of the effect of stress and showed some expressions governing the stress effect on various types of magnetization. Generally two factors are important in the stress effect; a reversible change in susceptibility and a reversible change in natural remanent magnetization (NRM). It is theoretically known that a reversible change results from the rotation of spontaneous magnetization within magnetic domains. On the other hand, an irreversible change which is representative of soft magnetization such as IRM (isothermal remanent magnetization) and PRM (piezo-remanent magnetization) is caused by irreversible displacement of 90° domain walls.

If the stress under consideration is within about 10^2 bars, as is the case for usual tectonomagnetic events, changes in magnetization (ΔJ) of various types of rocks can be expressed as

$$\Delta J_{/\!/} = -(\chi_0\beta' H + J_R\beta'')\sigma$$
$$\Delta J_{\perp} = \frac{1}{2}(\chi_0\beta' H + J_R\beta'')\sigma \tag{7.1}$$

for components parallel to and perpendicular to the direction of the applied stress (σ), respectively (e.g. NAGATA, 1969). χ_0 and J_R denote the initial susceptibility and remanent magnetization, respectively. H is the intensity of the applied magnetic field. β' and β'' are called the stress sensitivity and their values are given approximately by

$$\beta' = (0.5 \sim 5.0) \times 10^{-4}\ \mathrm{cm^2/kg}$$
$$\beta'' = (0.3 \sim 1.2) \times 10^{-4}\ \mathrm{cm^2/kg} \tag{7.2}$$

for igneous rocks (NAGATA, 1969).

STACEY and JOHNSTON (1972) also investigated theoretically changes in rock magnetization due to stress. They calculated the stress sensitivity of susceptibility and remanent magnetization, respectively, for titanomagnetite-bearing rocks and showed that both kinds of sensitivity increase with an increasing proportion of titanomagnetite. Therefore, they pointed out that a value of $2 \times 10^{-4}\ \mathrm{bar}^{-1}$ is more suitable for most igneous rocks rather than the $1 \times 10^{-4}\ \mathrm{bar}^{-1}$, which is assigned to pure magnetite.

However, changes in susceptibility and thermoremanent magnetization (TRM), in most cases the main constituent of NRM, seem to depend on the size of magnetic grains, as shown in the experiment made by KEAN *et al.* (1976). In view of the fact that natural rocks contain grains of various sizes, the stress effect would not be a simple manifestation of the rotation of spontaneous magnetization only. It was also pointed out by KEAN *et al.* (1976) that the morphology of grains and the composition of titanomagnetite are important factors in the stress response of rock magnetization.

The nature of dilatancy has attracted much attention in connection with anomalous phenomena prior to earthquake occurrences (e.g. SCHOLZ *et al.*, 1973). BRACE *et al.* (1966) showed experimentally that many rocks become dilatant as a result of creation of new cracks at one-third or two-thirds of the fracture stress. The relationship between dilatancy and magnetization changes was investigated experimentally for IRM and NRM by MARTIN and WYSS (1975) and MARTIN *et al.* (1978). At the initial elastic stage the magnetization changes linearly with stress. As the applied stress increases, a dilatancy stage is realized and at this stage non-linear behavior becomes evident. The non-linear change during dilatancy is generally characterized by a rotation of the magnetization vector as well as a slight change in its intensity. However, the details of magnetization change are different from one sample to another, presumably reflecting the effect of grain size. Such experimental results may provoke a new interpretation with consideration of a local stress state.

In fact, the non-linear behavior during dilatancy was interpreted in a speculative way by WYSS and MARTIN (1978) as resulting from local stress concentration at the tips of newly created cracks. Suppose that the elastic behavior still continues during the dilatancy stage, then the non-linear behavior turns out to be due to demagnetization during dilatancy. Provided that the distribution of the local stress field is random in the dilatant region, the rotation of magnetization will also be random, thus resulting in net demagnetization.

MARTIN (1980) investigated changes in susceptibility during the dilatancy stage in addition to the examination of the relationship between remanent magnetization and dilatancy. Non-linear behavior was again observed, apparently in association with dilatancy. However, the dependence of susceptibility change on stress turned out to be nearly the same for different confining pressures, which implies no correlation between non-linear behavior and dilatancy in view of strong dependence of dilatancy onset on confining pressure. Thus it was confirmed that a susceptibility change is a stress-dependent property. MARTIN (1980) also examined the dependence of remanent magnetization change on confining pressure and found a tendency similar to the susceptibility change as described above. Therefore, contrary to the interpretation given in the preceding paragraph, it cannot be concluded reasonably that remanent magnetization is strongly affected by dilatancy.

7.3 *Theories of Seismomagnetic and Volcanomagnetic Effects*

Changes in the geomagnetic field which will be observed on the earth's surface can be calculated on the basis of the expressions for rock magnetization changes due to stress, as given in the preceding section, if the distribution of the stress field is known. In the case of a seismomagnetic effect, that is an abrupt change in the geomagnetic field at the time of earthquake occurrence, the stress distribution could be derived from simple dislocation models (SHAMSI and STACEY, 1969; TALWANI and KOVACH, 1972). The results of calculation indicate only slight changes, a few gammas at maximum, for a reasonable assumed crustal magnetization, 10^{-3} emu/cm^3 (1 A/m).

Recent rapid progress in computer facilities and improvement in techniques of numerical calculation made it possible to evaluate numerically the stress distribution due to fault displacement, on the basis of analytical expressions derived from dislocation theory. For instance, OHSHIMAN (1980) made a numerical evaluation of stress in the directions of the principal axes. Then, expressions (7.1) could be used to calculate magnetization changes in the crust, and consequently magnetic field changes at the earth's surface could be derived on the assumption that the initial magnetization was uniform in the crust.

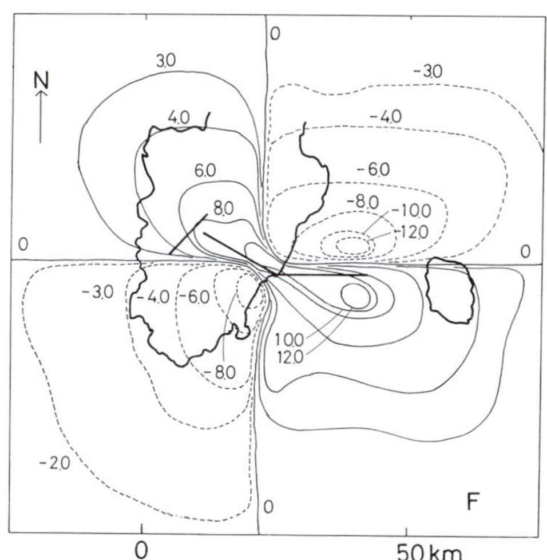

FIG. 7.7. Coseismic changes to be expected from seismomagnetic calculations for displacements of a main fault and two sub-faults as shown by straight lines. Contours are given in gammas (OHSHIMAN, 1980).

Figure 7.7 shows one of the examples of changes in the total intensity calculated for fault displacement (OHSHIMAN, 1980; RIKITAKE *et al.*, 1980).

The faults shown in the figure are supposed to have actually displaced at the time of the 1978 Izu-Oshima Kinkai earthquake of magnitude 7.0. Various parameters describing fault displacement had been determined for the major fault from fault mechanism solutions and they were used in the calculation. It is of interest to compare the calculated result with that observed, as shown in Fig. 7.4(a), which presumably indicates co-seismic changes. The overall pattern of changes seems to be in good agreement.

Analytical expressions for geomagnetic changes on the earth's surface were derived by SASAI (1980) on the basis of analytical expressions deduced from the elastic theory of dislocation (MARUYAMA, 1964). In Sasai's treatment, expressions (7.1) are extended to a general form as

$$\Delta J_i \boldsymbol{e}'_i = \beta J_i \left(\frac{\sigma_j + \sigma_k}{2} - \sigma_i \right) \boldsymbol{e}'_i \qquad (i, j, k = 1, 2, 3;\ i \neq j \neq k \neq i) \tag{7.3}$$

where i, j and k denote the axes in the directions of principal axes. J_i is the component of magnetization, either remanent magnetization (in this case $\beta = \beta''$) or induced magnetization ($\beta = \beta'$), along the i axis. $\boldsymbol{e}'_i$ indicates the unit vector along the i principal axis. It is troublesome, however, to calculate the directions of principal stress. This difficulty can be avoided by deriving a more general form in the original Cartesian coordinates (x, y, z). Suppose that the principal axes be (x', y', z') and cosines of angles between pairs of axes be a_{ij} $(i, j = 1, 2, 3)$, that is

$$\begin{pmatrix} x' \\ y' \\ z' \end{pmatrix} = \begin{pmatrix} a_{11} & a_{12} & a_{13} \\ a_{21} & a_{22} & a_{23} \\ a_{31} & a_{32} & a_{33} \end{pmatrix} \begin{pmatrix} x \\ y \\ z \end{pmatrix}. \tag{7.4}$$

The unit vectors $\boldsymbol{e}'_i (i = 1, 2, 3)$ can be expressed as a linear combination of the unit vectors $\boldsymbol{e}_j (j = 1, 2, 3)$ along the original Cartesian axes:

$$\boldsymbol{e}'_i = \sum_{j=1}^{3} a_{ji} \boldsymbol{e}_j \qquad (i = 1, 2, 3) \tag{7.5}$$

where 1, 2, 3 correspond to x, y, z or x', y', z', respectively.

Now, we consider a magnetization vector with its components J_p in the p direction. If this vector is projected onto each principal axis, its components in the x', y', z' directions are given by $J_p a_{ij} (j = 1, 2, 3)$ where $i = 1, 2, 3$ for $p = x, y, z$, respectively. Then (7.3) can be used to derive magnetization changes in the x', y', z' directions as $\beta \boldsymbol{J}_p a_{ij} c_j (j = 1, 2, 3)$ where

$$c_j = \frac{1}{2}(\sigma_1 + \sigma_2 + \sigma_3) - \frac{3}{2}\sigma_j. \tag{7.6}$$

Then the vector of magnetization change $\Delta \boldsymbol{J}$ is expressed in terms of $\boldsymbol{e}'_j (j = 1, 2, 3)$ as

$$\Delta \boldsymbol{J} = \beta J_p \sum_{j=1}^{3} a_{ij} c_j \boldsymbol{e}'_j. \tag{7.7}$$

From (7.5) we have

$$\begin{aligned} \Delta \boldsymbol{J} &= \beta J_p \sum_{j=1}^{3} a_{ij} c_j \sum_{k=1}^{3} a_{kj} \boldsymbol{e}_k \\ &= \beta J_p \sum_{k=1}^{3} s_{ik} \boldsymbol{e}_k \end{aligned} \tag{7.8}$$

where

$$s_{ik} = \sum_{j=1}^{3} c_j a_{ij} a_{kj}. \tag{7.9}$$

The relation between $\sigma_i (i = 1, 2, 3)$ and elements of the stress tensor T, which is written as

$$T = \begin{pmatrix} \tau_{xx} & \tau_{xy} & \tau_{xz} \\ \tau_{yx} & \tau_{yy} & \tau_{yz} \\ \tau_{zx} & \tau_{zy} & \tau_{zz} \end{pmatrix} \tag{7.10}$$

can be obtained as follows. Since T is symmetric ($\tau_{yx} = \tau_{xy}$, $\tau_{zx} = \tau_{xz}$, $\tau_{zy} = \tau_{yz}$), the matrix (7.10) can be transformed into a diagonal matrix by using a matrix P, which is orthogonal and normalized, as

$${}^tPTP = \begin{pmatrix} \sigma_1 & 0 & 0 \\ 0 & \sigma_2 & 0 \\ 0 & 0 & \sigma_3 \end{pmatrix}. \tag{7.11}$$

In the present case, P corresponds to the matrix in (7.5). (7.11) yields

$$\begin{aligned} a_{11}\tau_{xx} + a_{21}\tau_{xy} + a_{31}\tau_{xz} &= a_{11}\sigma_1 \\ a_{12}\tau_{xx} + a_{22}\tau_{xy} + a_{32}\tau_{xz} &= a_{22}\sigma_2 \\ a_{13}\tau_{xx} + a_{23}\tau_{xy} + a_{33}\tau_{xz} &= a_{33}\sigma_3. \end{aligned} \tag{7.12}$$

Since a_{11}, etc. are considered as direction cosines, the following relations hold:

$$\begin{aligned} a_{11}^2 + a_{12}^2 + a_{13}^2 &= 1 \\ a_{21}^2 + a_{22}^2 + a_{23}^2 &= 1 \\ a_{31}^2 + a_{32}^2 + a_{33}^2 &= 1 \end{aligned} \tag{7.13}$$

$$\begin{aligned} a_{11}a_{21} + a_{12}a_{22} + a_{13}a_{23} &= 0 \\ a_{21}a_{31} + a_{22}a_{32} + a_{23}a_{33} &= 0 \\ a_{31}a_{11} + a_{32}a_{12} + a_{33}a_{13} &= 0. \end{aligned} \tag{7.14}$$

From (7.6), (7.9), (7.12), (7.13), and (7.14), we obtain

$$s_{ik} = \frac{\delta_{ik}}{2}(\tau_{xx} + \tau_{yy} + \tau_{zz}) - \frac{3}{2}\tau_{ik} \qquad (i,k = x,y,z). \tag{7.15}$$

Thus we arrive at the following expression:

$$\begin{pmatrix} \Delta J_x \\ \Delta J_y \\ \Delta J_z \end{pmatrix} = \beta \begin{pmatrix} \dfrac{\tau_{yy} + \tau_{zz}}{2} - \tau_{xx} & -\dfrac{3}{2}\tau_{xy} & -\dfrac{3}{2}\tau_{xz} \\ -\dfrac{3}{2}\tau_{xy} & \dfrac{\tau_{xx} + \tau_{zz}}{2} - \tau_{yy} & -\dfrac{3}{2}\tau_{yz} \\ -\dfrac{3}{2}\tau_{xz} & -\dfrac{3}{2}\tau_{yz} & \dfrac{\tau_{xx} + \tau_{yy}}{2} - \tau_{zz} \end{pmatrix} \begin{pmatrix} J_x \\ J_y \\ J_z \end{pmatrix} \tag{7.16}$$

The magnetic potential at a point (x, y, z) on the earth's surface due to a change in magnetization J_x at a point (x', y', z') in the crust can be expressed as

$$W_x(x,y,z) = \iiint \beta J_x\{(x - x')s_{xx} + (y - y')s_{xy} + (z - z')s_{xz}\} \frac{dx'dy'dz'}{r'^3} \tag{7.17}$$

where r' denotes the distance between (x, y, z) and (x', y', z'). Similarly, for changes in magnetization J_y and J_z we obtain

$$W_y(x,y,z) = \iiint \beta J_y\{(x - x')s_{yx} + (y - y')s_{yy} + (z - z')s_{yz}\} \frac{dx'dy'dz'}{r'^3} \tag{7.18}$$

$$W_z(x,y,z) = \iiint \beta J_z\{(x - x')s_{zx} + (y - y')s_{zy} + (z - z')s_{zz}\} \frac{dx'dy'dz'}{r'^3}. \tag{7.19}$$

The distribution of stress released at the time of earthquake occurrence must be known in order to calculate the magnetic changes associated with an earthquake. An earthquake occurrence is a phenomenon closely related to a sudden displacement of a fault and stress release associated with fault displacement has been evaluated analytically by MARUYAMA (1964) on the basis of the elastic theory of dislocation. According to the theory, a stress element τ_{mn} at a point $Q(x', y', z')$ in the elastic medium caused by a dislocation at a point $P(\xi, \eta, \zeta)$ can be expressed as

$$\tau_{mn}(Q) = \iint_{\Sigma} \Delta u_k(P) H_{kl}^{mn} n_l(P) \mathrm{d}\Sigma \qquad (k, l = 1, 2, 3; \quad m, n = x, y, z) \tag{7.20}$$

where Δu_k denotes the discontinuity in displacements across the dislocation surface Σ, and n_l is the component of a unit vector normal to the surface element $\mathrm{d}\Sigma$. In (7.20) the Einstein convention is used and hence summation with respect to k and l is implicitly understood. H_{kl}^{mn} indicates a stress component corresponding to an elementary dislocation specified by a pair of k and l (e.g. MARUYAMA, 1964).

The expressions for magnetic potential are now written as

$$\begin{aligned} W_x &= \beta J_x \iint_{\Sigma} \Delta u_k(P) w_{kl}^x n_l(P) \mathrm{d}\Sigma \\ W_y &= \beta J_y \iint_{\Sigma} \Delta u_k(P) w_{kl}^y n_l(P) \mathrm{d}\Sigma \\ W_z &= \beta J_z \iint_{\Sigma} \Delta u_k(P) w_{kl}^z n_l(P) \mathrm{d}\Sigma \end{aligned} \tag{7.21}$$

w_{kl}^x, etc. are given by

$$\begin{aligned} w_{kl}^x &= \int_0^H \mathrm{d}z' \int_{-\infty}^{\infty} \int_{-\infty}^{\infty} \{(x - x')S_{kl}^{xx} + (y - y')S_{kl}^{xy} + (z - z')S_{kl}^{xz}\} \frac{\mathrm{d}x'\mathrm{d}y'}{r'^3} \\ w_{kl}^y &= \int_0^H \mathrm{d}z' \int_{-\infty}^{\infty} \int_{-\infty}^{\infty} \{(x - x')S_{kl}^{yx} + (y - y')S_{kl}^{yy} + (z - z')S_{kl}^{yz}\} \frac{\mathrm{d}x'\mathrm{d}y'}{r'^3} \\ w_{kl}^z &= \int_0^H \mathrm{d}z' \int_{-\infty}^{\infty} \int_{-\infty}^{\infty} \{(x - x')S_{kl}^{zx} + (y - y')S_{kl}^{zy} + (z - z')S_{kl}^{zz}\} \frac{\mathrm{d}x'\mathrm{d}y'}{r'^3} \end{aligned} \tag{7.22}$$

where

$$S_{kl}^{mn} = \frac{1}{2}\delta_{mn} H_{kl}^{jj} - \frac{3}{2} H_{kl}^{mn} \tag{7.23}$$

and H denotes the depth to the Curie point isotherm.

SASAI (1980) called w_{kl}^x, etc. the piezomagnetic potential due to an elementary dislocation represented by (k, l) and derived analytical expressions for them by making use of the Fourier transform theorem of convolution integrals.

Geomagnetic changes due to changes in the magnetization of crustal rocks were also examined by YUKUTAKE and TACHINAKA (1967) for the stress field caused by pressure build-up in the earth. The model considered by them is an infinitely long cylinder embedded in the elastic half-space. Pressure build-up was approximated by a hydrostatic pressure of 100 bars inside the cylinder. Magnetic field changes calculated for a cylinder of 5 km radius lying at a depth of 5 km amount to some gammas and are of the same order as those to be expected from fault displacement models.

A more realistic model simulating the volcanomagnetic effect should involve pressure build-up in a three-dimensional magma reservoir beneath a volcano body. MOGI (1958) approximated a magma reservoir by a sphere and examined crustal deformation due to pressure build-up in a sphere with application to crustal deformation at the time of volcanic eruption. Hence, such a sphere model has sometimes been called the "Mogi model."

On the basis of the Mogi model, geomagnetic changes were estimated numerically by DAVIS (1976) with special reference to the volcanomagnetic effect over Kilauea Volcano, Hawaii. On the other hand, SASAI (1979) evaluated analytically the changes in the geomagnetic field to be expected from the Mogi model. Sasai's treatment is based on the analytical expressions for the displacement field and the strain function for dilatation at a point (e.g. MINDLIN and CHENG, 1950; YAMAKAWA, 1955).

The stress components are given in cylindrical coordinates for a strain nucleus located at $(0, 0, D)$ as

$$\begin{aligned}
\sigma_{rr} &= C\left\{\frac{2}{R_1^3} - \frac{3(z-D)^2}{R_1^5} + \frac{2(2\lambda+3\mu)}{\lambda+\mu}\frac{1}{R_2^3} - \frac{3(z+D)(11z+3D)}{R_2^5} + \frac{30z(z+D)^3}{R_2^7}\right\} \\
\sigma_{\phi\phi} &= C\left\{-\frac{1}{R_1^3} + \frac{\lambda-3\mu}{\lambda+\mu}\frac{1}{R_2^3} + \frac{1}{\lambda+\mu}\frac{6(z+D)(\mu z-\lambda D)}{R_2^5}\right\} \\
\sigma_{zz} &= C\left\{-\frac{1}{R_1^3} + \frac{3(z-D)^2}{R_1^5} + \frac{1}{R_2^3} + \frac{3(z+D)(5z-D)}{R_2^5} - \frac{30z(z+D)^3}{R_2^7}\right\} \\
\sigma_{zr} &= C3r\left\{\frac{z-D}{R_1^5} + \frac{3z+D}{R_2^5} - \frac{10z(z+D)^2}{R_2^7}\right\} \\
\sigma_{r\phi} &= 0 \\
\sigma_{\phi z} &= 0
\end{aligned} \tag{7.24}$$

where λ and μ are Lame's constants. R_1 and R_2 denote the distance of a point (x, y, z) from $(0, 0, D)$ and $(0, 0, -D)$, respectively. By making use of a matrix P for the transformation of the cylindrical (r, ϕ, z) to Cartesian (x, y, z)

coordinates, that is,

$$P = \begin{pmatrix} \cos\phi & -\sin\phi & 0 \\ \sin\phi & \cos\phi & 0 \\ 0 & 0 & 1 \end{pmatrix} \tag{7.25}$$

the stress tensor T denoted by (7.10) can be derived from the tensor T', the elements of which are given in (7.24), as

$$T = P^{-1}T'P$$
$$= \begin{pmatrix} \sigma_{rr}\cos^2\phi + \sigma_{\phi\phi}\sin^2\phi & (\sigma_{rr} - \sigma_{\phi\phi})\sin\phi\cos\phi & \sigma_{zr}\cos\phi \\ (\sigma_{rr} - \sigma_{\phi\phi})\sin\phi\cos\phi & \sigma_{rr}\sin^2\phi + \sigma_{\phi\phi}\cos^2\phi & \sigma_{zr}\sin\phi \\ \sigma_{zr}\cos\phi & \sigma_{zr}\sin\phi & \sigma_{zz} \end{pmatrix} \tag{7.26}$$

Substitution of (7.26) into (7.15) yields

$$\begin{aligned}
s_{xx} &= \frac{1}{2}(\sigma_{rr} + \sigma_{zz} - 2\sigma_{\phi\phi}) - \frac{3}{2}(\sigma_{rr} - \sigma_{\phi\phi})\cos^2\phi \\
s_{xy} &= -\frac{3}{2}(\sigma_{rr} - \sigma_{\phi\phi})\sin\phi\cos\phi \\
s_{xz} &= -\frac{3}{2}\sigma_{zr}\cos\phi \\
s_{yy} &= \frac{1}{2}(\sigma_{\phi\phi} + \sigma_{zz} - 2\sigma_{rr}) + \frac{3}{2}(\sigma_{rr} - \sigma_{\phi\phi})\cos^2\phi \\
s_{yz} &= -\frac{3}{2}\sigma_{zr}\sin\phi \\
s_{zz} &= \frac{1}{2}(\sigma_{rr} + \sigma_{\phi\phi} - 2\sigma_{zz}) \\
s_{xy} &= s_{yx}, \quad s_{zx} = s_{xz}, \quad s_{zy} = s_{yz}.
\end{aligned} \tag{7.27}$$

From (7.17), (7.18) and (7.19), the magnetic potential on the earth's surface due to magnetization changes caused by hydrostatic pressure increase in a magma reservoir can be evaluated in a way similar to the case of the general seismomagnetic effect.

SASAI (1979) derived analytical expressions of magnetic potential for the Mogi model and estimated possible changes in the geomagnetic field in the case of Volcano Mihara where geomagnetic changes have been observed in association with its eruptions, as described in Subsection 7.1.2. In contrast to the interpretation in terms of thermal demagnetization (e.g. RIKITAKE, 1951b),

SASAI (1979) pointed out the possibility that the piezomagnetic effect is responsible, at least partly, for the observed volcanomagnetic effect. In fact, geomagnetic changes expected from pressure build-up inside the reservoir are very similar to those due to thermal demagnetization.

7.4 Magnetic Field Generation due to the Electrokinetic Effect

In relation to an important role played by water prior to earthquake occurrences as claimed in the dilatancy-diffusion hypothesis (e.g. SCHOLZ *et al.*, 1973), MIZUTANI *et al.* (1976) pointed out the possibility of generation of electric currents due to underground water flow through the cracks and pores commonly present in the earth's crust. Such a phenomenon would be a result of the electrokinetic effect which can be expressed as

$$\begin{aligned} -\boldsymbol{I} &= \eta\sigma\nabla\phi - \frac{\eta\varepsilon\zeta}{\mu}\nabla P \\ -\boldsymbol{J} &= -\frac{\eta\varepsilon\zeta}{\mu}\nabla\phi + \frac{k}{\mu}\nabla P \end{aligned} \tag{7.28}$$

where $\boldsymbol{I}$ and $\boldsymbol{J}$ denote the electric current density and the flux of fluid flow, respectively. P and ϕ are the pressure of the fluid and the electric potential. ζ is the so-called zeta-potential. Other parameters appearing in (7.28) are as follows. η, porosity; k, bulk permeability; σ, electrical conductivity; ε, dielectric constant; μ, viscosity of fluid. In deriving (7.28), water flow paths, usually consisting of cracks and pores, are approximated by a bundle of straight capillaries. Expressions for the more general case of tortuous capillaries are given in ISHIDO and MIZUTANI (1981).

In the case of a homogeneous medium, the current density is obtained as

$$-\boldsymbol{I} = \eta\sigma\nabla\left(\phi - \frac{\varepsilon\zeta}{\sigma\mu}P\right). \tag{7.29}$$

The term $\varepsilon\zeta/\sigma\mu$ is called the streaming potential coefficient. In a stationary case with ΔP fixed, the coefficient corresponds to the ratio of the streaming potential ($\Delta\phi$) to the driving pressure (ΔP), since in this case the electric current vanishes.

The streaming potential coefficient depends on various parameters of the medium. From experimental results obtained for values of parameters which are reasonable for the earth's crust, ISHIDO and MIZUTANI (1981) estimated the coefficient as -80 mV/bar for granites, -420 mV/bar for sandstones, and -470 mV/bar for porous rocks.

With the above electrokinetic effect in mind, MIZUTANI and ISHIDO (1976) pointed out that the stage of anomalous variations in the geomagnetic field during the Matsushiro earthquake swarm (see Subsection 7.1.1) coincides well with the stage of maximum water outflow in the epicentral area. The density of water flow could be estimated as 10^{-7} $m^3/(m^2 \cdot sec)$ from data on water flow from wells and springs in the epicentral area. Then, the total electric current induced by the water flow was calculated from the expressions for the electrokinetic effect. The calculated total current is as strong as 4×10^2 amperes. MIZUTANI and ISHIDO (1976) considered a very effective circuit and showed that the estimated total current can indeed generate a magnetic field of several gammas. Although the intensity of the magnetic field to be observed on the earth's surface depends strongly on circuit configuration, it must be concluded that the electrokinetic effect may not be ruled out as a possible cause of anomalous geomagnetic changes associated with earthquakes.

FITTERMAN (1978) attempted to estimate the magnitude of a possible anomaly in electric potential by applying the above theory to a model of the crust. It is known that the bulk conductivity of a porous medium is essentially controlled by the conductivity of fluid contained in the pores and the degree of their interconnection. In the present case we are dealing with fluid flow and hence pores must be interconnected to each other to some degree. If pores are effectively interconnected, the bulk conductivity can be approximated by the product of the fluid conductivity and the porosity. In this case, (7.29) is written as

$$-\boldsymbol{I} = \sigma \nabla (\phi - CP) \tag{7.30}$$

where σ is the bulk conductivity and C the streaming potential coefficient. Provided that no electric current source is available,

$$\nabla \cdot \boldsymbol{I} = 0 \tag{7.31}$$

and consequently

$$\nabla^2 \psi = 0 \tag{7.32}$$

in homogeneous regions. ψ is the total electric potential defined by

$$\psi = \phi - CP. \tag{7.33}$$

FITTERMAN (1978) treated a two layer model of the crust, in which the pressure distribution is specified as spherical due to a source located in the lower layer. As one of the conclusions, it was shown that no electric potential anomaly would be expected unless the following conditions are satisfied; there must be a boundary separating two regions, each having a different value of the streaming potential coefficient, and there must be pressure gradient in the direction parallel to the boundary. However, no magnetic field anomaly would be expected at the earth's surface for any pressure gradient in the crust, so long as the structure is horizontally uniform.

In seismically active regions, however, laterally non-uniform crustal structure is not unusual and the possibility cannot be ruled out that anomalous magnetic fields might appear as a result of electric current generation due to the electrokinetic effect. In fact, FITTERMAN (1979) showed that magnetic field anomalies, although not very remarkable, would be expected to appear in an area close to a fault. In the model for which the above conclusion was derived, the fault line is considered to be a boundary of the streaming potential coefficient and the pressure source is assumed to exist along the fault.

7.5 Methods of Detecting Tectonomagnetic Changes

There are many kinds of geomagnetic variations arising from sources in the ionosphere and magnetosphere (see Chapter 8) as well as those in the earth's core. It seems, then, almost impossible to detect slight changes associated with tectonic events, such as earthquakes and volcanic eruptions, from observation of the geomagnetic field at the particular location under consideration. Under certain conditions, however, it may be possible to eliminate geomagnetic variations of non-tectonic origin. For this purpose, comparison is usually made between data sets obtained simultaneously at two or more locations. This procedure is based upon the presumption that geomagnetic variations of non-tectonic origin can be considered to be spatially uniform at least for not-too-distant pairs of observation sites, whereas the spatial extent of an anomalous variation of tectonic origin would be limited to the neighborhood of an anomaly source.

The simplest data reduction technique is a simple difference method in which data at a remote reference station are simply subtracted from those at the site under consideration. If the reference station is far from the tectonomagnetic research area, any anomalous changes in the difference would be ascribed to an anomalous change at the field site. Many of the reported changes in the geomagnetic field were derived using this simple method (e.g. HONKURA, 1981).

However, geomagnetic variations of local nature yet of non-tectonic origin are quite often observed, for instance, local anomalies of short-period geomagnetic variations (see Chapter 12). Spatial non-uniformity is also considerable for daily variations (MORI and YOSHINO, 1970). Moreover, RIKITAKE (1966b) pointed out that day-to-day fluctuations in the night-time level, which is free from the effect of daily variation, are not necessarily the same at two stations.

As a method superior to the simple difference method, RIKITAKE (1966b) proposed a weighted difference method. Suppose that night-time values of the geomagnetic total intensity are given by F_A and F_B at two stations A and B. Their difference $F_A - F_B$ would be separated into the fields originating in the earth's core on one hand and in the ionosphere and magnetosphere on the other hand, that is

$$F_A - F_B = (F_A^c - F_B^c) + (F_A^e - F_B^e) \tag{7.34}$$

where superscripts c and e denote the core and external origins, respectively. Since the field of core origin is likely to be the same at both the stations, fluctuation in $F_A - F_B$ is governed by the field of external origin. In the weighted difference method, the ratio of F_A^e to F_B^e is determined empirically from sufficiently large data sets as

$$\alpha = F_A^e/F_B^e. \tag{7.35}$$

Now using α as the weighting factor for the weighted difference, we have

$$F_A - \alpha F_B = (F_A^c - F_B^c) + (1 - \alpha)F_B^c. \tag{7.36}$$

If the field of core origin is constant or, less likely, its variation follows the same relation as (7.35), the weighted difference method would turn out to be effective.

In general, neither condition is satisfied in a strict sense, because secular variations of core origin (see Chapter 2) have been observed and their nature is quite different from that of variations of external origin. Hence the weighted difference method may cause an erroneous result as pointed out by FUJITA (1973). However, the time scale of secular variation is fairly long compared with other variations and the weighted difference method would still be more effective than the simple difference method, as far as short-term tectonomagnetic phenomena are concerned.

The theoretical basis for the weighted difference method is derived by taking into account the three components of the geomagnetic field instead of the total intensity. Suppose that varying fields are separated from the stationary one peculiar to a location for the various components as

$$\begin{aligned} X &= X_0 + \Delta X \\ Y &= Y_0 + \Delta Y \\ Z &= Z_0 + \Delta Z \end{aligned} \tag{7.37}$$

where X, Y, and Z are the northward, eastward, and downward components, respectively. The total intensity F is derived as

$$F = \{(X_0 + \Delta X)^2 + (Y_0 + \Delta Y)^2 + (Z_0 + \Delta Z)^2\}^{1/2} \tag{7.38}$$

ΔX, etc. being much smaller than X_0, etc. By expanding (7.38) in a Taylor series and neglecting higher terms, we obtain

$$F \approx F_0 + \alpha \Delta X + \beta \Delta Y + \gamma \Delta Z \tag{7.39}$$

where

$$F_0 = (X_0^2 + Y_0^2 + Z_0^2)^{1/2} \tag{7.40}$$

$$\begin{aligned} \alpha &= X_0/F_0 = \cos I_0 \cos D_0 \\ \beta &= Y_0/F_0 = \cos I_0 \sin D_0 \\ \gamma &= Z_0/F_0 = \sin I_0. \end{aligned} \tag{7.41}$$

Even if ΔX, ΔY, and ΔZ at a station A are all equal to the corresponding values at another station B, $F_A - F_B$ depends on α, β and γ and hence on the dip and declination at the stations. The important thing is that the weighted difference method must be applied to all of the three components. The effect of a non-uniform permeability distribution would be treated in a similar way. In this case, however, α, β and γ must be determined empirically from the data sets (Davis *et al*., 1980)

If geomagnetic variations arising from electromagnetic induction are involved, the above procedure is not applicable because of a frequency-dependent response. In such a case, the use of transfer functions turns out to be powerful. In general, a total intensity variation is written as

$$T(t) \approx \int_{-\infty}^{\infty} [a(\tau)\Delta X(t-\tau) + b(\tau)\Delta Y(t-\tau) + c(\tau)\Delta Z(t-\tau)]\,d\tau \tag{7.42}$$

where a, b and c are convolution filters. (7.42) is transformed into the frequency domain as

$$T(f) \approx A(f)\Delta X(f) + B(f)\Delta Y(f) + C(f)\Delta Z(f) \tag{7.43}$$

where A, B and C are transfer functions.

Poehls and Jackson (1978) applied the above method to a data set and demonstrated that the transfer function method is more effective than the method based on (7.39). In the analysis of Beahn (1976), a similar transfer function method had been applied to the D component only. In his case, the difference in the total intensity between a pair of stations was well correlated with the D variation. The method of Beahn (1976) was further extended by Ware (1979) to a more general case; data reduction was made by successively applying the transfer function method first to H, then to D, and finally to Z.

A more general method based on multichannel Wiener filtering was proposed by Davis *et al.* (1981). In this method, the total intensity is also included as one of the input data sets. Supposing that there are some stations for total intensity measurements, then the difference to be predicted in the total intensity between two stations is expressed in a convolution form as

$$\delta_p \Delta T(t) \approx A(t) * \Delta X(t-\tau) + B(t) * \Delta Y(t-\tau) + C(t) * \Delta Z(t-\tau) + D(t) * \Delta T(t-\tau) + \ldots\ldots. \qquad (7.44)$$

The difference between $\delta \Delta T(t)$ (observed) and $\delta_p \Delta T(t)$ (predicted) should be a cleaned difference.

CHAPTER 8

CHAPTER 8

GEOMAGNETIC VARIATION OF EXTERNAL ORIGIN AND ELECTROMAGNETIC INDUCTION

The distribution of electrical conductivity within the earth can be researched using the phenomenon of electromagnetic induction, as will be described in the following chapters. In the sense that the nature of the inducing magnetic field must be known, at least approximately, for this purpose, geomagnetic variations that can be used for induction studies are those of external origin, arising from electromagnetic phenomena in the ionosphere and magnetosphere. In this chapter, some typical geomagnetic variations of external origin will be described briefly so as to serve as an introduction to the following chapters.

Geomagnetic variations are defined as deviations from the mean field, which would be considered the earth's main field, and can be expressed in terms of magnetic potential W as

$$W = a \sum_{n=1}^{\infty} \sum_{m=0}^{n} [(r/a)^n e_n^m \cos(m\phi + \varepsilon_n^m) + (r/a)^{-n-1} i_n^m \cos(m\phi + \iota_n^m)] P_n^m(\cos\theta) \tag{8.1}$$

where a is the earth's radius. e_n^m and i_n^m denote coefficients of magnetic potential of external and internal origins, respectively.

8.1 Solar Daily Variation

The geomagnetic field in mid- and low-latitudes usually undergoes a rather regular daily variation. A daily variation without any severe variations of short period is usually called the solar quiet daily variation and denoted by *Sq*. Figure 8.1 shows a typical *Sq* observed at the Kakioka Magnetic Observatory. A similar feature is to be found everywhere along the same latitude as Kakioka, if the variation is expressed in local time in place of ϕ in (8.1).

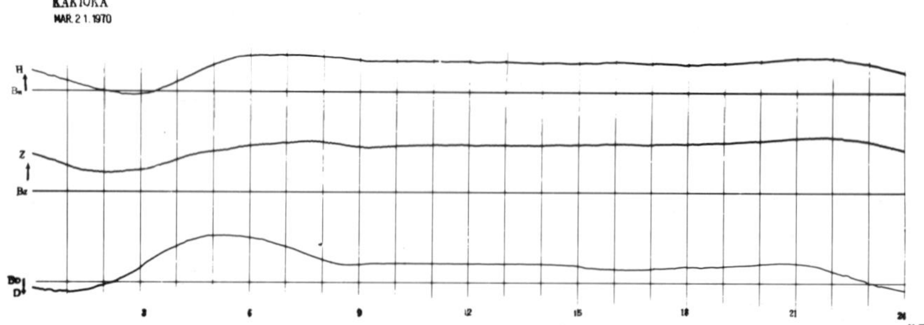

FIG. 8.1. A typical example of *Sq* at the Kakioka Magnetic Observatory. Time marks are given in universal time (U.T.), from which local time is derived by adding nine hours.

TABLE 8.1. i_n^m/e_n^m and $\iota_n^m-\varepsilon_n^m$ for *Sq*.

n	m	Chapman	Benkova	Hasegawa & Ota	Matsushita & Maeda	Suzuki
				i/e		
2	1	0.36	0.43	0.43	0.36	0.33
3	2	0.45	0.43	0.42	0.43	0.36
4	3	0.40		0.45	0.45	0.40
				$\iota-\varepsilon$		
2	1	13°	5°	9	13°	17°
3	2	18	5	10	13	19
4	3	21		14	12	22

The *Sq* field is analyzed in most cases by making use of spherical harmonic expansion as expressed by (8.1) (e.g. CHAPMAN, 1919; BENKOVA, 1940; HASEGAWA and OTA, 1950; MATSUSHITA and MAEDA, 1965a; SUZUKI, 1973). Detailed results of *Sq* analysis are also given in review papers by MATSUSHITA (1967, 1975). Such a method is not suitable for examining details of *Sq* fields, since a considerable number of modes of spherical harmonics would be required in (8.1). The surface integral method may be preferrable in such a case (HASEGAWA, 1936; VESTINE, 1941; PRICE and WILKINS, 1963).

Various results on the ratio of the internal to external coefficients (i_n^m/e_n^m) and the phase difference $\iota_n^m - \varepsilon_n^m$, which are often used in studies of electromagnetic induction on a global scale (see Chapters 9 and 10), are shown in Table 8.1. It should be noticed that the *Sq* field depends on the season of the year and also on solar activity. The results of Chapman indicate the values averaged over seasons in the epochs of solar activity minimum (1902) and maximum (1905). The results of Benkova were derived from the data obtained

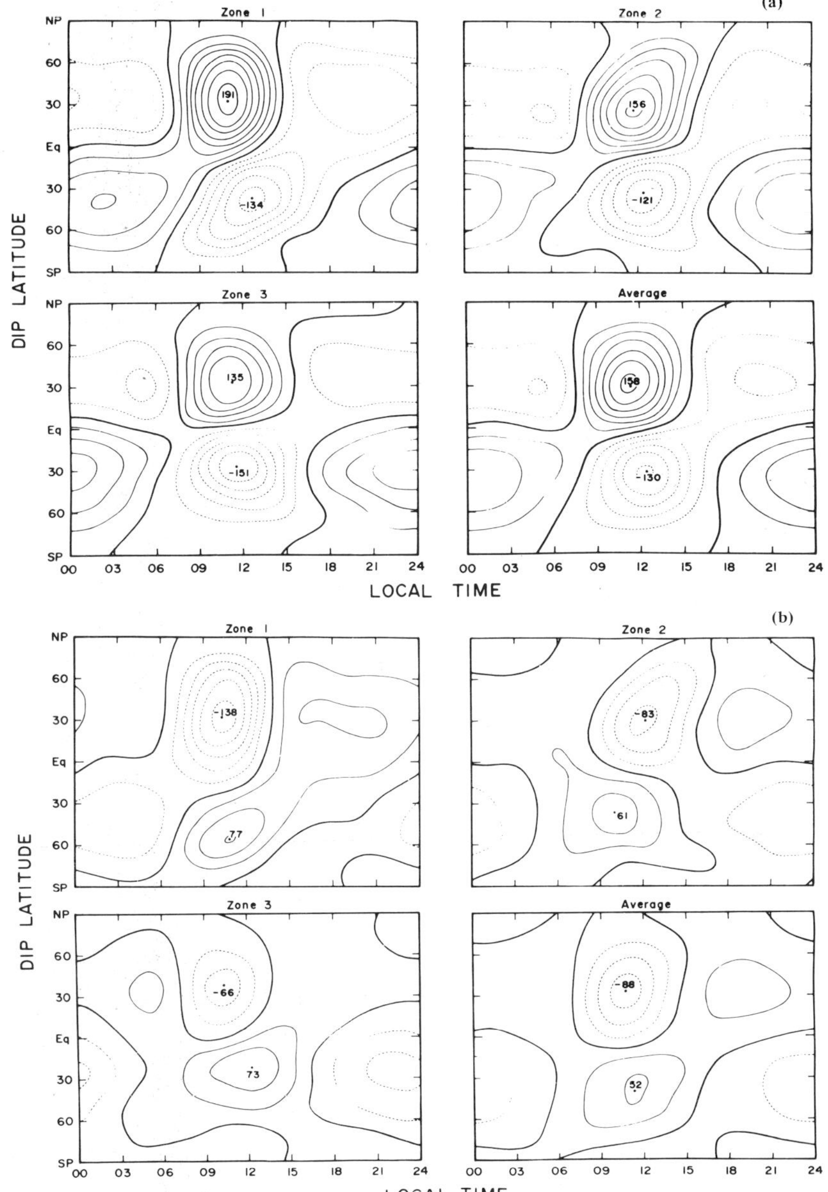

FIG. 8.2. External (a) and internal (b) current systems for the yearly average of *Sq* field. Solid and dotted lines indicate clockwise and counterclockwise current flows, respectively, while thick lines denote the zero lines. Numerals given at the centers of current vortices represent the total current intensity of the vortices in units of 10^3 amperes (SUZUKI, 1973).

in the northern hemisphere in the summer of 1933. The results of Hasegawa and Ota are the average over the summer seasons during 1932–33.

The results of Matsushita and Maeda are based on the data obtained at 68 stations over the world during the International Geophysical Year (IGY). These stations are located between the dip latitudes of $\pm 60°$. The coefficients in (8.1) were obtained for three zones dividing the world, zone 1 (Europe-Africa), zone 2 (Asia-Australia), and zone 3 (North-South America), and for three groups of months, D (November-February), J (May-August), and E (other months), respectively. The values shown in Table 8.1 are their averages for various pairs of (n, m). The results of Suzuki were derived from the same data sets as were used by MATSUSHITA and MAEDA (1965a). In Suzuki's analysis, however, the effect of the equatorial electroject is removed on the grounds that it would contaminate coefficients of spherical harmonics of low degree.

The mechanism of *Sq* field generation has been considered to be as follows. The electrical conductivity of the ionosphere is high, as it contains many charge carriers such as ions and electrons. Particles in the conducting ionosphere are set in motion by a tidal force and/or other forces. Then a dynamo action is generated under the influence of the earth's magnetic field. Generated electric currents flow in the ionosphere, at a level located approximately 100 km above the earth's surface. The electric current system responsible for the external part of the *Sq* field is fixed relative to the direction of the Sun.

The earth's rotation, in the above circumstances, results in electromagnetic induction within a conducting earth and hence the internal field as expressed by i_n^m arises. The *Sq* field observed on the earth's surface is certainly the superposition of the external and internal parts. The external and internal current systems can be calculated from the sets of external and internal coefficients, respectively, as shown in Fig. 8.2 (SUZUKI, 1973).

MATSUSHITA *et al.* (1973) examined the *Sq* current system in the polar regions using geomagnetic data at 40 stations located in and around the northern polar cap region. If this current system is included in the overall current system, the current intensity in mid-latitudes is expected to be enhanced more than previously estimated.

8.2 Lunar Daily Variation

Daily geomagnetic variations include the contribution from a dynamo action in the ionosphere due to lunar tidal forces. This portion is controlled by the lunar time and is called the lunar daily variation (L) (e.g. MATSUSHITA, 1975). However, its amplitude is usually much smaller than that of *Sq* and a

TABLE 8.2. ι_n^m/e_n^m and $\iota_n^m-\varepsilon_n^m$ for L.

n	m	Chapman		Matsushita & Maeda	
		i/e	$\iota-\varepsilon$	i/e	$\iota-\varepsilon$
2	2	0.53	6°	0.45	6°
3	2	0.56	29	0.31	6

long sequence of data is required for detailed studies of L (SCHLAPP, 1978; SHIRAKI, 1981).

Separation of the L field into external and internal parts is also possible. Table 8.2 shows the ratio of internal to external parts and the phase difference as determined by CHAPMAN (1919) and MATSUSHITA and MAEDA (1965b). It is found that the response of the earth to L is not very different from that to Sq.

8.3 Geomagnetic Storms

The geomagnetic field sometimes undergoes violent disturbances called a geomagnetic storm. In mid- and low-latitudes fairly systematic features can be recognized in storm-time variations. Figure 8.3 shows typical storm-time variations observed at the Kakioka Magnetic Observatory. Storm-time variations are characterized mostly by the behavior of the horizontal component. At the beginning of storm the horizontal component suddenly increases as shown clearly in Fig. 8.3. This event is called the sudden storm commencement (ssc). The field remains at the enhanced level for some hours, with some minor fluctuations. This stage is called the initial phase of the storm. The main phase follows this initial phase. During the main phase, the horizontal field is reduced to a considerably lower level, usually a few hundred

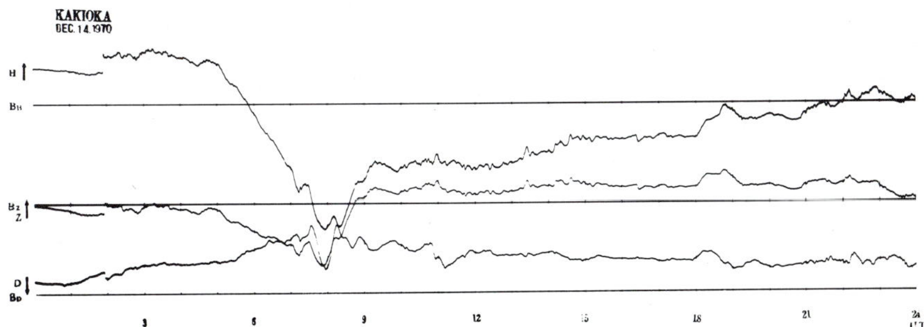

FIG. 8.3. Magnetograms at the Kakioka Magnetic Observatory showing a typical geomagnetic storm.

gammas or less. Then the field tends to recover gradually, still with some fluctuations. This recovery phase usually lasts for a few days. The declination and the vertical component also change markedly during a storm, although their variations are not so typically characterized as the horizontal component.

The earth's main field is confined in a cavity by a plasma stream, called the solar wind, ejected from the sun. This cavity is called the magnetosphere of the earth. Hampered by this stream, the magnetosphere cannot extend beyond about 10 earth-radii on the side facing the sun, while on the opposite side it extends as far as 100 earth-radii or more, as shown in Fig. 8.4.

High energy plasma, ejected at the time of explosions on the sun's surface, collides with the magnetosphere and compresses it. This effect is manifested as an enhancement of magnetic flux density at the earth's surface. This is considered to be the mechanism of ssc. Severe fluctuations during a storm are generated through interactions between high energy plasma and the magnetosphere.

Variations can be classified statistically into two types; *Dst* (storm time variation) and *DS* (disturbance daily-variation). More popularly, the latter has been called the polar substorm (e.g. MATSUSHITA, 1975). Although *Dst* and *DS* fields are not completely separable, the latter will be treated in the next section and the former will further be discussed below.

The main and recovery phases of a storm are supposed to result from the formation and subsequent decay of an equatorial ring current system located at 5–6 earth-radii. Although various mechanisms with three-dimensional configuration have been proposed for the formation of an electric current system in the magnetosphere and ionosphere during a storm (FUKUSHIMA and

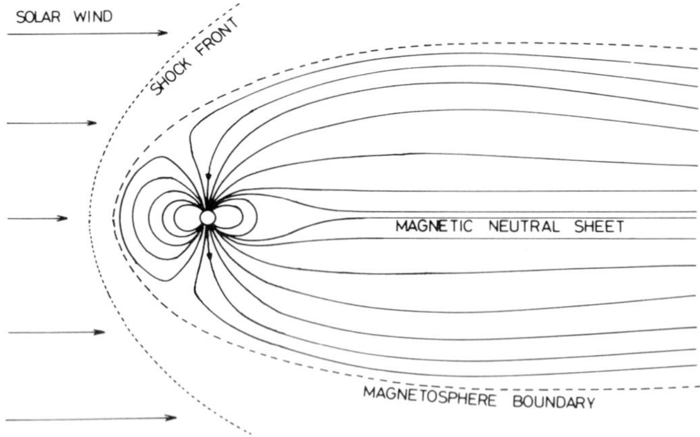

FIG. 8.4. Schematic representation of the earth's magnetosphere.

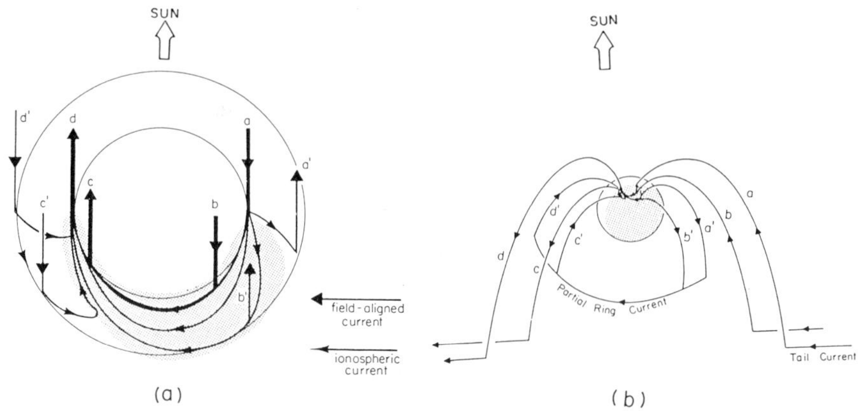

FIG. 8.5. (a) A model of an ionospheric and field-aligned current system in the auroral zone for a magnetospheric substorm. (b) Schematic representation of a three-dimensional current system in the ionosphere and magnetosphere for a magnetospheric substorm (KAMIDE *et al.*, 1976).

KAMIDE, 1973), it is likely that the equatorial ring current is not a complete ring but an asymmetric or partial ring and the ring current system is connected with the electrojet in the auroral zones through field-aligned currents, that is, electric currents flowing along magnetic lines of force. Such a current system is well illustrated in Fig. 8.5 (KAMIDE and FUKUSHIMA, 1972).

RIKITAKE (1950b) analyzed magnetograms of ssc obtained at a number of stations over the world and pointed out that the magnetic potential W can be expressed approximately as

$$W = a[(r/a)e_1 + (a/r)^2 i_1]\, P_1(\cos\theta) \tag{8.2}$$

More detailed analyses by MAEDA *et al.* (1965) yielded $i_1/e_1 \approx 0.5$, which implies that, for a short-period variation, the earth responds as a highly conducting sphere. It was also pointed out that the distribution of Z variation is fairly irregular, although on the average it can be accounted for by (8.2). The irregular distribution of Z variations is now interpreted as reflecting local anomalies of short-period geomagnetic variations.

Dst fields have also been analyzed to examine the earth's response to long-period variations (e.g. CHAPMAN and PRICE, 1930; RIKITAKE and SATO, 1957; TAKEUCHI and SAITO, 1963; ANDERSSEN and SENETA, 1969). The magnetic potential of *Dst* is also expressed approximately by (8.2) as theoretically expected from the equatorial ring current system. For *Dst* fields the ratio i_1/e_1 was obtained as 0.37–0.39 (RIKITAKE and SATO, 1957).

8.4 Substorm and Bay-Type Variation

A considerable variation having a duration of a few hours often occurs in high-latitudes. From studies of the three components of geomagnetic variation, the electric current system responsible for the variation has been identified as a strong westward current (electroject) in the auroral zone. This type of magnetic disturbance is called the polar magnetic substorm (e.g. ROSTOKER, 1972). Studies of substorms revealed that the westward and eastward electrojets are connected with the ring-current and tail-current systems through field-aligned currents and large-scale circuits are formed in the ionosphere and magnetosphere (e.g. FUKUSHIMA and KAMIDE, 1973; KISABETH, 1975). A typical model is shown in Fig. 8.5 (KAMIDE and FUKUSHIMA, 1972). In this sense, a substorm may be called a magnetospheric substorm.

The influence of a polar substorm in mid- and low-latitudes is manifested as a bay-type variation. KAMIDE and FUKUSHIMA (1972) attributed it to return currents in the ionosphere, while KISABETH (1975) pointed out that it can be ascribed to field-aligned currents.

BANNISTER and GOUGH (1977, 1978) and GOUGH and BANNISTER (1978) studied electric current systems responsible for some substorm events by analyzing magnetograms obtained at an array of stations beneath the auroral oval in western Canada. In particular, ionospheric current systems in the auroral zone were examined in detail. According to their conclusions, a combination of westward and eastward electrojets is required to account for the distribution of the observed geomagnetic variations during a substorm.

According to the result of spherical harmonic analysis applied to bay-type variations observed in mid- and low-latitudes, i/e amounts to 0.45 or so for spherical harmonics of low degree (RIKITAKE, 1950a, b, c).

8.5 Micropulsations

On many occasions, the geomagnetic field exhibits variations of very short period and small amplitudes, usually a few tens of gammas at largest. These variations are called micropulsations and have been classified into several types according to their morphology. The term *Pc* is assigned to regular and continuous variations, while *Pi* is used for rather irregular variations. *Pc* and *Pi* are further divided according to their periods as shown in Table 8.3 (JACOBS, 1970). Amplitudes representative of *Pc* 1, *Pc* 2–*Pc* 4, and *Pc* 5 are 0.05–0.1, 0.1–1, and 1–10 gammas, respectively.

Generation of micropulsations has something to do with interactions between solar wind plasma and the earth's magnetosphere and the periods of

TABLE 8.3. Classification of *Pc* and *Pi*.

	Period (sec)
Pc 1	0.2–5
2	5–10
3	10–45
4	45–150
5	150–600
Pi 1	1–40
2	40–150

micropulsations are related to the dimensions of the magnetosphere and its resonance cavity (e.g. JACOBS, 1970). Thus studies of micropulsations are expected to provide information on the structure of the magnetosphere, although this has little bearing on solid earth geomagnetism.

From the viewpoint of electromagnetic induction within the earth, micropulsations can be used to investigate the surface conductivity structure, particularly in connection with crustal conductivity anomalies. In fact, crustal anomalies have recently been drawing much attention because of their potential relation with studies of geothermal energy and earthquake prediction (see Chapter 12).

8.6 *Geomagnetic Variations of Long Period*

In order to determine the electrical conductivity in the deep mantle from electromagnetic induction studies, we must essentially rely on geomagnetic variations that can penetrate deep into the earth. In view of this, geomagnetic variations of periods longer than *Dst* have sometimes been studied. RIKITAKE (1951a) analyzed a variation of 27 day period, which is presumably associated with the sun's rotation. YUKUTAKE (1965) examined the contribution of the solar cycle to geomagnetic variations and derived a variation of 11 year period. Similar studies of 11 year period variation were undertaken by ECKHART *et al.* (1963) and CURRIE (1966). The above variations are characterized by $P_1(\cos\theta)$ in spherical harmonic representation. Analyses of these variations yielded $i_1/e_1 \approx 0.25$, 0.13–0.16 for the 27 day (RIKITAKE, 1951a) and 11 year (YUKUTAKE, 1965) periods, respectively.

Continuum spectra of geomagnetic variation have also been brought to light by spectral analysis of a long sequence of magnetic data. CURRIE (1966) examined the geomagnetic spectrum for periods ranging from 40 days to 5.5 years. Two line spectra, semi-annual and annual, were evident and Currie

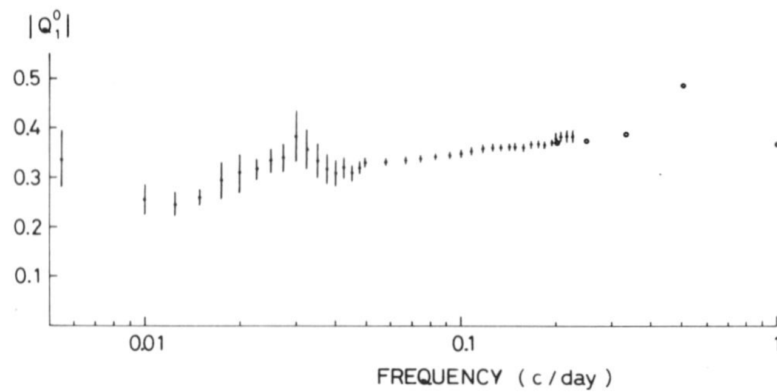

FIG. 8.6. Modulus of response function Q_1^0 corresponding to i_1/e_1. Bars indicate standard errors. Open circles denote estimates derived from *Dst* data by TAKEUCHI and SAITO (1963) (BANKS, 1972).

pointed out that the semi-annual variation has its origin in the equatorial ring current in the magnetosphere, while the annual one originates in the ionosphere; the latter is due to ionospheric winds blowing from the summer to winter hemispheres. Other continuum spectra for the period range 40 days–3.7 days were attributed to modulations of the *Dst* field. CURRIE (1973) also examined geomagnetic line spectra for periods of 2–70 years.

Similarly, BANKS (1969, 1972) derived continuum spectra for 4.4–100 day periods, in addition to annual and semi-annual lines. Except for the annual variation, which can be approximated by P_2 in spherical harmonic representation, and *Sq*, long-period variations of external origin are expressed in terms of P_1, reflecting, most probably, fluctuations in the equatorial ring current. Figure 8.6 shows values of i_1/e_1 plotted against frequency (BANKS, 1972).

CHAPTER 9

CHAPTER 9

ELECTROMAGNETIC INDUCTION IN A SPHERICAL CONDUCTOR

It has been shown in the previous chapter that the internal part of *Sq*, *L* and other temporal geomagnetic variations can be interpreted as the magnetic field produced by the electric currents induced within the earth by the time-dependent, external part of those variations. In order to test whether such an interpretation is correct, it is useful to estimate the induced electric current and magnetic field on the basis of the theory of electromagnetic induction in a sphere, such a theory having been developed by CHAPMAN (1919), CHAPMAN and PRICE (1930), LAHIRI and PRICE (1939). RIKITAKE (1966a) and PRICE (1967) have presented summary reviews of the problem.

9.1 Fundamental Equations

We denote magnetic field, electric current density, electric displacement, electric field, magnetic induction, volume electric charge and time by $\boldsymbol{H}, \boldsymbol{i}, \boldsymbol{D}$, $\boldsymbol{E}, \boldsymbol{B}, \rho$ and t, respectively, so that Maxwell's equations in a stationary medium are written as

$$\operatorname{curl} \boldsymbol{H} = 4\pi \boldsymbol{i} + \partial \boldsymbol{D}/\partial t \tag{9.1}$$

$$\operatorname{curl} \boldsymbol{E} = -\partial \boldsymbol{B}/\partial t \tag{9.2}$$

$$\left.\begin{aligned} \operatorname{div} \boldsymbol{B} &= 0 \\ \operatorname{div} \boldsymbol{D} &= 4\pi\rho \end{aligned}\right\} \tag{9.3}$$

in electromagnetic units.

Assuming an isotropic medium, we have relations such as

$$\boldsymbol{D} = \varepsilon \boldsymbol{E}, \quad \boldsymbol{B} = \mu \boldsymbol{H}, \quad \boldsymbol{i} = \sigma \boldsymbol{E} \tag{9.4}$$

where ε, μ and σ are the dielectric constant, magnetic permeability and electrical conductivity, respectively.

Taking the divergence of both sides of Eq. (9.1), we obtain

$$4\pi \operatorname{div} \boldsymbol{i} + \operatorname{div} \partial \boldsymbol{D}/\partial t = 0$$

which can be rewritten as

$$\operatorname{div} (\sigma/\varepsilon \boldsymbol{D}) = -\partial \rho/\partial t$$

by virtue of (9.4). Accordingly, we obtain

$$\partial \rho/\partial t + 4\pi\sigma\rho/\varepsilon = (\sigma/\varepsilon)\boldsymbol{i} \cdot \operatorname{grad} (\varepsilon/\sigma). \tag{9.5}$$

We therefore see that the right-hand side of (9.5) vanishes either when ε/σ is constant or when the electric current flows in a direction perpendicular to grad (ε/σ), so that electric charge ρ is independent of electric and magnetic fields in such cases. Even if ρ is assumed to be finite, it should decay exponentially with time.

Except for the above two cases, electric charge is not independent of electric and magnetic fields, so that the general theory of electromagnetic induction becomes very difficult to handle.

Assuming ε and μ are constant, we further assume either that σ is constant or that the electric currents flow in a layer in which σ is constant. In such cases, we may put $\rho = 0$, as justified above and so, from (9.3), we can write

$$\operatorname{div} \boldsymbol{E} = 0 \tag{9.6}$$

without loss of generality. Taking the curl of (9.1), we obtain

$$\nabla^2 \boldsymbol{H} = \mu(4\pi\sigma\partial/\partial t + \varepsilon\partial^2/\partial t^2)\boldsymbol{H} \tag{9.7}$$

with the aid of relation (9.6) and (grad $\sigma \cdot \boldsymbol{E}) = 0$. We also see that an equation similar to (9.7) holds good for $\boldsymbol{E}$.

When the change in $\boldsymbol{H}$ is not too rapid, we may assume that $\varepsilon\partial \boldsymbol{H}/\partial t \ll 4\pi\sigma \boldsymbol{H}$ holds in a conductor. In that case, (9.7) becomes

$$\nabla^2 \boldsymbol{H} = 4\pi\sigma\mu\partial \boldsymbol{H}/\partial t. \tag{9.8}$$

On the other hand, the first term of the right-hand side of (9.7) can be ignored in a dielectric or an insulator, so that we have

$$\nabla^2 \boldsymbol{H} = \mu\varepsilon\partial^2 \boldsymbol{H}/\partial t^2 \tag{9.9}$$

which is a wave equation indicating that an electromagnetic disturbance propagates with a velocity of $1/\sqrt{\mu\varepsilon}$. When we discuss relatively slow field variations, the time of propagation through any portion of the dielectric can be regarded as negligible compared with the time taken for the field changes at a point to become effective. In this case, we may ignore the right-hand side of (9.9) obtaining

$$\nabla^2 \boldsymbol{H} = 0. \tag{9.10}$$

In the more general latter case, we can define a magnetic potential W in a dielectric (insulator). If the displacement current is ignored in (9.1) as discussed above, the potential is related to the magnetic field by the equation

$$\boldsymbol{H} = -\operatorname{grad} W. \tag{9.11}$$

In the conductor, we may define a vector potential $\boldsymbol{A}$ by

$$\operatorname{curl} \boldsymbol{A} = \boldsymbol{B}. \tag{9.12}$$

In this case, (9.2) leads to

$$\boldsymbol{E} = -\partial \boldsymbol{A}/\partial t - \operatorname{grad} \varphi \tag{9.13}$$

where φ is an arbitrary function.

It is shown that, if

$$\operatorname{div} \boldsymbol{A} + 4\pi\sigma\mu\varphi = 0 \tag{9.14}$$

the fundamental equation (9.8) can be satisfied. It is therefore possible to take $\varphi = 0$, which is consistent with (9.4), (9.12) and (9.14) because we see that

$$\operatorname{div} \boldsymbol{A} = 0 \tag{9.15}$$

holds, from (9.6). Equation (9.13) then reduces to

$$\boldsymbol{E} = -\partial \boldsymbol{A}/\partial t \tag{9.16}$$

while $\boldsymbol{A}$ satisfies

$$\nabla^2 \boldsymbol{A} = 4\pi\sigma\mu\partial \boldsymbol{A}/\partial t. \tag{9.17}$$

9.2 *Conductor of Spherical Symmetry*

Let us assume a spherical conductor in which the conductivity is a function of radius. We put

$$\sigma = \sigma(\rho) \tag{9.18}$$

where

$$\rho = r/a \tag{9.19}$$

a being the radius of the sphere. The origin of spherical polar coordinates (r, θ, ϕ) is taken at the center of the sphere.

LAHIRI and PRICE (1939) showed that a solution of (9.17) appropriate for the present problem is given by

$$\boldsymbol{A} = \boldsymbol{r} \times \operatorname{grad} u \tag{9.20}$$

where $\boldsymbol{r}$ denotes the radius vector and u satisfies

$$\nabla^2 u = 4\pi\sigma\mu\partial u/\partial t. \tag{9.21}$$

It is evident that $\boldsymbol{A}$ as given by (9.20) satisfies div $\boldsymbol{A} = 0$ and that $\boldsymbol{A}$ is a vector of toroidal type. We may also have another vector of poloidal type as a solution of (9.17), but the magnetic field for such a vector potential vanishes outside the conductor, so that it is not involved with electromagnetic induction by an external magnetic field.

A typical solution of (9.21) is obtained as

$$u_n = a\, f_n(p, \rho)\, S_n \tag{9.22}$$

where S_n is a surface spherical function having degree n and f_n is a solution of a differential equation

$$(\mathrm{d}/\mathrm{d}\rho)(\rho^2 \mathrm{d} f_n/\mathrm{d}\rho) = [n(n+1) + 4\pi\mu a^2 \rho^2 \sigma(\rho) p]\, f_n \tag{9.23}$$

in which p stands for $\partial/\partial t$.

Corresponding to the typical solution for the vector potential, the r, θ and ϕ components of magnetic induction can be obtained as

$$\begin{aligned} B_r &= -\rho^{-1} n(n+1) f_n S_n \\ B_\theta &= -\rho^{-1} \mathrm{d}(\rho f_n)/\mathrm{d}\rho \cdot \partial S_n/\partial\theta \\ B_\phi &= -\rho^{-1} \mathrm{d}(\rho f_n)/\mathrm{d}\rho \cdot \partial S_n/\sin\theta\,\partial\phi. \end{aligned} \tag{9.24}$$

On the other hand, a typical solution for the magnetic potential, satisfying

$$\nabla^2 W = 0 \tag{9.25}$$

outside the conductor ($\sigma = 0$, $\mu = 1$) is given by

$$W_n = a(e_n \rho^n + i_n \rho^{-n-1}) S_n \tag{9.26}$$

in which e_n and i_n are the coefficients of magnetic potential originating outside and inside the conductor, respectively. The magnetic field components are then obtained as

$$\begin{aligned} H_r &= -[n e_n \rho^{n-1} - (n+1) i_n \rho^{-n-2}] S_n \\ H_\theta &= -(e_n \rho^{n-1} + i_n \rho^{-n-2}) \partial S_n/\partial\theta \\ H_\phi &= -(e_n \rho^{n-1} + i_n \rho^{-n-2}) \partial S_n/\sin\theta\,\partial\phi. \end{aligned} \tag{9.27}$$

When the surface of the spherical conductor is identified by $r = qa$ or $\rho = q$, the continuity conditions for B_r, H_θ and H_ϕ lead to

$$\left.\begin{aligned} n(n+1)(\rho^{-1} f_n)_{\rho=q} &= ne_n q^{n-1} - (n+1) i_n q^{-n-2} \\ \mu^{-1}[\rho^{-1}\mathrm{d}(\rho f_n)/\mathrm{d}\rho]_{\rho=q} &= e_n q^{n-1} + i_n q^{-n-2}. \end{aligned}\right\} \tag{9.28}$$

9.2.1 Uniform sphere

When σ is constant throughout the sphere, differential equation (9.23) reduces to

$$\mathrm{d}^2 f_n/\mathrm{d}\rho^2 + 2\rho^{-1}\mathrm{d} f_n/\mathrm{d}\rho - [n(n+1)\rho^{-2} + k^2a^2] f_n = 0 \tag{9.29}$$

where

$$k^2 = 4\pi\mu\sigma p. \tag{9.30}$$

A solution of (9.29), which is finite at $\rho = 0$, is readily obtained as

$$f_n = C_n \rho^{-1/2} I_{n+1/2}(ka\rho) \tag{9.31}$$

where $I_{n+1/2}$ is a modified Bessel function and C_n is a constant. Putting (9.31) into (9.28) and solving i_n and C_n, we obtain

$$\frac{i_n}{e_n} = \frac{n}{n+1} q^{2n+1}\left[1 - \mu\left\{\frac{kqa}{2n+1}\frac{I_{n-1/2}(kqa)}{I_{n+1/2}(kqa)} + \frac{n(\mu-1)}{2n+1}\right\}^{-1}\right] \tag{9.32}$$

$$\frac{C_n}{e_n} = \frac{\mu(2n+1)}{n+1} q^{n+1/2}[kqaI_{n-1/2}(kqa) + n(\mu-1)I_{n+1/2}(kqa)]^{-1}. \tag{9.33}$$

Supposing that the coefficient of magnetic potential of external origin e_n is given, we see from (9.32) and (9.33) that i_n and C_n can be determined, so that magnetic and electric fields and electric currents arising from electromagnetic induction can be obtained in terms of e_n.

When the change in the external field is purely periodic, we may put

$$p = i\alpha, \quad \alpha = 2\pi/T, \quad i = \sqrt{-1}. \tag{9.34}$$

In this case, i_n/e_n and C_n/e_n become complex quantities. The amplitude ratio and phase difference between the external and internal fields are given by mod i_n/e_n and arg i_n/e_n, respectively.

When $|kqa| \gg 1$, asymptotic expansions of modified Bessel functions are available. In such a case, we obtain

$$\frac{I_{n-1/2}(kqa)}{I_{n+1/2}(kqa)} \simeq 1 + \frac{n}{kqa} + \frac{n(n+1)}{2(kqa)^2} + \frac{n(n+1)}{2(kqa)^3} + \cdots \tag{9.35}$$

which makes numerical computation of (9.32) easy.

CHAPMAN (1919), who applied a uniform sphere model to his results of analyses of *Sq*, obtained $q = 0.96$ and $\sigma/\mu = 3.6 \times 10^{-13}$ e.m.u. As no ferromagnetism is likely to be present in the deep interior of the earth, $\sigma = 3.6 \times 10^{-13}$ e.m.u. (3.6×10^{-2} S/m) is obtained for the conductivity putting $\mu = 1$. A combination $q = 0.94$ and $\sigma = 5 \times 10^{-1}$ S/m has been obtained on the basis of more recent analyses of *Sq* (RIKITAKE, 1950a).

When the external inducing field is aperiodic, (9.32) becomes an operational equation in which e_n is a function of time. LAHIRI and PRICE (1939) argued that the relationship between external and internal fields during the main phase of a magnetic storm (*Dst*) cannot be explained by a uniform sphere model which is compatible with that of *Sq*. As the *Dst* study required a higher conductivity, it has been surmised that the conductivity in the deeper part of the earth must be higher than that of the shallower part because the electric currents induced by a *Dst*, which changes more slowly than *Sq*, penetrate deeper than those by *Sq*.

9.2.2 Non-uniform sphere

LAHIRI and PRICE (1939) developed a theory of electromagnetic induction in a non-uniform sphere in which the conductivity is given by

$$\sigma = \sigma_0 \varrho^{-l} \tag{9.36}$$

where σ_0 and l are constants.

The development of high-speed computers has enabled us to calculate the electromagnetic response of a sphere in which the conductivity is distributed arbitrarily (TAKEUCHI and SAITO, 1963). ECKHARDT (1963) presented a method of obtaining i_n/e_n for any distribution of conductivity as long as spherical symmetry holds.

We shall, in the following, develop a theory of electromagnetic induction within a non-uniform sphere in a manner which is suitable for applying to periodic, as well as non-periodic, cases to be handled by computer.

Let us measure the length in units of a and the time in units of $4\pi\mu\sigma_0 a^2$ where a and σ_0 are typical length and conductivity of the system concerned, respectively.

In that case, (9.17) becomes

$$\partial \boldsymbol{A}/\partial t = K \nabla^2 \boldsymbol{A} \tag{9.37}$$

in which

$$K = \sigma_0/\sigma. \tag{9.38}$$

We take a typical solution as given by (9.22) where, for the sake of convenience, we assume

$$u_n(t, \rho) = \rho^n y_n(t, \rho). \tag{9.39}$$

From (9.37), we then see that y_n should satisfy

$$\partial y_n/\partial t = K(\rho)[\rho^2 \partial^2 y_n/\partial \rho^2 + 2(n+1)\partial y_n/\partial \rho]. \tag{9.40}$$

Taking relations (9.24) and (9.27) into account, the boundary conditions at $\rho = 1$ become

$$\begin{aligned} n(n+1)(y_n)_{\rho=1} &= n e_n - (n+1) i_n \\ \mu^{-1}[\rho^{-1}\partial(\rho^{n+1} y_n)/\partial \rho]_{\rho=1} &= e_n + i_n \end{aligned} \tag{9.41}$$

in place of (9.28).

(1) Periodic variation

Putting

$$\eta_n = i_n/e_n \tag{9.42}$$

ECKHARDT (1963) showed that η_n satisfies

$$\frac{\mathrm{d}\eta_n}{\mathrm{d}\rho} = \frac{i4\pi\mu\sigma\omega(n+1)}{(2n+1)n}\left(\eta_n - \frac{n}{n+1}\right)^2 - \frac{2n+1}{\rho}\eta_n \tag{9.43}$$

where ω is the angular frequency of the variation.

At a depth well below the skin depth, the induced current intensity is practically zero, so that $\eta_n = n/(n+1)$, which holds in a perfect conductor, may be assumed. Starting from that, η_n at $\rho = 1$ can be calculated for any distribution of σ by means of numerical integration.

Once η_n at $\rho = 1$ is obtained, it is possible to get at y_n and $\partial y_n/\partial \rho$ at $\rho = 1$ from (9.41). Then, y_n and consequently the electric current density can be calculated by numerical methods from Eq. (9.40).

(2) Aperiodic variation

In order to study electromagnetic induction by a transient variation in a spherical body with an arbitrary conductivity distribution, Eq. (9.40) has to be converted into a difference equation. The range from $\rho = 0$ to 1 is divided into M equal parts. If $y_{n\nu}$, σ_ν and K_ν are the values of y_n, σ and K at the ν-th point of division, (9.40) is converted to a set of difference equations such as

$$\begin{aligned} y_\nu(t+\Delta t) = y_\nu(t) + \Delta t K_\nu \rho_\nu M[&(\rho_\nu M + n + 1) y_{\nu+1}(t) \\ &- 2\rho_\nu M y_\nu(t) + (\rho_\nu M - n - 1) y_{\nu-1}(t)] \\ &\text{for } \nu = 1, 2, \ldots, M \end{aligned} \tag{9.44}$$

from which subscript n is dropped.

Equation (9.44) holds for $\nu = 1, 2, \ldots, M$ or $\rho = 1/M, 2/M, \ldots, 1$. For $\nu = 0$ or $\rho = 0$, it is obvious that y_ν assumes a constant value which may be taken as zero. On the other hand, we see that y_{M+1} or y_ν at $\rho = 1 + (1/M)$ have no physical existence. Since, however, the boundary conditions given by (9.41) imply

$$\begin{aligned} M(n+1)(y_{M+1} - y_{M-1})/2 + (n+1)(2n+1)y_M &= (2n+1)e_n \\ Mn(y_{M+1} - y_{M-1})/2 &= (2n+1)i_n \end{aligned} \tag{9.45}$$

provided $\mu = 1$ is assumed, y_{M+1} can be obtained from the first equation of (9.45) as

$$y_{M+1} = [2(2n+1)e_n - 2(n+1)(2n+1)y_M + M(n+1)y_{M-1}]/M(n+1) \tag{9.46}$$

i_n is also calculated as

$$i_n = ne_n/(n+1) - ny_M \tag{9.47}$$

which serves for estimating the induced field.

Putting (9.46) into one of Eq. (9.44) for $\nu = M$, we obtain

$$\begin{aligned} y_M(t+\Delta t) = y_M(t) + \Delta t K_M M\,[(M+n+1)\{&2(2n+1)e_n/M(n+1) \\ &- 2(2n+1)y_M(t)/M + y_{M-1}(t)\} \\ &- 2My_M(t) + (M-n-1)y_{M-1}(t)] \end{aligned} \tag{9.48}$$

which should be used for $\nu = M$ in place of (9.44).

Let us first assume a step function inducing field. In this case, we assume $e_n = 1$ for $t = 0$ and that e_n for other values of t is always zero. In other words, the induced currents are excited at $t = 0$ only. They are subjected to simple dissipation thereafter.

Starting from $y_1, y_2, \ldots, y_M$ at $t = 0$, which are all zero, it is now possible to estimate $y_1, y_2, \ldots, y_M$ at $t = \Delta t$ from (9.44) and (9.48). The y_ν at $t = 2\Delta t$, $3\Delta t$, ... are calculated successively in a similar way. The i_n for the corresponding times are estimated from (9.47). Care should be taken to use an appropriate Δt in order that the successive calculation converges.

Let us write $y_\nu(t)$ thus obtained as $A_\nu(t)$. If $e_n(t)$ is an arbitrary function of time, a theorem of operational calculus leads to

$$y_\nu(t) = \frac{d}{dt}\int_0^t A_\nu(t-\tau)e_n(\tau)d\tau \tag{9.49}$$

which is transformed to

$$y_\nu(t) = A_\nu(t)e_n(0) + \int_0^t A_\nu(t-\tau)e'_n(\tau)d\tau. \tag{9.50}$$

It is therefore seen that induced currents proportional to the increment of e_n are excited at each instant. The currents thus excited are then subjected to natural decay. The y_ν are the sum of those currents excited previously.

9.3 *Nearly Spherical Conductor*

No general theory of electromagnetic induction in a conductor having an irregular shape has ever been put forward because of various difficulties. However, RIKITAKE (1964b) tackled a problem of electromagnetic induction in a perfectly conducting body with a shape slightly different from a sphere.

Let us take a conductor, with a surface described by

$$r = q_0 a[1 + \varepsilon u(\theta)] \tag{9.51}$$

in spherical coordinates; q_0, a and ε are constants.

Equation (9.51) is rewritten as

$$\rho = q_0[1 + \varepsilon u(\theta)] \tag{9.52}$$

where ρ is defined by (9.19).

When the external field is uniform and parallel to the $\theta = 0$ axis, its potential is given as

$$W_e = a e_1 \rho P_1(\cos\theta). \tag{9.53}$$

Consequently, the r and θ components of the magnetic field become

$$H_{re} = -e_1 P_1,\ H_{\theta e} = -e_1 \mathrm{d}P_1/\mathrm{d}\theta. \tag{9.54}$$

The magnetic potential of the internal field produced by the induced currents is given by

$$W_i = a \sum_n i_n \rho^{-n-1} P_n(\cos\theta) \tag{9.55}$$

and so the magnetic field components are given by

$$H_{ri} = \sum_n (n+1) i_n \rho^{-n-2} P_n,\ H_{\theta i} = -\sum_n i_n \rho^{-n-2} \mathrm{d}P_n/\mathrm{d}\theta. \tag{9.56}$$

From the boundary condition that the magnetic field normal to the surface of a perfect conductor should vanish, we have

$$(H_{re} + H_{ri})\cos\alpha - (H_{\theta e} + H_{\theta i})\sin\alpha = 0 \tag{9.57}$$

over the whole surface of conductor. In (9.57), α is the angle between the normal to the surface at a point and the straight line connecting that point to the origin of the coordinates. $\cos\alpha$ and $\sin\alpha$ are given by

$$\cos \alpha = 1/\{1 + (f'/r)^2\}^{1/2}, \ \sin \alpha = (f'/r)/\{1 + (f'/r)^2\}^{1/2} \tag{9.58}$$

where f denotes the right-hand side of (9.51) and $f' = \mathrm{d}f/\mathrm{d}\theta$.

Accordingly, (9.57) can be rewritten as

$$H_{re} + H_{ri} - (H_{\theta e} + H_{\theta i})\varepsilon u'(\theta)/[1 + \varepsilon u(\theta)] = 0. \tag{9.59}$$

Substituting (9.54) and (9.56) into (9.59) we obtain

$$\sum_n (n+1) i_n \rho^{-n-2} P_n + q_0 \varepsilon \sum_n i_n \rho^{-n-3} \frac{\mathrm{d}P_n}{\mathrm{d}\theta} u'(\theta) = e_1 \left[P_1 - \varepsilon \frac{\mathrm{d}P_1}{\mathrm{d}\theta} u'(\theta)/\{1 + \varepsilon u(\theta)\} \right] \tag{9.60}$$

in which ρ is a function of θ as defined by (9.52).

Multiplying both the sides of (9.60) by $P_N \sin\theta$ and integrating it with respect to θ from 0 to π, we obtain

$$\sum_n [(n+1)\alpha_{nN} + \varepsilon \beta_{nN}] q_0^{-n-2} i_n = (\xi_{1N} + \varepsilon \eta_{1N}) e_1 \tag{9.61}$$

where

$$\begin{aligned} \alpha_{nN} &= \int_0^\pi \frac{P_n P_N \sin\theta \, \mathrm{d}\theta}{[1 + \varepsilon u(\theta)]^{n+2}} \\ \beta_{nN} &= \int_0^\pi \frac{(\mathrm{d}P_n/\mathrm{d}\theta) P_N u'(\theta) \sin\theta \, \mathrm{d}\theta}{[1 + \varepsilon u(\theta)]^{n+3}} \\ \xi_{1N} &= \int_0^\pi P_1 P_N \sin\theta \, \mathrm{d}\theta \\ \eta_{1N} &= -\int_0^\pi \frac{(\mathrm{d}P_1/\mathrm{d}\theta) P_N u'(\theta) \sin\theta \, \mathrm{d}\theta}{1 + \varepsilon u(\theta)} \end{aligned} \tag{9.62}$$

(9.61) is a set of simultaneous equations in i_n.

If we assume $n, N \leqq 12$ and

$$q_0 = 0.8837, \ \varepsilon = 0.1, \ u(\theta) = \sin\theta \tag{9.63}$$

we obtain i_n as given in Table 9.1, where we see that i_n vanishes for even values of n. When ε becomes large, it becomes difficult to solve the simultaneous equations with sufficient accuracy, so that the present approach cannot be applied to a conductor having an undulation of large amplitude.

Neglecting i_n for $n \geqq 3$, which are much smaller than i_1, in Tabel 9.1, let us try to account for the amplitude ratio as represented by i_1/e_1 by a spherical model with perfect conductivity. In this case, (9.32) shows that

TABLE 9.1. i_n in units of e_1.

n	i_n
1	0.4466
3	−0.0252
5	−0.0027
7	−0.0012
9	0.0007
11	−0.0006

$$i_n/e_n = \frac{n}{n+1} q^{2n+1}$$

should hold. We therefore have $q = 0.963$.

On the other hand, the average radius calculated from (9.63) amounts to only $0.94a$. It should therefore be noticed that the radius of the conductor obtained only from i_1/e_1 does not agree with the average radius. It is evident that the estimate is greatly affected by the uplifted portions of the conductor, leading to an overestimate of the radius.

9.4 Thin Spherical Shell

9.4.1 General theory of electromagnetic induction in a thin sheet

When the thickness of a conductor is extremely small compared to the spatial extent of the magnetic field, it is possible to treat the conductor as a thin sheet. Generally speaking, it is difficult to discuss electromagnetic induction in a non-uniform conductor because no separation of magnetic and electric fields from electric charge is possible. However, we can get rid of such a difficulty in the case of a thin sheet, so that mathematical treatment becomes feasible. In geophysics, electromagnetic induction in thin sheets can be applied to problems related to the ionosphere, the oceans, the superficial layers of the earth's crust and so on.

From the Maxwell equations, we can deduce (e.g. RIKITAKE, 1966a, pp. 154–156) that

$$\rho \operatorname{div}(\boldsymbol{H}_+ - \boldsymbol{H}_-) + \operatorname{grad} \rho \cdot (\boldsymbol{H}_+ - \boldsymbol{H}_-) = -4\pi \partial H_n/\partial t \tag{9.64}$$

holds for a thin sheet, where ρ is the reciprocal of the integrated conductivity as given by

$$1/\rho = \int_0^d \sigma \mathrm{d}\zeta \tag{9.65}$$

in which d is the thickness of the sheet. $\boldsymbol{H}_+$ and $\boldsymbol{H}_-$ denote the magnetic vectors immediately outside and inside the thin sheet, respectively. H_n is the component of magnetic field normal to the sheet. The magnetic permeability is assumed to be 1, as before. t denotes the time.

We denote the magnetic potential of the field arising from origins outside the sheet by W_e, while W_i is that arising from the electric currents induced in the sheet by the electromagnetic induction. The total magnetic potential is then given by $W_e + W_i$. As the external field is continuous at the sheet, (9.64) can be rewritten as

$$\begin{aligned}&\rho \operatorname{div} \operatorname{grad}(W_{i+} - W_{i-}) + \operatorname{grad} \rho \cdot \operatorname{grad}(W_{i+} - W_{i-}) \\ &= -4\pi\partial(\partial W_e/\partial n + \partial W_i/\partial n)/\partial t\end{aligned} \tag{9.66}$$

where n is the normal to the sheet.

Consequently , electromagnetic induction by a time-dependent field in a thin sheet, with a magnetic potential given by W_e, can be reduced to the problem of solving

$$\nabla^2 W_i = 0 \tag{9.67}$$

under the conditions that W_i becomes zero at infinity and that W_i satisfies (9.66) at the sheet.

It is sometimes convenient to use a current function Ψ to describe the electric currents in the sheet. Ψ is defined by

$$\Psi = (1/4\pi)(W_+ - W_-). \tag{9.68}$$

As W_e is continuous at the sheet, W in (9.68) can be replaced by W_i.

9.4.2 *Electromagnetic induction in a spherical sheet*

Equation (9.66) can be rewritten in spherical polar coordinates as

$$\begin{aligned}&a^{-2}\left[\frac{\partial \rho}{\partial \theta}\frac{\partial}{\partial \theta} + \frac{1}{\sin^2\theta}\frac{\partial \rho}{\partial \phi}\frac{\partial}{\partial \phi} + \frac{\rho}{\sin\theta}\frac{\partial}{\partial \theta}\left(\sin\theta\frac{\partial}{\partial \theta}\right) + \frac{\rho}{\sin^2\theta}\frac{\partial^2}{\partial \phi^2}\right](W_{i+} - W_{i-}) \\ &= -4\pi\frac{\partial^2}{\partial t\partial r}(W_e + W_i)\end{aligned} \tag{9.69}$$

where a is the radius of the spherical shell. Equation (9.69) is the boundary condition to be satisfied at $r = a$.

When we use the current function Ψ, as defined in (9.68), (9.69) is rewritten as

$$\left[\frac{\partial\rho}{\partial\theta}\frac{\partial}{\partial\theta}+\frac{1}{\sin^2\theta}\frac{\partial\rho}{\partial\phi}\frac{\partial}{\partial\phi}+\frac{\rho}{\sin\theta}\frac{\partial}{\partial\theta}\left(\sin\theta\frac{\partial}{\partial\theta}\right)+\frac{\rho}{\sin^2\theta}\frac{\partial^2}{\partial\phi^2}\right]\Psi$$
$$= -a^2\frac{\partial^2}{\partial t\partial r}(W_e+W_i) \qquad r=a. \tag{9.70}$$

9.4.3 Shielding by a uniform spherical sheet

When the sheet is uniform, (9.70) becomes

$$\rho\left[\frac{1}{\sin\theta}\frac{\partial}{\partial\theta}\left(\sin\theta\frac{\partial}{\partial\theta}\right)+\frac{1}{\sin^2\theta}\frac{\partial^2}{\partial\phi^2}\right]\Psi = a^2\frac{\partial H_r}{\partial t} \qquad r=a. \tag{9.71}$$

The components of electric current in the sheet are given as

$$i_{s,\theta} = \partial\Psi/(a\sin\theta\,\partial\phi), \quad i_{s,\phi} = -\partial\Psi/(a\partial\theta). \tag{9.72}$$

Denoting the magnetic potentials outside and inside the sheet by W and W' respectively, we have the expressions

$$\begin{aligned} W_n^m &= a[e_n^m(r/a)^n + i_n^m(r/a)^{-n-1}]S_n^m \qquad & r>a \\ W_n^{m'} &= ae_n^{m'}(r/a)^n S_n^m & r<a \end{aligned} \tag{9.73}$$

where S_n^m is the spherical surface harmonic. The meaning of each coefficient is self-evident.

Writing a typical term of the current function as

$$\Psi_n^m = K_n^m S_n^m \tag{9.74}$$

the magnetic potentials arising from the current in the sheet become

$$\begin{aligned} w_n^m &= 4\pi n(2n+1)^{-1}K_n^m(r/a)^{-n-1}S_n^m \qquad & r>a \\ w_n^{m'} &= -4\pi(n+1)(2n+1)^{-1}K_n^m(r/a)^n S_n^m & r<a \end{aligned} \tag{9.75}$$

outside and inside the sheet, respectively.

The magnetic potential arising from origins outside the sheet is continuous, so we must have

$$W' - w' = W - w \tag{9.76}$$

where suffices m and n are omitted for the sake of simplicity. Accordingly, we have

$$\begin{aligned} e_n^{m'} - e_n^m &= \frac{4\pi}{a}\frac{n+1}{2n+1}K_n^m \\ i_n^m &= \frac{4\pi}{a}\frac{n}{2n+1}K_n^m. \end{aligned} \tag{9.77}$$

On the other hand, (9.71) leads to

$$K_n^m = \frac{a^2 p}{\rho n(n+1)}[ne_n^m - (n+1)i_n^m] \tag{9.78}$$

and we have

$$\begin{aligned} e_n^{m'} &= e_n^m - \frac{4\pi a p}{\rho n(2n+1)}[ne_n^m - (n+1)i_n^m] \\ i_n^m &= \frac{4\pi a p}{\rho(n+1)(2n+1)}[ne_n^m - (n+1)i_n^m]. \end{aligned} \tag{9.79}$$

When the inducing field is uniform, we have $n = 1, m = 0$. It then follows that

$$e' = (1 + i\omega C)^{-1} e \tag{9.80}$$

where

$$C = \frac{4\pi a}{3\rho}. \tag{9.81}$$

A periodic change having an angular frequency ω is assumed here.

Putting

$$v = \frac{4\pi a\omega}{3\rho} \tag{9.82}$$

the amplitude ratio, which is defined by the absolute value of e'/e, and the phase difference are given by

$$f = \frac{1}{\sqrt{1+v^2}}, \quad \delta = -\tan^{-1} v \tag{9.83}$$

respectively. f is the quantity which describes the decrease of the magnetic field in the spherical sheet or, in other words, f specifies the extent of the shielding effect.

Table 9.2 shows how f and δ change with the change in v. When v becomes larger, the shielding effect becomes more marked. In order to have a large v, a large radius, a high frequency and a high conductivity are required.

When the earth's surface is covered by a sea having a thickness of 1,000 m, it becomes clear that f takes on a value quite close to zero for a magnetic change having a period of 1 min. In such a case, therefore, the external field is shielded almost completely.

TABLE 9.2. The shielding effect of a uniform spherical shell. f and δ denote the amplitude ratio and phase difference, respectively.

v	f	δ
0	1.0000	0.0°
1	0.7072	45.0
2	0.4472	63.4
3	0.3163	71.6
4	0.2425	76.0
5	0.1961	78.7
6	0.1644	80.5
7	0.1414	81.9
8	0.1240	82.9
9	0.1104	83.7
10	0.0995	84.3
15	0.0665	86.2
20	0.0499	87.1

9.4.4 *Electromagnetic induction in a non-uniform spherical sheet*

PRICE (1949) was the first to present a general theory of electromagnetic induction in non-uniform thin sheets. The theory was applied by ASHOUR and PRICE (1948) to electromagnetic induction in an ionosphere having a non-uniform distribution of electrical conductivity. It then became clear that the induced electric currents decay slowly in highly-conducting portions of the sheet, while they decay rapidly in less conducting portions.

RIKITAKE (1967b) developed a method for computing electric currents in non-uniform plane and spherical sheets, the method having been formulated in such a way that it is suitable for computer programming.

Let us assume that a non-uniform spherical sheet is denoted by $r = a$, while the electrical resistance in the sheet or ρ as defined by (9.65) is a function of θ and ϕ, spherical polar coordinates (r, θ, ϕ) being used throughout the following discussion.

The magnetic potential of an inducing field is assumed to be

$$W_e = ae_k^l\left(\frac{r}{a}\right)^k P_k^l(\cos\theta)\cos(l\phi + \varepsilon_k^l) \quad (r > a) \tag{9.84}$$

where $P_k^l(\cos\theta)$ is an associated Legendre function having degree k and order l as defined by Neumann.

The induced potential may be of the following form;

$$W_i = a\sum_n\sum_m (a/r)^{n+1} P_n^m(\cos\theta)(i_n^{mC}\cos m\phi + i_n^{mS}\sin m\phi) \quad (r > a). \tag{9.85}$$

If the current function for the electric currents induced in the sheet is denoted by

$$\Psi = \sum_n \sum_m P_n^m(\cos\theta)(K_n^{mC}\cos m\phi + K_n^{mS}\sin m\phi) \tag{9.86}$$

the coefficients of the induced potential are correlated with those of the current function by

$$ai_n^{mC} = 4\pi\frac{n}{2n+1}K_n^{mC}, \quad ai_n^{mS} = 4\pi\frac{n}{2n+1}K_n^{mS}. \tag{9.87}$$

According to the theory of electromagnetic induction in a non-uniform sheet, equation (9.70) should hold at the sheet.

Putting (9.84) and (9.85) into (9.70) and taking (9.87) into account, we obtain

$$\begin{aligned}
&\sum_n \sum_m \frac{\partial\rho_1}{\partial\theta}\frac{\mathrm{d}P_n^m}{\mathrm{d}\theta}(K_n^{mC}\cos m\phi + K_n^{mS}\sin m\phi)\\
&\quad + \sum_n \sum_m \frac{m}{\sin^2\theta}\frac{\partial\rho_1}{\partial\phi}P_n^m(-K_n^{mC}\sin m\phi + K_n^{mS}\cos m\phi)\\
&\quad - \sum_n \sum_m n(n+1)\rho_1 P_n^m(K_n^{mC}\cos m\phi + K_n^{mS}\sin m\phi)\\
&\quad = -\beta\Big[ake_k^l P_k^l\cos(l\phi + \varepsilon_k^l)\\
&\quad - 4\pi\sum_n \sum_m \frac{n(n+1)}{2n+1}P_n^m(K_n^{mC}\cos m\phi + K_n^{mS}\sin m\phi)\Big]
\end{aligned} \tag{9.88}$$

where, putting the time operator as $\partial/\partial t = p$, we write

$$\beta = ap/\rho_0, \quad \rho_1 = \rho/\rho_0. \tag{9.89}$$

ρ_0 is a constant having the dimension of resistance in the sheet, while β and ρ_1 are dimensionless.

If we multiply both sides of (9.88) by $P_N^M \sin\theta\cos M\phi$ and integrate with respect to θ and ϕ, respectively, from $\theta = 0$ to $\theta = \pi$ and from $\phi = 0$ to $\phi = 2\pi$, we obtain

$$\begin{aligned}
&\sum_n \sum_m [(A_N^{MC} - C_N^{MC} - E_N^{MC})K_n^{mC} + (B_N^{MC} + D_N^{MC} - F_N^{MC})K_n^{mS}]\\
&\quad \Bigg| -\beta\Big[aNe_N^M\cos\varepsilon_N^M\frac{2\pi}{2N+1}\frac{(N+M)!}{(N-M)!}\\
&\quad \Bigg| -8\pi^2\frac{N(N+1)}{(2N+1)^2}\frac{(N+M)!}{(N-M)!}K_N^{MC}\Big]
\end{aligned}$$

$$= \begin{cases} \text{for } k = N, l = M \\ 8\beta\pi^2 \dfrac{N(N+1)}{(2N+1)^2}\dfrac{(N+M)!}{(N-M)!} K_N^{MC} \\ \text{for other combinations of } N \text{ and } M \end{cases} \tag{9.90}$$

where

$$\begin{aligned}
A_N^{MC} &= \int_0^{\pi}\int_0^{2\pi} \frac{\partial \rho_1}{\partial \theta}\frac{\mathrm{d}P_n^m}{\mathrm{d}\theta} P_N^M \sin\theta\cos m\phi\cos M\phi \mathrm{d}\theta\mathrm{d}\phi \\
B_N^{MC} &= \int_0^{\pi}\int_0^{2\pi} \frac{\partial \rho_1}{\partial \theta}\frac{\mathrm{d}P_n^m}{\mathrm{d}\theta} P_N^M \sin\theta\sin m\phi\cos M\phi \mathrm{d}\theta\mathrm{d}\phi \\
C_N^{MC} &= m\int_0^{\pi}\int_0^{2\pi} \frac{\partial \rho_1}{\partial \phi}\frac{P_n^m P_N^M}{\sin\theta} \sin m\phi\cos M\phi \mathrm{d}\theta\mathrm{d}\phi \\
D_N^{MC} &= m\int_0^{\pi}\int_0^{2\pi} \frac{\partial \rho_1}{\partial \phi}\frac{P_n^m P_N^M}{\sin\theta} \cos m\phi\cos M\phi \mathrm{d}\theta\mathrm{d}\phi \\
E_N^{MC} &= n(n+1)\int_0^{\pi}\int_0^{2\pi} \rho_1 P_n^m P_N^M \sin\theta\cos m\phi\cos M\phi \mathrm{d}\theta\mathrm{d}\phi \\
F_N^{MC} &= n(n+1)\int_0^{\pi}\int_0^{2\pi} \rho_1 P_n^m P_N^M \sin\theta\sin m\phi\cos M\phi \mathrm{d}\theta\mathrm{d}\phi .
\end{aligned} \tag{9.91}$$

Similarly, multiplying by $P_N^M \sin\theta \sin M\phi$ leads to

$$\sum_n \sum_m [(A_N^{MS} - C_N^{MS} - E_N^{MS})K_n^{mC} + (B_N^{MS} + D_N^{MS} - F_N^{MS})K_n^{mS}]$$

$$= \begin{cases} \beta\left[aNe_N^M \sin\varepsilon_N^M \dfrac{2\pi}{2N+1}\dfrac{(N+M)!}{(N-M)!} + 8\pi^2 \dfrac{N(N+1)}{(2N+1)^2}\dfrac{(N+M)!}{(N-M)!} K_N^{MS}\right] \\ \text{for } k = N, \quad l = M \\ 8\beta\pi^2 \dfrac{N(N+1)}{(2N+1)^2}\dfrac{(N+M)!}{(N-M)!} K_N^{MS} \\ \text{for other combinations of } N \text{ and } M \end{cases} \tag{9.92}$$

where

$$\begin{aligned}
A_N^{MS} &= \int_0^{\pi}\int_0^{2\pi} \frac{\partial \rho_1}{\partial \theta}\frac{\mathrm{d}P_n^m}{\mathrm{d}\theta} P_N^M \sin\theta\cos m\phi\sin M\phi \mathrm{d}\theta\mathrm{d}\phi \\
B_N^{MS} &= \int_0^{\pi}\int_0^{2\pi} \frac{\partial \rho_1}{\partial \theta}\frac{\mathrm{d}P_n^m}{\mathrm{d}\theta} P_N^M \sin\theta\sin m\phi\sin M\phi \mathrm{d}\theta\mathrm{d}\phi
\end{aligned}$$

$$C_N^{MS} = m \int_0^\pi \int_0^{2\pi} \frac{\partial \rho_1}{\partial \phi} \frac{P_n^m P_N^M}{\sin\theta} \sin m\phi \sin M\phi \mathrm{d}\theta \mathrm{d}\phi$$

$$D_N^{MS} = m \int_0^\pi \int_0^{2\pi} \frac{\partial \rho_1}{\partial \phi} \frac{P_n^m P_N^M}{\sin\theta} \cos m\phi \sin M\phi \mathrm{d}\theta \mathrm{d}\phi \tag{9.93}$$

$$E_N^{MS} = n(n+1) \int_0^\pi \int_0^{2\pi} \rho_1 P_n^m P_N^M \sin\theta \cos m\phi \sin M\phi \mathrm{d}\theta \mathrm{d}\phi$$

$$F_N^{MS} = n(n+1) \int_0^\pi \int_0^{2\pi} \rho_1 P_n^m P_N^M \sin\theta \sin m\phi \sin M\phi \mathrm{d}\theta \mathrm{d}\phi.$$

Putting $N = 1$, $M = 0$; $N = 1$, $M = 1$; $N = 2$, $M = 0$; . . . , (9.90) and (9.92) provide a set of simultaneous equations which can be solved to determine the K_n^{mC} and K_n^{mS}.

When the inducing field is purely periodic, we put

$$p = i\alpha \ (\alpha = 2\pi/T, \ i = \sqrt{-1}) \tag{9.94}$$

where T is the period. In such a case, let us also put

$$\left.\begin{aligned} K_n^{mC} &= \bar{K}_n^{mC} + iK_n^{mC*} \\ K_n^{mS} &= \bar{K}_n^{mS} + iK_n^{mS*}. \end{aligned}\right\} \tag{9.95}$$

Introducing (9.95) into (9.90) and (9.92), we obtain

$$\begin{aligned} \sum_n \sum_m &[(A_N^{MC} - C_N^{MC} - E_N^{MC})\bar{K}_n^{mC} + (B_N^{MC} + D_N^{MC} - F_N^{MC})\bar{K}_n^{mS}] \\ &= -8\beta^*\pi^2 \frac{N(N+1)}{(2N+1)^2} \frac{(N+M)!}{(N-M)!} K_N^{MC*} \end{aligned} \tag{9.96}$$

$$\begin{aligned} \sum_n \sum_m &[(A_N^{MC} - C_N^{MC} - E_N^{MC})K_n^{mC*} + (B_N^{MC} + D_N^{MC} - F_N^{MC})K_n^{mS*}] \\ &= \begin{cases} -\beta^*\left[aNe_N^M \cos\varepsilon_N^M \dfrac{2\pi}{2N+1} \dfrac{(N+M)!}{(N-M)!} - 8\pi^2 \dfrac{N(N+1)}{(2N+1)^2} \dfrac{(N+M)!}{(N-M)!} \bar{K}_N^{MC}\right] \\ \text{for } k = N, \ l = M \\ 8\beta^*\pi^2 \dfrac{N(N+1)}{(2N+1)^2} \dfrac{(N+M)!}{(N-M)!} \bar{K}_N^{MC} \\ \text{for other combinations of } k \text{ and } l \end{cases} \end{aligned} \tag{9.97}$$

$$\begin{aligned} \sum_n \sum_m &[(A_N^{MS} - C_N^{MS} - E_N^{MS})\bar{K}_n^{mC} + (B_N^{MS} + D_N^{MS} - F_N^{MS})\bar{K}_n^{mS}] \\ &= -8\beta^*\pi^2 \frac{N(N+1)}{(2N+1)^2} \frac{(N+M)!}{(N-M)!} K_N^{MS*} \end{aligned} \tag{9.98}$$

$$\sum_n \sum_m [(A_N^{MS} - C_N^{MS} - E_N^{MS}) K_n^{mC*} + (B_N^{MS} + D_N^{MS} - F_N^{MS}) K_n^{mS*}]$$

$$= \begin{cases} \beta^* \left[a N e_N^M \sin \varepsilon_N^M \dfrac{2\pi}{2N+1} \dfrac{(N+M)!}{(N-M)!} + 8\pi^2 \dfrac{N(N+1)}{(2N+1)^2} \dfrac{(N+M)!}{(N-M)!} \bar{K}_N^{MS} \right] \\ \text{for } k = N, l = M \\ 8\beta^* \pi^2 \dfrac{N(N+1)}{(2N+1)^2} \dfrac{(N+M)!}{(N-M)!} \bar{K}_N^{MS} \\ \text{for other combinations of } k \text{ and } l \end{cases} \tag{9.99}$$

where

$$\beta^* = a\alpha/\rho_0. \tag{9.100}$$

Equations (9.96), (9.97), (9.98) and (9.99) provide a set of simultaneous equations which can be solved to determine the $K_n^{\bar{m}C}$, K_n^{mC*}, $K_n^{\bar{m}S}$, and K_n^{mS*}.

In a similar fashion, a theory of electromagnetic induction by an aperiodic variation can be effected as follows.

Going back to (9.90), the difference form is written as

$$\sum_n \sum_m [(A_N^{MC} - C_N^{MC} - E_N^{MC}) K_n^{mC}(t) + (B_N^{MC} + D_N^{MC} - F_N^{MC}) K_n^{mS}(t)]$$

$$= \begin{cases} -\gamma \left[a N \{e_N^M(t+\Delta t) - e_N^M(t)\} \cos \varepsilon_N^M \dfrac{2\pi}{2N+1} \dfrac{(N+M)!}{(N-M)!} \right. \\ \left. - 8\pi^2 \dfrac{N(N+1)}{(2N+1)^2} \dfrac{(N+M)!}{(N-M)!} \{K_N^{MC}(t+\Delta t) - K_N^{MC}(t)\} \right] \\ \text{for } k = N, l = M \\ 8\gamma\pi^2 \dfrac{N(N+1)}{(2N+1)^2} \dfrac{(N+M)!}{(N-M)!} \{K_N^{MC}(t+\Delta t) - K_N^{MC}(t)\} \\ \text{for other combinations of } k \text{ and } l \end{cases} \tag{9.101}$$

where

$$\gamma = a/\rho_0 \Delta t. \tag{9.102}$$

Equation (9.101) can be rewritten as

$$K_N^{MC}(t+\Delta t) = K_N^{MC}(t) + \frac{1}{8\gamma\pi^2} \frac{(2N+1)^2}{N(N+1)} \frac{(N-M)!}{(N+M)!}$$

$$\times \sum_n \sum_m [(A_N^{MC} - C_N^{MC} - E_N^{MC}) K_n^{mC}(t) + (B_N^{MC} + D_N^{MC} - F_N^{MC}) K_n^{mS}(t)]$$

$$+ \lambda \frac{a}{4\pi} \frac{2N+1}{N+1} \{e_N^M(t+\Delta t) - e_N^M(t)\} \cos \varepsilon_N^M \tag{9.103}$$

in which

$$\left.\begin{array}{l}\lambda = 1 \text{ for } k = N,\, l = M \\ \lambda = 0 \text{ for other combinations of } k \text{ and } l.\end{array}\right\} \tag{9.104}$$

Similarly, the following holds:

$$K_N^{MS}(t + \Delta t) = K_N^{MS}(t) + \frac{1}{8\gamma\pi^2}\frac{(2N+1)^2}{N(N+1)}\frac{(N-M)!}{(N+M)!}$$
$$\times \sum_n \sum_m [(A_N^{MS} - C_N^{MS} - E_N^{MS})K_n^{mC}(t) + (B_N^{MS} + D_N^{MS} - F_N^{MS})K_n^{mS}(t)]$$
$$- \lambda\frac{a}{4\pi}\frac{2N+1}{N+1}\{e_N^M(t + \Delta t) - e_N^M(t)\} \sin \varepsilon_N^M. \tag{9.105}$$

It is therefore possible to estimate the coefficients of the current function at $t + \Delta t$ from those at t provided that Δt is sufficiently small. The accuracy of the present method may be discussed in a fashion similar to that described by RIKITAKE (1967b) for the case of non-uniform plane sheets.

RIKITAKE (1967b) applied the above theory to the problem of electromagnetic induction by *Sq* and *ssc* (sudden storm commencement) in a non-uniform spherical sheet, which modelled the surface layers of the earth, representing the actual distribution of land and sea. Assuming a step-function type jump of the external magnetic field parallel to the $\theta = 0$ axis, the distributions of the current function for various epochs can be calculated on the basis of the above formulae for an aperiodic variation. In Fig. 9.1, for instance, the current pattern 300 s after the jump is illustrated. The details of conductivity distribution are given in the original paper (RIKITAKE, 1967b). Looking at Fig. 9.1, we see that circulation of induced currents in the highly conducting oceans, especially in the South Pacific, becomes apparent.

The above estimates were made by truncating the series of spherical functions at $n = m = 3$, so that more detailed features of induced currents cannot be described, although it is believed that the large-scale effects of land-and-sea distribution should be reflected. The theory was extended to electromagnetic induction by a hypothetical *ssc* in an earth model consisting of a non-uniform surface sheet representing the oceans and an inner, uniform, spherical conductor (RIKITAKE, 1968b). In general, agreement between the expected and observed magnetograms is poor. It is concluded that most of such disagreements are caused by local irregularities in the electrical conductivity which cannot be described by the large-scale conductivity contrast represented by the model.

An important improvement in the theory of electromagnetic induction was made by BULLARD and PARKER (1970) who introduced an integro-

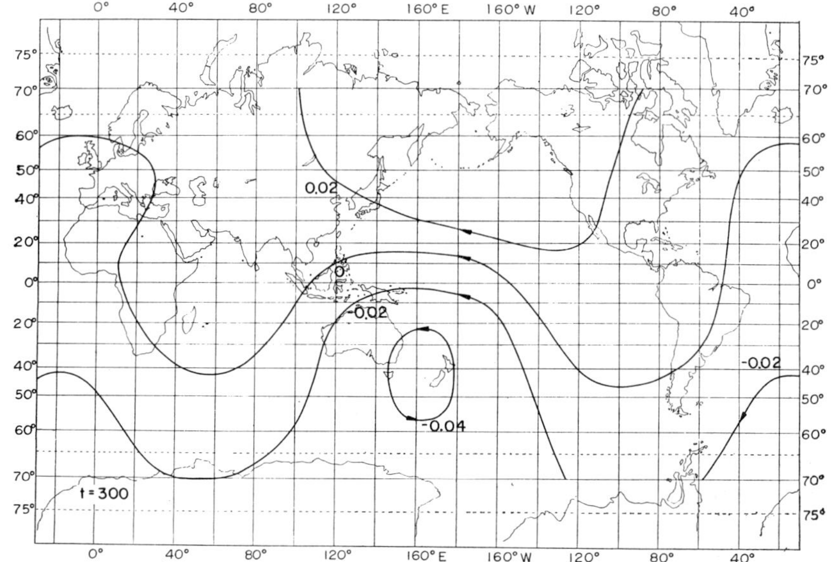

FIG. 9.1. The distribution of electric current in a non-uniform spherical sheet modelling the actual land and sea distribution 300 s after a jump of the magnetic field parallel to the earth's rotation axis (RIKITAKE, 1967b).

differential equation

$$\operatorname{div}(\rho \operatorname{grad} \Psi) = \frac{\partial}{\partial t}\left\{H_n^e - \frac{1}{4\pi}\int_S \left(\operatorname{grad}\frac{1}{r}\cdot \operatorname{grad}\Psi\right)\mathrm{d}S\right\} \tag{9.106}$$

in place of (9.66) or (9.70). The first term in parenthesis on the right-hand side of (9.106) is the normal component of the inducing field, while the second is that of the field arising from the electric currents induced in the sheet. In a sense, (9.106) is similar to the equation of induction in a thin circular sheet as obtained by ASHOUR (1950) who also introduced an integro-differential equation.

BULLARD and PARKER (1970) tried to solve Eq. (9.106) for a land-and-sea model of the earth by an iterative method originally proposed by PRICE (1949). First of all, the second term in parenthesis on the right-hand side, which is nothing but the effect of self-induction, is ignored. The current function, which is denoted by Ψ_0, is then obtained by solving the equation. In the next place, Ψ_0 is substituted in Ψ in the integral resulting from the equation for Ψ_1. It was proved that such an iterative procedure converges for periods longer than 7 hours or so. Using this approach, electric current vortices induced in the oceans by *Sq* were demonstrated very well.

The iterative method of estimating induced currents in a non-uniform spherical sheet has been further improved by a number of British researchers in ways that can be applied even in the case of short-period variations. The number of relevant papers published is so large that only a few representative references are quoted in the following.

HOBBS and PRICE (1970), HOBBS (1971, 1975a) and HOBBS and BRIGNALL (1976) developed a theory representing geophysical induction effects using surface integrals. With the aid of such surface integral representations, Hobbs and his associates arrived at a very successful iterative method that converges for any period. The method was identified by them as an analytic continuation.

HUTSON *et al.* (1972, 1973) and HEWSON-BROWNE *et al.* (1973) proposed another approach to iterative methods for oceanic induction problems. Their work can be formally presented in the following way (KENDALL, 1978). If f is an unknown quantity such as the current function Ψ in a thin ocean, an equation for f may be obtained in linear operator form

$$(\lambda + L)f = g. \tag{9.107}$$

where g is known from the inducing magnetic field, λ is a constant composed of several physical constants such as frequency, depth and conductivity. L is a generalized linear operator with a spectrum. When f is identified as the current density, Lf is composed of the induced field, expressible as an integral of the current density together with the surface gradient of the corresponding scalar electric potential.

HUTSON *et al.* (1972, 1973) observed that the iterative scheme

$$\lambda f_{n+1} = g - Lf_n \tag{9.108}$$

converges when $|\lambda| > R_L$, where R_L is the spectral radius. This is equivalent to Price's iteration for low frequency.

For $|\lambda| < R_L$, it is possible to choose a constant β such that the scheme

$$(\lambda - \beta)f_{n+1} = g - (\beta + L)f_n \tag{9.109}$$

converges to the solution of (9.107).

It is convenient to choose $f_{-1} = 0$, so that $f_0 = (\lambda - \beta)^{-1}g$. Then the solution of (9.107) is

$$f = \left\{1 + \sum_{n=1}^{\infty} \left(\frac{\beta + L}{\beta - \lambda}\right)^n\right\} f_0 \tag{9.110}$$

provided that the series converges.

It was proved by KENDALL (1978) that the analytic continuation procedure proposed by HOBBS and BRIGNALL (1976) is equivalent to the method of shifting the spectrum for a linear operator L.

Similar problems have also been studied by Soviet workers such as M. N. Berdichevsky, M. S. Zhdanov, B. Sh. Zinger, L. P. Lagutinskaya, V. G. Dubrowsky and E. B. Fainberg. Although no detailed account of this work is presented here, there is an excellent review of the problem by FAINBERG (1980).

It is the authors' feeling, however, that most of the above work has concerned the theoretical side of the problem, so that oceans represented by a hemispherical shell or a spherical cap have usually been involved.

9.4.5 Electromagnetic induction in a hemispherical sheet

Electromagnetic induction in a hemispherical thin sheet has been the target of mathematical and experimental studies by which it is hoped to examine the effect of a vast ocean such as the Pacific.

RIKITAKE and YOKOYAMA (1955b) were the first to tackle the electromagnetic induction in a hemispherical thin sheet by a uniform field with a method which is essentially equivalent to that of RIKITAKE (1967b), as given in the last subsection. As the induced electric current and magnetic field are approximated by a sum of spherical harmonic functions of finite number, no accurate results are obtained at the edge of conductor where the current density becomes infinitely large. However, the general features of the induced magnetic field do agree with the results of experiment (NAGATA *et al.*, 1955).

An essentially equivalent approach to electromagnetic induction in a hypothetical ocean, represented by a hemispherical uniform sheet, by *Sq* was attempted by RIKITAKE (1960). The study was later extended to a hemispherical sheet with an inner spherical conductor (RIKITAKE, 1961b, 1962a).

ASHOUR (1965a, 1965b) succeeded in obtaining an exact solution for the problem of electromagnetic induction in a perfectly-conducting hemispherical shell by a uniform magnetic field by making use of a suitable transformation of Bessel functions appearing in the potential theory. Taking the radius of the shell as a, we assume that the shell occupies the portion $0 \leqq \theta \leqq \pi/2$. It is shown by Ashour that the current function Ψ, ϕ component of current density i_ϕ $(i_\theta = 0)$, and normal and tangential components of the magnetic field produced by the currents Z_i and H_i are given as

$$\Psi = \begin{cases} (-H_0 a/4\pi^2)\{(3\cos\theta + 1)\tan^{-1}(\cos\theta)^{1/2} + 3(\cos\theta)^{1/2}\} & 0 \leq \theta \leq \pi/2 \\ 0 & \pi/2 < \theta \leq \pi \end{cases} \quad (9.111)$$

$$i_\phi = \begin{cases} (-H_0/4\pi^2)\sin\theta\{3\tan^{-1}(\cos\theta)^{1/2} + 2(\sec\theta)^{1/2} \\ \qquad + (\cos\theta)^{1/2}/(1+\cos\theta)\} & 0 \le \theta \le \pi/2 \\ 0 & \pi/2 < \theta \le \pi \end{cases} \tag{9.112}$$

$$Z_i = \begin{cases} -H_0\cos\theta & 0 \le \theta < \pi/2 \\ -H_0\cos\theta + (H_0/\pi)\{2\cos\theta\tan^{-1}(-\cos\theta)^{1/2} \\ \qquad -2(-\cos\theta)^{1/2} + (-\sec\theta)^{1/2}\} & \pi/2 < \theta \le \pi \end{cases} \tag{9.113}$$

$$H_{i\pm} = \begin{cases} \pm 2\pi i + (H_0/4)\sin\theta & 0 \le \theta < \pi/2 \\ (H_0/2\pi)\{\pi/2 - (-\cos\theta)^{1/2}/(1-\cos\theta) \\ \qquad - \tan^{-1}(-\cos\theta)^{1/2}\}\sin\theta & \pi/2 < \theta \le \pi. \end{cases} \tag{9.114}$$

The behavior of the magnetic lines of force around the shell is shown in Fig. 9.2. Meanwhile, the change in the current density with θ is shown in Fig. 9.3 in which we see that the current density increases enormously near the edge of the shell, where it becomes infinitely large. The normal component of the magnetic field is distributed as shown in Fig. 9.4, in which we clearly see that the inducing field is completely cancelled by the induced field over the shell and that the field becomes infinitely large at the edge.

When the inducing field is parallel to the plane denoted by $\theta = \pi/2$, we obtain

$$\Psi = \begin{cases} -(H_0a/4\pi^2)[3\sin\theta\{\tan^{-1}(\cos\theta)^{1/2} + (\cos\theta)^{1/2}/(1+\cos\theta)\} \\ \qquad + 2(\pi+2)^{-1}\{2\cot(\theta/2)\tan^{-1}(\cos\theta)^{1/2} \\ \qquad -\pi\,\mathrm{cosec}\,\theta(\cos\theta)^{1/2}\}]\cos\phi & 0 \le \theta \le \pi/2 \\ 0 & \pi/2 < \theta \le \pi \end{cases} \tag{9.115}$$

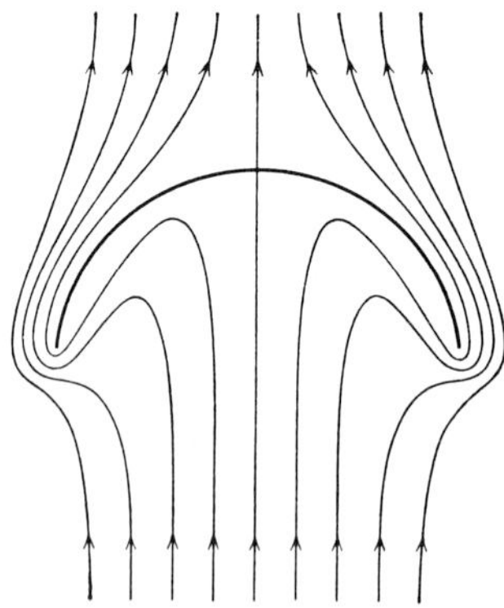

FIG. 9.2. Magnetic lines of force around a perfectly-conducting hemispherical sheet. The field is parallel to the $\theta = 0$ axis at a distance (ASHOUR, 1965b).

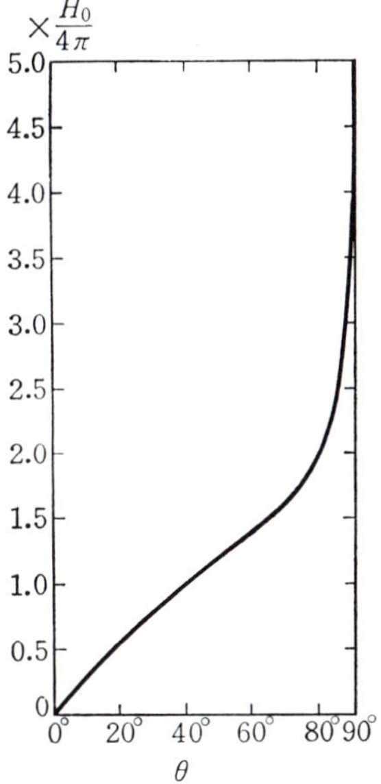

FIG. 9.3. The distribution of electric current corresponding to Fig. 9.2 (ASHOUR, 1965b).

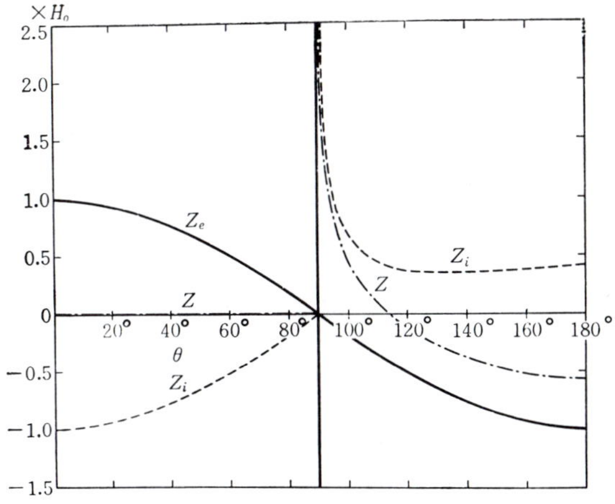

FIG. 9.4. The distributions of the magnetic fields normal to the sheet corresponding to Fig. 9.2 (ASHOUR, 1965b).

$$\left.\begin{aligned} i_\theta &= -(1/a)\operatorname{cosec}\theta\tan\phi\Psi(\theta,\phi) \\ i_\phi &= (H_0/4\pi^2)[3(\cos\theta)^{1/2} - 3(\sec\theta)^{1/2}/(1+\cos\theta) \\ &\quad + (\pi+2)^{-1}\{2\pi(\cos\theta)^{3/2}\operatorname{cosec}^2\theta + (\pi-2)(\sec\theta)^{1/2}\} \\ &\quad + \{3\cos\theta - 4(\pi+2)^{-1}(1-\cos\theta)^{-1}\} \\ &\quad \times \tan^{-1}(\cos\theta)^{1/2}]\cos\phi \end{aligned}\right\} \quad 0 \le \theta < \pi/2 \tag{9.116}$$

$$i_\theta = i_\phi = 0 \qquad \pi/2 < \theta \le \pi \tag{9.117}$$

$$Z_i = \begin{cases} -H_0\sin\theta\cos\phi & 0 \le \theta < \pi/2 \\ -H_0\sin\theta\cos\phi \\ \quad +(H_0/\pi)[2\sin\theta\{\tan^{-1}(-\cos\theta)^{1/2} + (-\sec\theta)^{1/2}\} \\ \quad -\pi(\pi+2)^{-1}(1+\cos\theta)^{1/2}(\cos^2\theta - \cos\theta)^{-1/2}]\cos\phi & \pi/2 < \theta \le \pi \end{cases} \tag{9.118}$$

$$H_{i,\phi\pm} = \begin{cases} \mp 2\pi i_\theta + (H_0/2)\{1 - 4(\pi+2)^{-1}\operatorname{cosec}\theta\tan(\theta/2)\}\sin\phi & 0 \le \theta < \pi/2 \\ -(H_0/\pi)\operatorname{cosec}\theta\sin\phi[\{\sin\theta - 4(\pi+2)^{-1}\tan(\theta/2)\}\tan^{-1}(-\sec\theta)^{1/2} \\ \quad - \{4(\pi+2)^{-1}\operatorname{cosec}\theta - \tan(\theta/2)\}(-\cos\theta)^{1/2}] & \pi/2 < \theta \le \pi \end{cases} \tag{9.119}$$

$$H_{i,\theta\pm} = \begin{cases} \pm 2\pi i_\phi - (H_0/2)\{\cos\theta - 2(\pi+2)^{-1}\sec^2(\theta/2)\}\cos\phi & 0 \le \theta < \pi/2 \\ -(H/\pi)[\{\cos\theta - 2(\pi+2)^{-1}\sec^2(\theta/2)\}\tan^{-1}(-\sec\theta)^{1/2} \\ \quad + \{1 + \sin^2\theta - (\pi-2)(\pi+2)^{-1}\cos\theta\} \\ \quad \times \operatorname{cosec}^2\theta(-\cos\theta)^{1/2}]\cos\phi & \pi/2 \le \theta < \pi. \end{cases} \tag{9.120}$$

The magnetic lines of force for this case are shown in Fig. 9.5. The distribution of induced electric currents is illustrated in Figs. 9.6 and 9.7, while that of the normal magnetic field is shown in Fig. 9.8. It is thus made clear that the magnetic field becomes anomalously large at the edge of a hemispherical sheet, such an effect being important for interpreting the anomalous magnetic field variations at an edge of ocean.

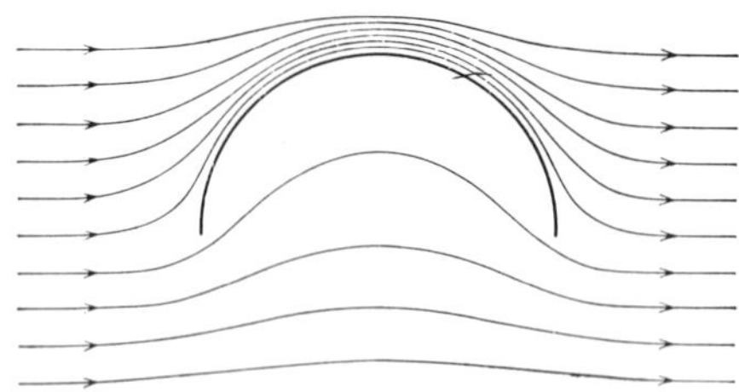

FIG. 9.5. Magnetic lines of force around a perfectly-conducting hemispherical sheet. The field is parallel to the $\theta = \pi/2$ and $\phi = 0$ plane at a distance (ASHOUR, 1965b).

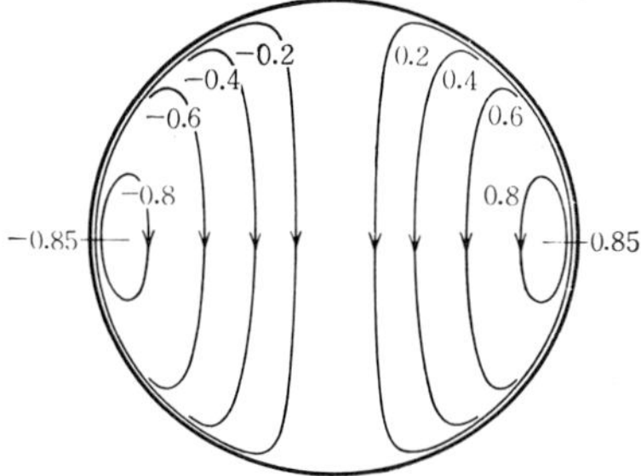

FIG. 9.6. Stream lines of induced electric current in units of $H_0a/4\pi$ corresponding to Fig. 9.5 (ASHOUR, 1965b).

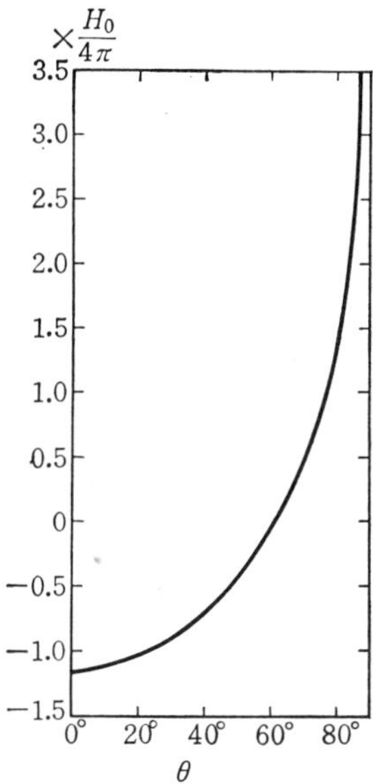

FIG. 9.7. The distribution of electric current $i_\phi \sec\phi$ corresponding to Fig. 9.5 (ASHOUR, 1965b).

DOSS and ASHOUR (1971) extended the above theory to the case of finite conductivity. An iterative procedure starting from thc solution for perfect conductivity as originally proposed by PRICE (1949) was adopted. As infinities of current density and magnetic field occur at the edge of shell, special caution should be taken during the iteration procedure to avoid the singularities.

9.4.6 *Hemispherical sheet overlying an inner spherical conductor*

RIKITAKE (1961a), who studied the electromagnetic induction in a

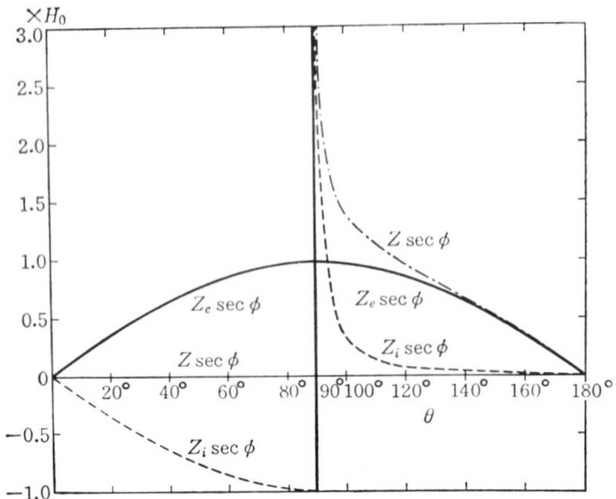

FIG. 9.8. The distribution of magnetic fields $Z_e\sec\phi$, $Z_i\sec\phi$ and $Z\sec\phi$ corresponding to Fig. 9.5 (ASHOUR, 1965b).

perfectly-conducting hemispherical sheet overlying an inner perfectly-conducting sphere, was the first to point out that the electromagnetic coupling between the two conductors considerably reduces the enhancement of anomalous current density and magnetic field at the edge of the overlying sheet. As the portion of the earth's mantle deeper than a few hundred kilometers is believed to be highly conducting, as will be shown in the following chapter, the electromagnetic coupling between the surface conducting layers and the conductors in the deep interior is important for discussing electromagnetic induction within the earth.

In view of the importance of the coupling, most studies referred to in the last two subsections have been made using a model with an inner spherical conductor.

In his study of electromagnetic induction in a hemispherical ocean by *Sq*, RIKITAKE (1960, 1961b, 1962a) made use of a transformation of the coordinate system, for the sake of mathematical convenience. Such a transformation was also adopted by HOBBS (1971) who tackled the same problem on the basis of a surface integral method developed by him. What follows is mostly due to Hobbs.

HOBBS (1971) takes a system of conductors as shown in Fig. 9.9 in which we have a hemispherical sheet at $r = a$, while the inner perfectly-conducting sphere is denoted by $r = b$. The spherical polar coordinate system (r, θ_0, ϕ_0) is such that the axis $\theta_0 = 0$ passes through the rim of the sheet, and the sheet occupies $0 \leqq \phi_0 \leqq \pi$, $0 \leqq \theta_0 \leqq \pi$.

The magnetic potential of the 24-hourly component of the *Sq* field is assumed to be

$$W=a(r/a)^2(e^1_{2,c}-ie^1_{2,s})\exp(i\omega t)(\cos\phi_0+i\sin\phi_0)P^1_2(\cos\theta_0) \tag{9.121}$$

which can be deduced on the assumption that the *Sq* depends solely on local time. ω is the angular frequency of the variation. According to Chapman's analysis (CHAPMAN, 1919),

$$e^1_{2,c}=7.1,\quad e^1_{2,s}=-3.1 \tag{9.122}$$

in units of gamma, for the mean *Sq* for 1905.

We now transform the coordinate system from (r, θ_0, ϕ_0) to (r, θ, ϕ) in which the axis $\theta=0$ passes through the center of the hemispherical sheet (see Fig. 9.9). In the new coordinate system, the inducing field given by (9.121) becomes

$$W=a(r/a)^2(e^1_{2,c}-ie^1_{2,s})\exp(i\omega t)\{-\cos\phi P^1_2(\cos\theta)+i\sin 2\phi P^2_2(\cos\theta)\}. \tag{9.123}$$

It is then possible to estimate the electromagnetic induction by the P^1_2 and P^2_2 terms separately. Adding the separate solutions and taking the real part, we can solve the problem.

Hobbs assumed the average depth of ocean represented by the sheet as 4 km and the conductivity of sea-water as 4 S/m, so that the integrated resistivity of the sheet amounts to $0.625\times10^{-4}\text{S}^{-1}$. $a=6.37\times10^8$ cm and $b/a=0.9$ are also assumed.

Figure 9.10 shows the currents induced in the hemispherical ocean by the 24-hourly component of *Sq*, the figures denoted by (a), (b), (c) and (d) representing the epochs $t=0, 3, 6$ and 9 hours, respectively. The current flow

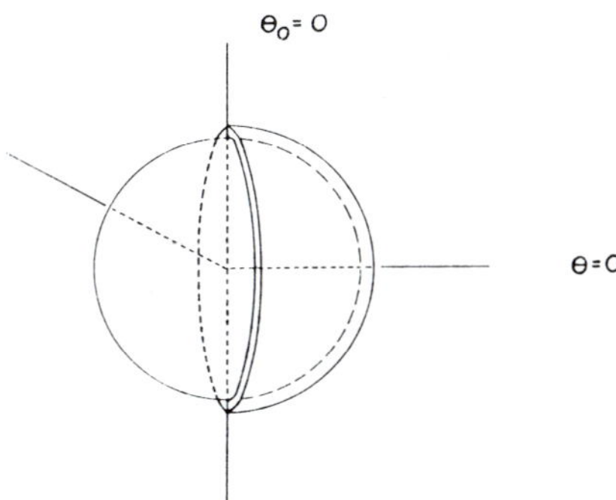

FIG. 9.9. System of conductors and the coordinates.

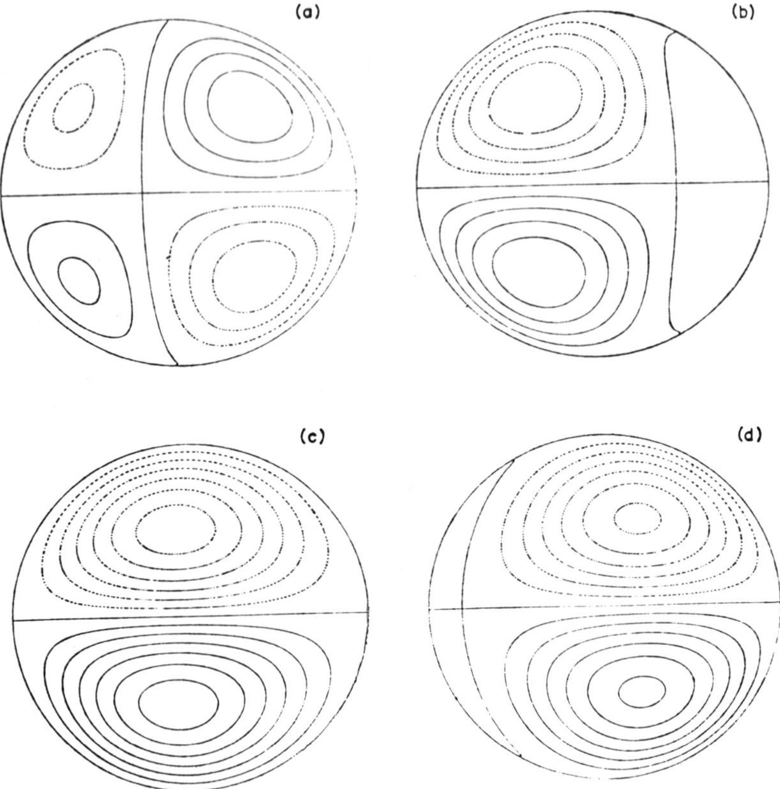

FIG. 9.10. Currents induced in the hemispherical ocean by the 24-hourly component of the *Sq* field. (a) $t = 0$, (b) $t = 3$, (c) $t = 6$, (d) $t = 9$ hours (HOBBS, 1971).

between adjacent lines is 1,000 amperes. The maximum value amounts to some 6,000 amperes. The hemispherical sheet is viewed from a point above the center on an azimuthal equidistant plot.

HOBBS (1971) also illustrated the vertical component of the induced magnetic field for the various epochs as reproduced in Fig. 9.11. The maximum value reached is estimated at just under 4 gammas.

9.5 Electromagnetic Induction by a Progressive Magnetic Field

In this section, electromagnetic induction by a progressive magnetic field will be studied. This may be applied to the interpretation of the interaction between the moon and an interplanetary magnetic field passing by it, as discussed by NESS (1969).

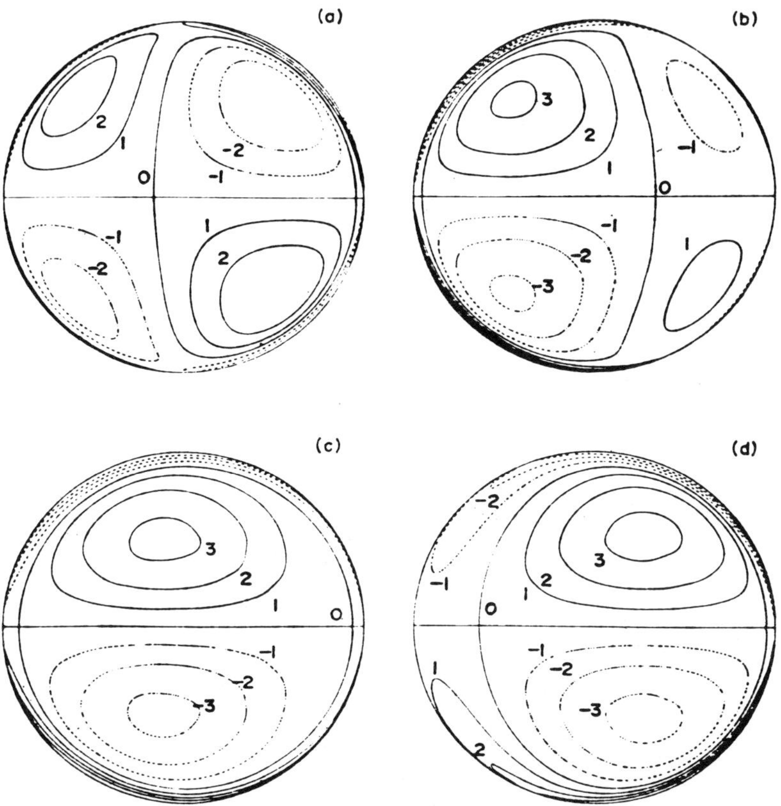

FIG. 9.11. Vertical field produced by the induced currents over the oceanic hemisphere for the 24-hourly component of the *Sq* field. (a) $t = 0$, (b) $t = 3$, (c) $t = 6$, (d) $t = 9$ hours (HOBBS, 1971).

A typical solution of (9.25), representing a progressive magnetic field, is given by

$$W_e = \begin{cases} A_k e^{-\frac{k\pi}{L}z} \sin \dfrac{k\pi}{L}(x - vt) & z \geq 0 \\ A_k e^{\frac{k\pi}{L}z} \sin \dfrac{k\pi}{L}(x - vt) & z \leq 0 \end{cases} \tag{9.124}$$

where v and L are the velocity of the progressive field and a typical length, respectively. The origin of the coordinate system is taken at the center of the moon, while the x-y plane coincides with the ecliptic plane.

An inducing field having a sharp change in the z direction followed by a gradual decrease, as observed by Explorer 35 (NESS, 1969), may be represented by summing up typical fields of different wave-length. A field normal to the ecliptic plane at $z = 0$, as shown in Fig. 9.12, may be expressed by

$$H_z(x) = \frac{2\alpha L}{\pi^2(1-\gamma)} \sum_{k=1}^{\infty} k^{-2} \sin k\pi\gamma \sin \frac{k\pi}{L}(x - vt) \tag{9.125}$$

where α is a constant specifying the amplitude and

$$\gamma = L'/L. \tag{9.126}$$

Taking 5, 20 and 50 terms for illustration, a demonstration of how well the series can approximate the original curve is presented in Fig. 9.13 which shows that 20 terms may be enough for practical use.

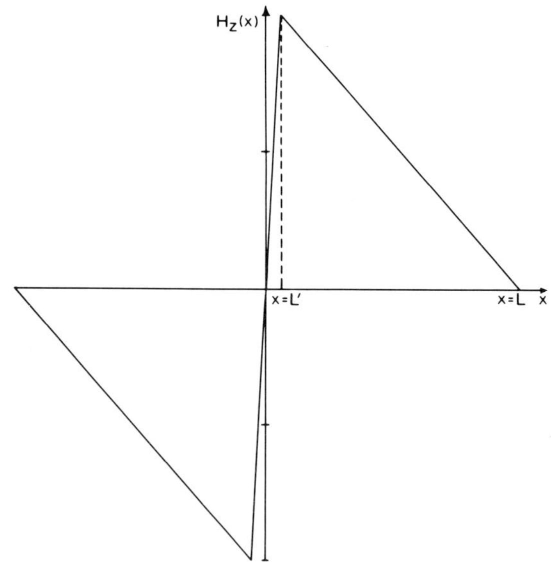

FIG. 9.12. A model of discontinuous interplanetary magnetic field.

The magnetic potential of the inducing field for $z \leq 0$ corresponding to (9.125) becomes

$$W_e = \sum_{k=1}^{\infty} A_k e^{\frac{k\pi}{L}z} \sin \frac{k\pi}{L}(x - vt) \tag{9.127}$$

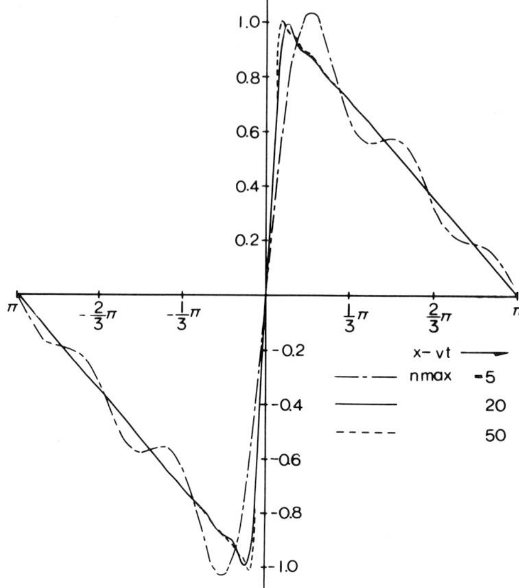

FIG. 9.13. Approximation by Fourier series of the discontinuous field. n is the number of terms.

in which

$$A_k = \frac{2\alpha L^2}{\pi^3(1-\gamma)} \frac{\sin k\pi\gamma}{k^3}. \tag{9.128}$$

Equation (9.127) can be rewritten as

$$W_e = \sum_{k=1}^{\infty} A_k e^{\frac{k\pi}{L}z}\left(\sin\frac{k\pi x}{L}\cos\frac{k\pi vt}{L} - \cos\frac{k\pi x}{L}\sin\frac{k\pi vt}{L}\right) \tag{9.129}$$

so that it is necessary to study inducing fields with spatial distributions specified by $e^{\frac{k\pi}{L}z}\sin(k\pi x/L)$ and $e^{\frac{k\pi}{L}z}\cos(k\pi x/L)$.

We have a formula in terms of Bessel functions such as

$$\sin(\lambda r\sin\theta\cos\phi) = 2\sum_{m=0}^{\infty}(-1)^m J_{2m+1}(\lambda r\sin\theta)\cos(2m+1)\phi. \tag{9.130}$$

On the other hand, the theory of spherical functions indicates the following relation:

$$e^{\lambda r \cos\theta} J_{2m+1}(\lambda r \sin\theta) = \sum_{n=0}^{\infty} \frac{(\lambda r)^{n+2m+1}}{(n+4m+2)!} P_{n+2m+1}^{2m+1}(\cos\theta). \tag{9.131}$$

We therefore obtain

$$e^{\frac{k\pi}{L}z} \sin\frac{k\pi x}{L} = 2 \sum_{n=0}^{\infty} \sum_{m=0}^{\infty} \frac{(-1)^m (k\pi r/L)^{n+2m+1}}{(n+4m+2)!} \times P_{n+2m+1}^{2m+1}(\cos\theta) \cos(2m+1)\phi \tag{9.132}$$

and, in a similar fashion,

$$e^{\frac{k\pi}{L}z} \cos\frac{k\pi x}{L} = \sum_{n=0}^{\infty} \sum_{m=0}^{\infty} \frac{\varepsilon_m (-1)^m (k\pi r/L)^{n+2m}}{(n+4m)!} P_{n+2m}^{2m}(\cos\theta) \cos 2m\phi \tag{9.133}$$

where

$$\varepsilon_0 = 1, \quad \varepsilon_m = 2 \quad (m = 1, 2, 3, \ldots). \tag{9.134}$$

Putting (9.132) and (9.133) into (9.129), the progressive inducing field can be rewritten as

$$W_e = a \sum_{k=1}^{\infty} \cos\frac{k\pi vt}{L} \sum_{n=0}^{\infty} \sum_{m=0}^{\infty} B_{k,n,m} \left(\frac{r}{a}\right)^{n+2m+1} P_{n+2m+1}^{2m+1}(\cos\theta)\cos(2m+1)\phi$$
$$+ a \sum_{k=1}^{\infty} \sin\frac{k\pi vt}{L} \sum_{n=0}^{\infty} \sum_{m=0}^{\infty} C_{k,n,m} \left(\frac{r}{a}\right)^{n+2m} P_{n+2m}^{2m}(\cos\theta)\cos 2m\phi \tag{9.135}$$

where

$$B_{k,n,m} = \frac{2A_k(-1)^m a^{n+2m} (k\pi/L)^{n+2m+1}}{(n+4m+2)!}$$
$$C_{k,n,m} = -\frac{A_k \varepsilon_m (-1)^m a^{n+2m-1} (k\pi/L)^{n+2m}}{(n+4m)!}. \tag{9.136}$$

We therefore see that the inducing field concerned is expressed by superposing elementary inducing fields, expressed in spherical polar coordinates.

The theory of electromagnetic induction in a spherical conductor in which the conductivity is distributed spherically symmetrically indicates that the magnetic potential induced by a field as given by (9.135) is expressed by

$$W_i = a \sum_{k=1}^{\infty} \sum_{n=0}^{\infty} \sum_{m=0}^{\infty} \bar{B}_{k,n,m} \cos\left(\frac{k\pi vt}{L} + \beta_{k,n,m}\right)\left(\frac{r}{a}\right)^{-n-2m-2}$$
$$\times P_{n+2m+1}^{2m+1}(\cos\theta)\cos(2m+1)\phi$$
$$+ a \sum_{k=1}^{\infty} \sum_{n=0}^{\infty} \sum_{m=0}^{\infty} \bar{C}_{k,n,m} \sin\left(\frac{k\pi vt}{L} + \gamma_{k,n,m}\right)\left(\frac{r}{a}\right)^{-n-2m-1}$$
$$\times P_{n+2m}^{2m}(\cos\theta)\cos 2m\phi \tag{9.137}$$

in which $\bar{B}_{k,n,m}$, $\bar{C}_{k,n,m}$, $\beta_{k,n,m}$ and $\gamma_{k,n,m}$ can be obtained from the theory of electromagnetic induction by a periodic variation.

A computer program is used to calculate the A_k, $B_{k,n,m}$ and $C_{k,n,m}$ for $k = 1, 2, \ldots, 20$ and $n, m \leq 6$, and $\bar{B}_{k,n,m}$, $\bar{C}_{k,n,m}$, $\beta_{k,n,m}$ and $\gamma_{k,n,m}$. The following parameters are assumed:

$$\begin{aligned} L &= 45{,}900 \text{ km} \\ v &= 500 \text{ km/sec} \\ \gamma &= 0.05555 \\ \alpha &= 1/L' \\ a &= 1{,}783 \text{ km}. \end{aligned} \tag{9.138}$$

After computing these quantities, a further program is used to sum series like (9.135) and (9.137). The program then serves to estimate magnetic fields around the moon.

Figure 9.14, 9.15 and 9.16 show the component perpendicular to

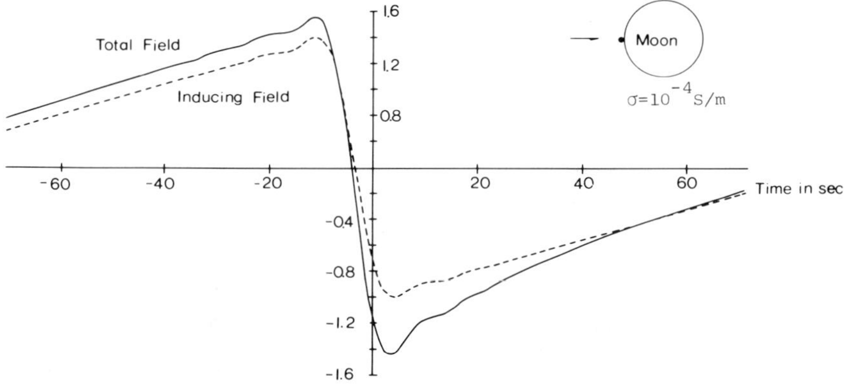

FIG. 9.14. The inducing and total fields observed by a magnetometer placed at a point on the equator shown by the small solid circle. The arrow indicates the direction of the progressive field.

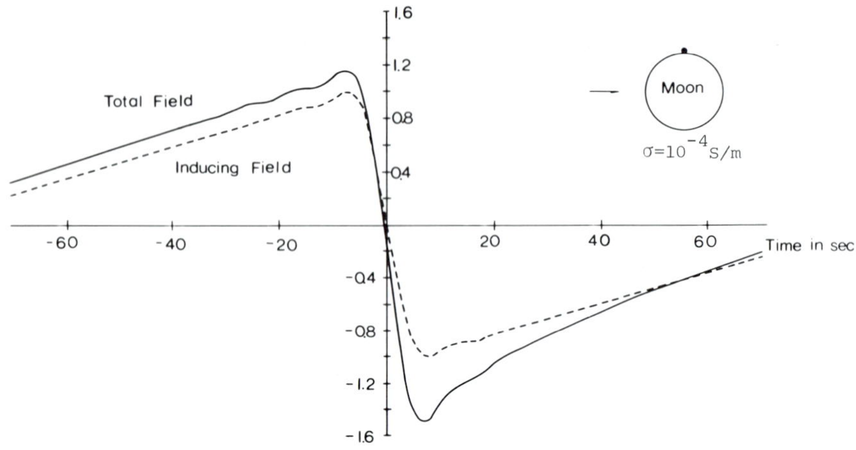

FIG. 9.15. As Fig. 9.14.

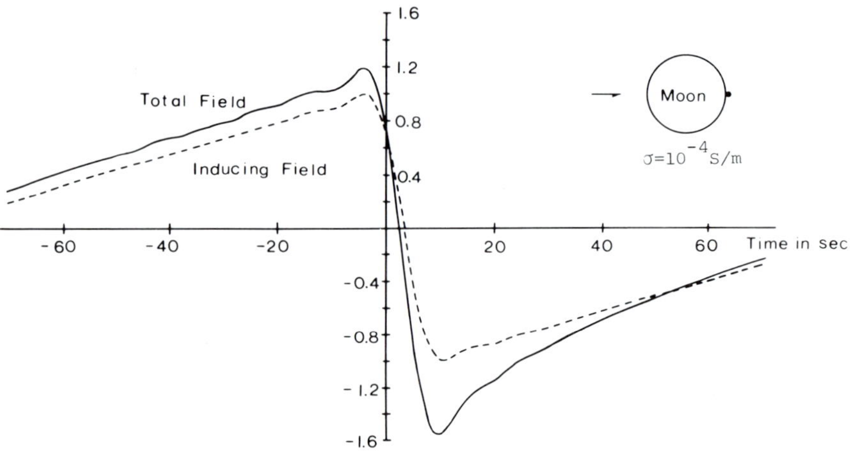

FIG. 9.16. As Fig. 9.14.

the ecliptic plane of the inducing and total (inducing plus induced) fields as would be observed by a magnetometer placed at the three locations on the moon's equator shown in the figures. A uniform moon in which the electrical conductivity amounts to 10^{-4} S/m is assumed. The time origin is taken at the instant when the center of the discontinuity passes the center of the moon.

It should be emphasized that the superposition of spherical magnetic fields as treated here presents a good approximation of the progressive field judging from the small fluctuations of the inducing field shown in the three figures. The fact also suggests that the approximation of the total fields as calculated by a similar superposition of spherical fields would be quite good.

It is felt that the following points should be noted. First of all, the total field exhibits asymmetry about the center of discontinuity. The amplitude of the field before the center of discontinuity passes the moon is always smaller than that after the center passes the moon. This is probably caused by the fact that induction by the rapidly changing portion of the progressive field affects the latter case more strongly. The second point to be noticed is the fact that the field amplitude, when it is measured behind the moon, is larger than that when measured ahead of it. This can be interpreted by the fact that some time is required for the induction currents to be fully developed.

In Figs. 9.17 and 9.18, the inducing and total fields are computed for the two positions of satellite, ahead and behind the moon, at a distance which is 1.89 times as large as the moon radius. In these cases, errors in numerical work seem to become fairly large, so that the undulations in the curves in the figures are meaningless. But we see that the effect of the moon becomes so small that no marked differences between the inducing and the total fields are observed. This should be the case for the discontinuity considered by Ness.

In order to have a clearer interpretation of the effect of a highly conducting sphere, a similar computation is made for a perfectly conducting moon. As can be seen in Figs. 9.19, 9.20 and 9.21 in which the fields at three points on the surface of the moon as studied before are shown, the amplitude of the total field becomes anti-symmetric about the center of discontinuity. No

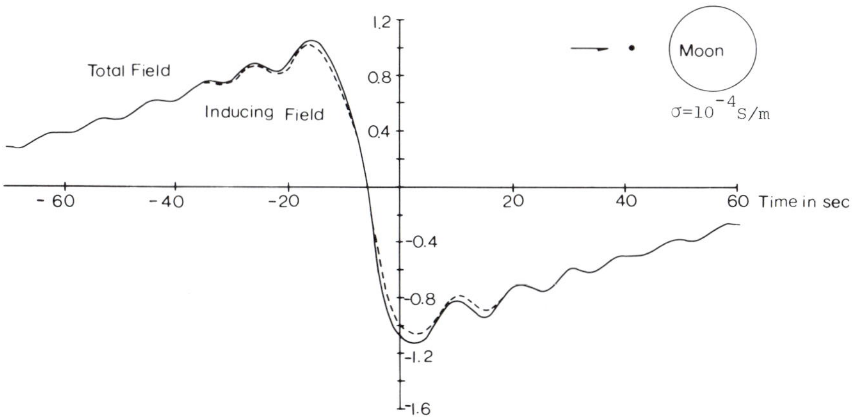

FIG. 9.17. As Fig. 9. 14, but the magnetometer is placed at a distance of 1.89 R_M(R_M: the moon's radius).

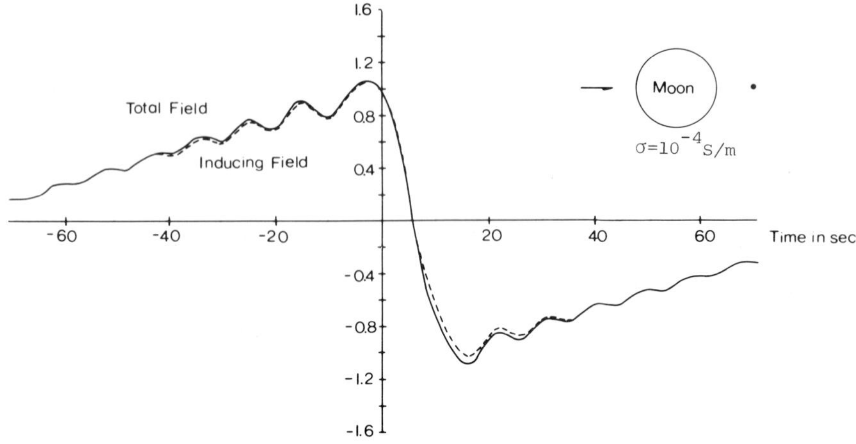

FIG. 9.18. As Fig. 9.17.

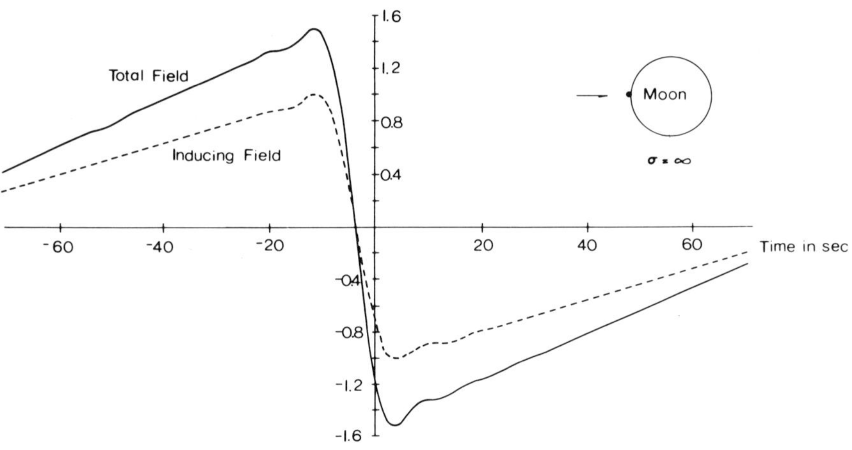

FIG. 9.19. As Fig. 9.14, but the conductivity of the moon is assumed as infinite.

difference in the amplitude between the three locations is observed. The amplitude of the total field is of course larger than that for the finite conductivity model.

No dilation of the discontinuity profile as suggested by NESS (1969) is deduced from the electromagnetic induction in a few moon models. Roughly speaking, the total field is approximately in phase with the inducing field for the models studied.

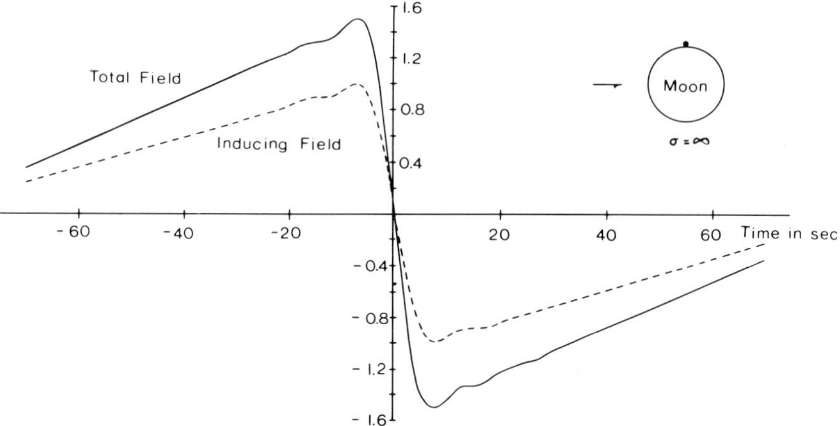

FIG. 9.20. As Fig. 9.15, but the conductivity of the moon is assumed as infinite.

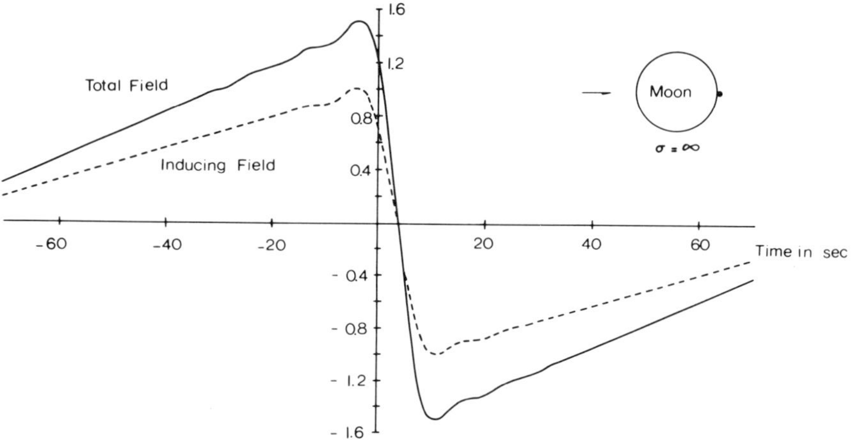

FIG. 9.21. As Fig. 9.16, but the conductivity of the moon is assumed as infinite.

It is disappointing that a dilation, as suggested by Ness, cannot be associated with electromagnetic induction by a progressive field, representing an interplanetary magnetic field discontinuity. Mathematically speaking, such a field with a wavelength far larger than the size of the moon is seen largely as a uniform field. Electromagnetic induction by a uniform field excites electric currents encircling an axis which is parallel to the field direction, so that the field due to the induced currents is dipolar. Such a field intensifies the inducing one at a point on the equatorial plane outside the conductor. No screening of the inducing field takes place in this case. As it is observed that the coefficient for P_1^0 is very much larger than those of other spherical harmonics, no screening effect is anticipated even for the progressive field considered.

As studied by JOHNSON and MIDGLEY (1968), a small number of magnetic lines of force, very near the center of a moon model of relatively low conductivity, can diffuse through the moon, being subjected to a considerable delay. Most of the magnetic lines of force, however, proceed by slipping around the moon, so that the delay in the above seems to be almost completely masked. No dilation is thus possible. If a progressive field, with a wavelength much smaller than the moon's size, is considered, however, the screening effect of the moon and accompanying dilation would become important.

In the light of the above result, it is not possible to interpret the dilation of the time profile as observed by Explorer 35 on the basis of the diffusion of the field through the moon. A more likely interpretation would be to assume that the velocity of the field-bearing solar plasma near the moon is smaller than that near Explorer 33, by a factor of about 5. The latter probe was far away from Explorer 35 which was located behind the moon. The conclusion about electrical conductivity presented by NESS (1969) does not seem quite reliable. The time constant discussed by Ness is that of the free decay of the induced currents, but what we observed with the satellite-borne magnetometer is the sum of the inducing and induced fields.

CHAPTER 10

CHAPTER 10

GLOBAL DISTRIBUTION OF ELECTRICAL CONDUCTIVITY IN THE EARTH AND MOON

It has been shown in the last chapter that the electromagnetic response of a spherical body to a time-dependent external magnetic field is controlled by the electrical conductivity of the body. It is also apparent that the longer the period of variation is, the deeper is the penetration depth of induced currents. Comparison between external and internal parts of a transient geomagnetic variation thus provides a means of inferring the electrical conductivity within the earth.

Uniform and non-uniform sphere models have been used in classical studies along these lines by SCHUSTER (1908), CHAPMAN (1919), CHAPMAN and PRICE (1930), LAHIRI and PRICE (1939), RIKITAKE (1950a, b, c) and others.

10.1 Uniform Sphere Models

Let us assume that the conducting interior of the earth is approximated by a uniform sphere having a radius qa (a: the earth's radius) and a conductivity σ. CHAPMAN (1919), who applied such a uniform sphere model to his results of analyses of *Sq*, obtained $q = 0.96$ and $\sigma = 3.6 \times 10^{-13}$ e.m.u. (3.6 $\times 10^{-2}$ S/m) on the assumption that the magnetic permeability is unity, in electromagnetic units. A combination $q = 0.94$ and $\sigma = 5 \times 10^{-1}$ S/m has been obtained by RIKITAKE (1950a) on the basis of later analyses of *Sq*.

RIKITAKE (1950a, b) obtained uniform sphere models which were compatible with the then-available analyses of *Sq*, *Dst* (main phase of magnetic storm), geomagnetic bay, s.f.e. (solar flare effect) and the like. He concluded that a superficial layer of poor conductor seems likely to overlie the earth's interior, represented by the spherical conductor, judging from the results of overall analyses of geomagnetic variations of short period. It was thus found, for all variations treated, that the conductivity jumps to a value about 10^{-1} S/m from 10^{-4} S/m at a depth of about 400 km. However, the conductivity values obtained for slower variations are larger than for those for

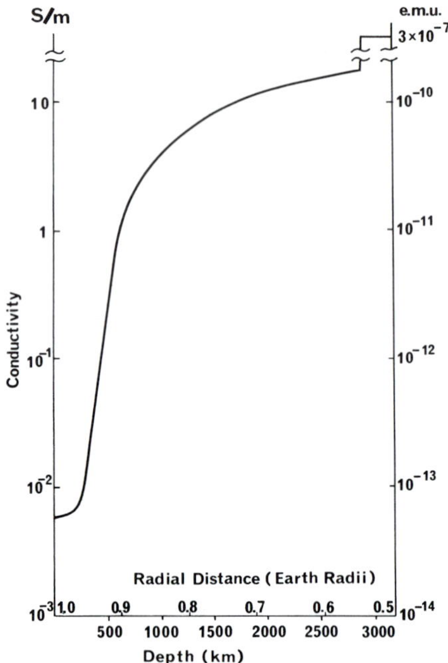

FIG. 10.1. Overall distribution of electrical conductivity in the earth's interior (RIKITAKE, 1973b).

more rapid variations. On the basis of a rough estimate of the depth, where the induced currents are most likely to contribute to the surface magnetic field, a radial distribution of the electrical conductivity was presented (RIKITAKE, 1950c). A gradual increase in the conductivity below a depth of 400 km was thus disclosed (see Fig. 10.1).

10.2 Non-Uniform Sphere Models

LAHIRI and PRICE (1939) were the first to apply a theory of electromagnetic induction in a non-uniform sphere in which the conductivity distribution is given in an analytical form such as (9.36). They concluded that the earth's conductivity increases sharply at a depth of several hundred kilometers.

In association with the development of high-speed computers, it has become possible to tackle the electromagnetic induction problem in a spherical model having an arbitrary conductivity distribution. Such a method was also applied to the study of the earth's conductivity (TAKEUCHI and SAITO, 1963; ECKHARDT, 1963).

Spectral analysis techniques have become widely used in geophysical analysis in recent years. Development of high-speed computers enables us to deal with a large set of data. According to spectral analysis of the horizontal component of magnetic variations in the frequency range 10^{-3} to 10 cycles · day^{-1} as recorded at Greenwich, England (BANKS, 1969), it is noticeable that the spectrum in the frequency range 10^{-3} to 0.5 cycles · day^{-1} consists of a number of lines, i.e. 1 and 2 cycles · yr^{-1}, 1/27, 2/27 and 3/27 cycles · day^{-1} and etc., superimposed on an approximately white continuum.

In order to investigate the radial distribution of the electrical conductivity within the earth, it is important to analyse geomagnetic variations covering a wide frequency range. BAILEY (1970) in fact proved that, if the electromagnetic response of the earth is known for all frequencies, the conductivity distribution within the earth can be uniquely determined under certain conditions.

BANKS (1969, 1972) examined the nature of the line spectra and continuum which he obtained. High coherence in the spectrum between widely spaced observatories indicates that the world-wide distribution of these variations is simple. For instance, the magnetic potential of the 27-day variation and continuum can be described approximately by the P_1 spherical harmonic only. Most likely, these variations are caused by fluctuations of the equatorial ring current. On the other hand, P_2 seems to fit the annual variation, though the semi-annual one seems likely to be expressed by P_1 (CURRIE, 1966).

A model which is compatible with the continuum response, and the response for the annual and semi-annual variations, as well as the 27-day variation and its harmonics, was obtained by BANKS (1969). The result agrees roughly with those of previous workers (RIKITAKE, 1950c; MCDONALD, 1957).

As the Banks model is derived from a frequency range 0.01 to 0.25 cycles · day^{-1}, nothing accurate can be said about the conductivity of the top 400 km of the earth. Any conductivity less than 10^{-1} S/m seems compatible with the observations. The situation is much the same for the previous models. BANKS (1972) improved his model by taking a better fit of phase relations into account, getting a slightly higher conductivity below a depth of 600 km.

10.3 *Inverse Problem*

PARKER (1971) advanced an inversion technique applicable to the induction problem and determined the conductivity on the basis of the continuum data due to Banks. The aim of Parker's method is to look for the distribution of conductivity σ under the condition that $\int_0^a (\delta\sigma/\sigma)^2 \mathrm{d}r$ is minimum, a and $\delta\sigma$ being the radius of the earth and a small deviation of σ. On applying a variational technique, the condition yields a set of simultaneous equations in which the unknowns are functions of radial function f_n as defined as a solution of (9.23) for different frequencies. Starting from an initial choice of conductivity distribution, a small deviation σ which tends to satisfy the above condition is obtained as a linear combination of the solutions of such simultaneous equations. The same procedure should be repeated until a convergent distribution is achieved.

Parker's method is so sophisticated that it is difficult to reproduce a full account here. It should be pointed out that the first choice should be sufficiently close to the true conductivity distribution for a linear approximation to be valid. Parker showed the distribution thus obtained, along with the two-dimensional uncertainties. Parker's distribution differs from Banks' in that it gives a near-surface conductivity one order of magnitude higher, although the cause of such a discrepancy is not quite clear. Logically, an inversion method is superior to the method of best fitting based on trials. Future development of the inversion technique applied to an extensive set of highly accurate geomagnetic data is highly desirable.

The work of BANKS (1969, 1972) and PARKER (1971) did not put much stress on analyses of *Sq* because *Sq* is thought to be contaminated by the effects of near-surface inhomogeneity in the electrical conductivity distribution. It appears to the authors, however, that the discrepancy between the observed and calculated amplitude ratios for the Banks and Parker models is not quite negligible. The Parker model gives rise to a set of responses which are too high. To avoid this, the conductivity values at depths of a few hundred kilometers should be considerably lowered. The Banks models also lead to slightly larger amplitude ratios especially for the P_2^1 response. None of the existing models seems to fit exactly in the P_1 and *Sq* responses at the same time, although it is not entirely clear whether this is caused by near-surface conductivity anomalies.

ANDERSSEN (1975), who reviewed existing inversion studies for global electromagnetic induction data, wrote that none has adequately examined the extent of the underlying non-uniqueness or made a systematic search for globally distinct solutions. He concluded that global electromagnetic induction data can only constrain qualitative conclusions about conductivity structure in the upper mantle, at least until a set of greatly improved data becomes available. The authors believe, however, that global electromagnetic induction is nevertheless one of the most important techniques for probing the earth's interior.

10.4 Geomagnetic Secular Variation and the Electrical Conductivity of the Lower Mantle

It has been shown that electric currents induced by magnetic field variations of external origin do not penetrate into the deep interior of the earth, so that the conductivity below a depth of 1,500 km or so cannot be determined. On the other hand, we have the core-origin secular variations, with periods which are predominantly longer than those of the magnetic variations of external origin so far considered. Actually, the variations at frequencies less than 10^{-3} cycles · day^{-1} are entirely due to the secular variation. As the secular variations are diffused through the mantle, they must be subjected to a screening by the conducting mantle. The degree of screening being dependent on the period and spatial distribution as well as the conductivity, it seems possible to estimate the conductivity in the lower mantle by comparing relative amplitudes of various harmonic constituents of secular variation. In such a study, it is necessary to assume details of the source of secular variation.

Studies along these lines were made by McDonald (1957) and Yukutake (1959), who assumed time-dependent dipoles at the core-mantle interface, deriving a conductivity amounting to 10^3 S/m or thereabouts in the lower mantle, immediately above the interface. McDonald (1957) incorporated his result in one of the distributions due to Lahiri and Price (1939). The theory of screening of the geomagnetic variation has been summarized by Rikitake (1966a, pp. 213–220).

We should not put too much emphasis on the conductivity distribution obtained from the studies of secular variation because the method adopted is equivalent to determining the characteristics of a filter only from the output, without knowing the details of the input.

10.5 Electrical Conductivity of the Earth's Core

No accurate estimate of the core's conductivity has been possible because physical conditions and composition in the core are very poorly known. It has been widely accepted that the core is mostly composed of molten iron with some silicate and/or sulphide. Physical discussion about the conductivity of such a material, under the temperatures and pressures prevailing, led to values amounting to $10^4 \sim 10^5$ S/m (Elsasser, 1946b; Bullard, 1948; Stacey, 1969).

The possibility of exciting MHD waves and their effect on seismic wave propagation has often been examined (Cagniard, 1952; Rikitake, 1952; Knopoff, 1955; Yukutake, 1957; Knopoff and MacDonald, 1958;

KRAUT, 1965; LILLEY and CARMICHAEL, 1968, 1970). It has been concluded, however, that seismic waves are not affected unless we have an incredibly large magnetic field within the core. Nothing definite can therefore be said about the core's conductivity from this kind of study.

10.6 Overall Distribution of the Electrical Conductivity within the Earth

Summarizing what has been stated in the foregoing sections, the most reliable model representing the electrical conductivity distribution in the mantle would seem to be the one due to BANKS (1972), although it is suspected that the model will shortly be modified by improved analyses of geomagnetic data. In particular, it is the authors' desire to find a model which achieves a better fit to the *Sq* data.

It is disappointing that no accurate determination of the conductivity in the top layer of the mantle is possible, even if geomagnetic variations of shorter period are analyzed. This is caused mainly by noise arising from lateral inhomogeneity of the conductivity.

Figure 10.1 is a smoothed version of Bank's distribution of conductivity in the mantle supplemented by his previous distribution (BANKS, 1969) for the top layer. The conductivity in the lower mantle seems to be one order of magnitude smaller than that of the McDonald model (MCDONALD, 1957).

As for the conductivity of the earth's core, no convincing method of estimation has been developed although there are a few studies (RIKITAKE, 1966a, chap. 16; STACEY, 1969, p. 150). We will take 3×10^4 S/m as a typical value of the conductivity in the core.

Although further improvement based on geomagnetic data of high quality and a proper theory of inversion is required, Fig. 10.1 seems to be the most reliable distribution of electrical conductivity throughout the earth, at the present stage of investigation.

10.7 Electrical Conductivity of the Moon

10.7.1 Interest in the electrical conductivity of the moon

When installation of various instruments on the lunar surface became a reality in the later half of 1960's, much debate about the internal structure and constitution within the moon was stimulated. The electrical conductivity was certainly one of the important subjects considered. For example, TOZER and WILSON (1967), ENGLAND *et al.* (1968) and others, who assumed probable temperature and pressure distributions, estimated the electrical conductivity of the moon's interior. According to these studies, it has been guessed that the moon consists of two parts; the outer layer, which is almost insulating, and the inner conducting core having a radius amounting to $0.5 \sim 0.7\ R_M$ (R_M: moon's radius). The conductivity in the inner core has been supposed to amount to $10^{-2} \sim 1$ S/m.

As mentioned in Subsection 5.8.4, it seems likely that a kind of unipolar generation takes place when the solar wind, bearing a magnetic field, encounters the moon, provided it is conducting. A magnetic field arising from such a generation process would form a shock front against the solar wind as observed for the earth's magnetosphere. According to the observations by Explorer 35, however, no such front was detected (COLBURN *et al.*, 1967; NESS *et al.*, 1967). It is therefore surmised either that there is no conducting core within the moon or that the core is small, if it exists at all.

BLANK and SILL (1969), NAGATA (1971) and DYAL *et al.* (1974) argued that the magnetic field produced by the electric currents induced in the moon's interior by the time-varying magnetic field, which is carried by the solar wind, is subjected to a compression by the highly-conducting plasma of solar wind, so that its magnetic lines of force would not leak outside the lunar surface facing the sun. It is supposed, however, that the situation is much the same as the electromagnetic induction in a conducting sphere placed in free space over the night side of the moon. This may also be the case when the moon is located at a suitable position in the geomagnetic tail. It is therefore surmised that electromagnetic induction in a sphere, placed in free space, by a time-dependent external magnetic field is important even in the case of the moon.

10.7.2 Actual response of the moon

(1) An interpretation at an early stage of investigation

NESS *et al.* (1967), who analysed the magnetic data obtained by Explorer 35, concluded that the mean electrical conductivity of the moon would be less than 10^{-5} S/m because the delay of the interplanetary magnetic field through the moon is small. JOHNSON and MIDGLEY (1968) reached a similar conclusion by investigating the shape of magnetic lines of force passing through the moon.

The rise-times of a discontinuous interplanetary magnetic field as observed by Explorers 33 and 35 amount to 10.2 and 56.2s, respectively (NESS, 1969). The distance between the two space probes was 56 R_E (R_E: the earth's radius). NESS (1969) concluded that such a delay of rise-time is caused by the time required for the magnetic lines of force to pass through the moon because Explorer 35 was located behind the moon. Ness further estimated the upper limit of the electrical conductivity of the moon's interior as 5×10^{-5} S/m on the basis of the time-constant for free decay of a magnetic field in a conducting sphere. As discussed in Subsection 9.4.7, it seems likely that most of the magnetic lines of force of a progressive field slip around the moon, so that the conductivity obtained by Ness may not be accurate. It seems more natural to

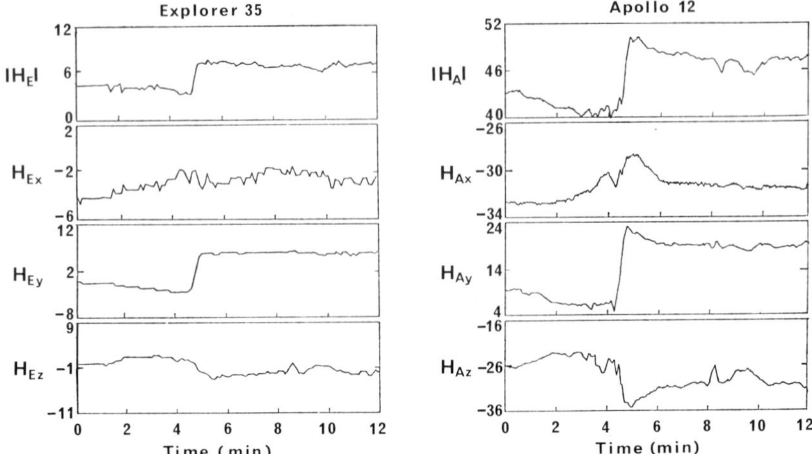

FIG. 10.2. An example of simultaneous observation of discontinuous interplanetary field, in units of gamma, by Explorer 35 (left) and the Apollo 12 LSM (DYAL *et al.*, 1970c).

presume that the velocity of solar wind is faster at the site of Explorer 33 by a factor 5 than that at that of Explorer 35.

(2) The response of the moon placed in the solar wind plasma

a) The data taken on the night side of the moon

Observations by means of the lunar surface magnetometers (LSM) since the Apollo 12 mission have made it possible for us to discuss the electromagnetic response of the moon in fair detail. Figure 10.2 (DYAL *et al.*, 1970c) shows a discontinuous change in the magnetic field as observed by the Apollo 12 LSM when it was in the dark area on the lunar surface. The change simultaneously observed by Explorer 35, which was located at a great distance from the moon, is also shown in the figure. Suffices x, y and z indicate the components in the radial, east and north directions on the moon's surface, respectively. It is clearly observed from the figure that, when the y component of the interplanetary field changes rapidly, the same component on the lunar surface is markedly enhanced.

DYAL *et al.* (1970c) presented a model of the distribution of conductivity (σ) in the moon by which a number of observed responses, including the above, can be accounted for. According to them, the model is given as follows;

$$\sigma < 10^{-7}\ \text{S/m for } r > 0.97\,R_M$$
$$\sigma = 1.7 \times 10^{-4}\ \text{S/m for } 0.97\,R_M > r > 0.67\,R_M$$
$$\sigma \geqq 10^{-2}\ \text{S/m for } 0.6\,R_M > r.$$

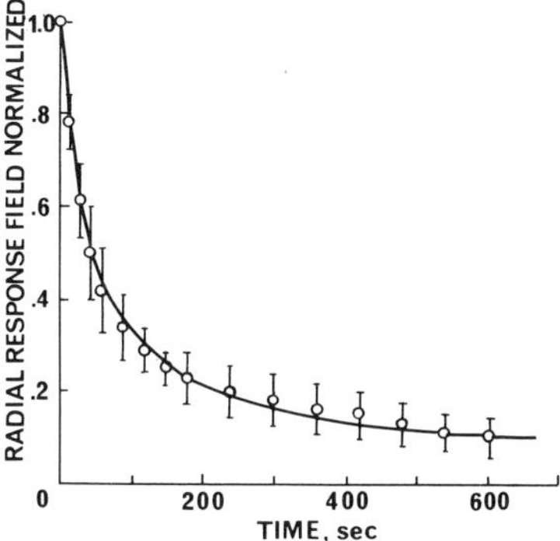

FIG. 10.3. The moon's transient response, observed on the dark side of the moon, in the solar wind plasma. The component of the induced magnetic field normal to the lunar surface is shown (DYAL *et al.*, 1974).

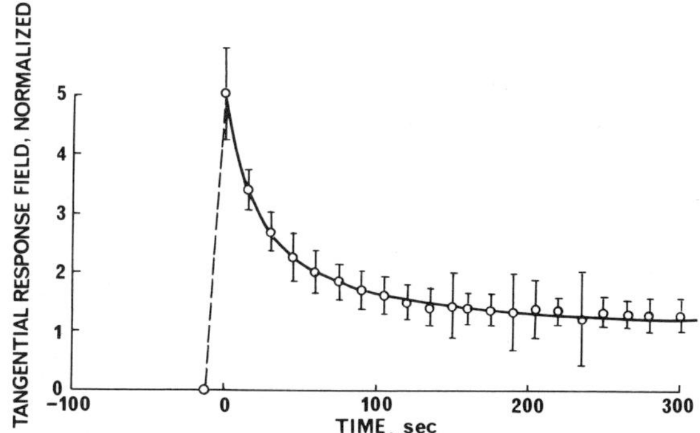

FIG. 10.4. The moon's transient response, observed on the daylight side of the moon, in the solar wind plasma. The component of the induced magnetic field tangential to the lunar surface is shown (DYAL *et al.*, 1974).

Accumulation of examples observed by LSM's in later years brought out the transient response of the moon fairly clearly. Figure 10.3 shows the transient behavior of the internal magnetic field component normal to the lunar surface when the external field changes stepwise with a rise-time shorter than 15s. The amplitude is normalized in such a way that the field becomes 1 at $t = 0$. The error bars in the figure indicate the standard deviations (DYAL *et al.*, 1974; DYAL and PARKIN, 1973). The response thus obtained may be explained by assuming that $\sigma = 1 \times 10^{-4} \sim 2 \times 10^{-3}$ S/m at a depth of 250 km and $\sigma = 2 \times 10^{-3} \sim 8 \times 10^{-2}$ S/m down to a depth of 1,000 km.

b) The data taken on the daylight side of the moon

When the magnetic field is measured by a LSM which is placed on the daylight side of the moon, the induced magnetic field is so compressed by the impinging solar wind plasma that almost all the normal fields are confined within the lunar interior. The normal component of the internal field is therefore almost zero at the lunar surface, while the tangential component is so enhanced that it sometimes becomes five times as large as that of the external field. According to DYAL *et al.* (1974), the response of the tangential component of the internal field is shown in Fig. 10.4. It is said that such a response can be obtained if a conducting core having a conductivity amounting to 10^{-3} S/m underlies an insulating layer of 0.1 R_M in depth.

The results of spectral analyses of magnetic fluctuations observed by LSM's can be compared to those observed by Explorer 35 at a great distance from the moon. On the basis of transfer functions obtained from such a comparison, SONETT *et al.* (1971, 1972) put forward a model in which they assumed a spiky, highly conducting zone at a depth of 1,500 km. Such a rather unnatural distribution was withdrawn some years later because of newly observed results.

(3) The data observed by LSM's when the moon is in the geomagnetic tail

When the moon is in the shadow zone of the earth or, in other words, in the geomagnetic tail, we can have data which are more appropriate for discussing the electromagnetic response than those taken on the daylight side of the moon. In this case, it is not necessary to pay much attention to the influence of fluctuations due to the compression of the permanent lunar magnetic field by the solar wind plasma, the diamagnetic effect of the plasma and the like.

Figure 10.5 (DYAL *et al.*, 1974) shows the magnetic fields observed by the Apollo 12 LSM and Explorer 35. The magnetic field which would be observed on the surface of the moon, within which the conductivity is distributed as shown in the insert of Fig. 10.6, is also shown. The actual observations were made when the moon was located north of the neutral plane in the geomagnetic tail. In the figure, the component of magnetic field normal to the lunar surface is shown.

Figure 10.6 (DYAL *et al.*, 1974) shows the observed and calculated responses of the moon along with the magnetic field observed by Explorer 35 when the moon is in the plasma sheet of the geomagnetic tail. It is thus shown that the observed transient response of the moon can be approximately accounted for by assuming a distribution of electrical conductivity such as shown in the insert of Fig. 10.6 although there is no guarantee that this is unique.

A magnetic disturbance having an anomalously large amplitude was observed on April 20, 1970. On that occasion, the amplitude observed by

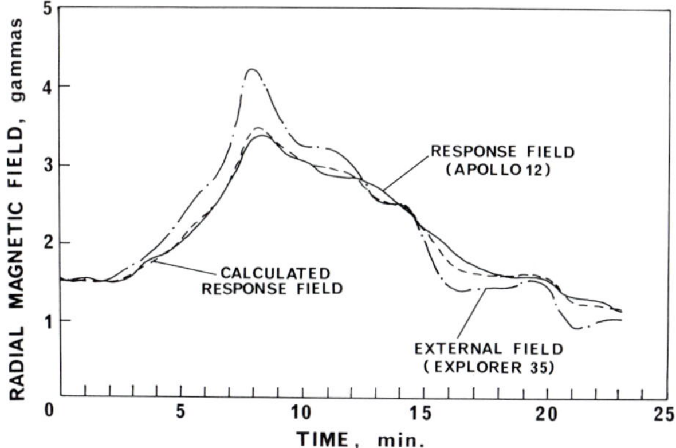

FIG. 10.5. An example of magnetic field fluctuation simultaneously observed by the Apollo 12 LSM and Explorer 35. The broken line shows the fluctuation that should be observed on the moon when electromagnetic induction in a model shown in the insert of Fig. 10.6 is taken into consideration. The observations were made when the moon was situated north of the neutral plane in the geomagnetic tail (DYAL *et al.*, 1974).

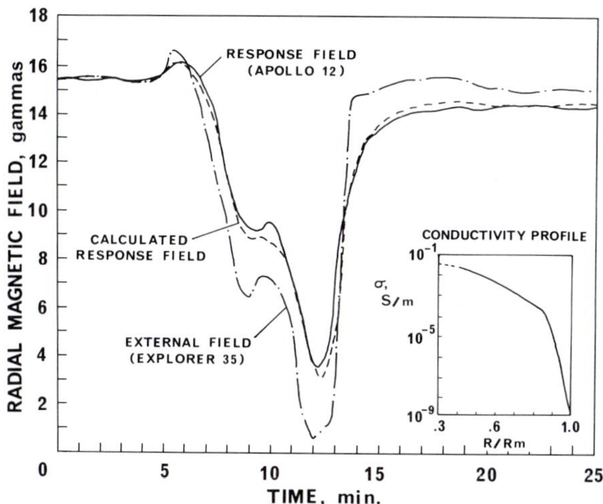

FIG. 10.6. The results of simultaneous observations of magnetic field by Apollo 12 LSM and Explorer 35 along with that expected for a model shown in the insert. The data were taken when the moon was in the plasma sheet of the geomagnetic tail (DYAL *et al.*, 1974).

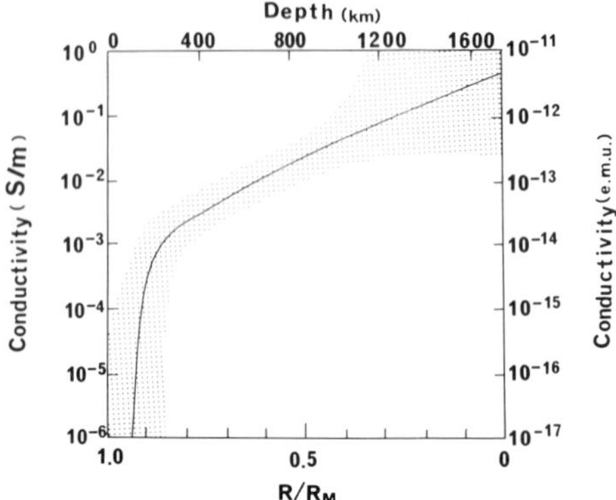

FIG. 10.7. Probable distribution of the electrical conductivity in the moon (DYAL *et al.*, 1976).

Explorer 35 amounted to 12–35 γ. The moon was in the geomagnetic tail at that time. DYAL *et al.* (1976) analysed the event on the basis of the observations by Explorer 35, Apollo 12, 15 and 16 LSM's, and Apollo 15 and 16 command modules. After testing more than 200 models, the most likely distribution of the electrical conductivity within the moon was obtained as shown by the solid line in Fig. 10.7 in which the allowances for probable errors are also indicated with the shadow. It thus becomes apparent that the conductivity in the moon is substantially lower than that in the earth. The moon's conductivity thus inferred provides a constraint for estimating the internal temperature of the moon.

10.8 Electrical Conductivity of Lunar Rock Samples

NAGATA *et al.* (1971) and NAGATA (1971) reported on the actual measurements of electrical conductivity of lunar rock samples. For example, the temperature change in the conductivity of the Apollo 11 igneous sample No. 10024–22 can be expressed as

$$\sigma = \sum_{i=1}^{2} \sigma_i e^{-A_i/kT}$$

in which

$$\sigma_1 = 7.9 \text{ S/m}, \ A_1 = 0.51 \text{ eV}$$
$$\sigma_2 = 3.1 \times 10^6 \text{ S/m}, \ A_2 = 1.25 \text{ eV}.$$

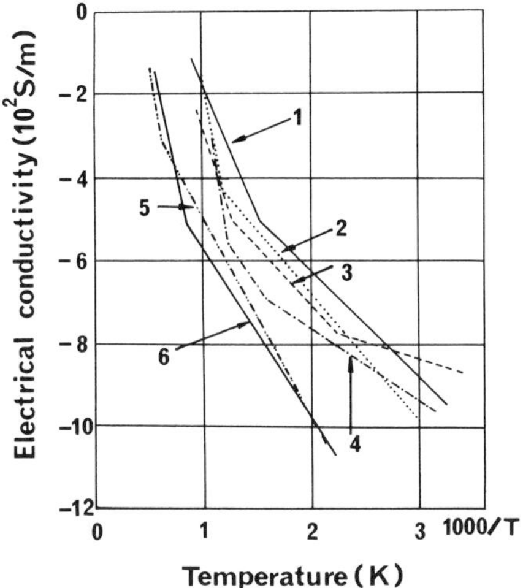

FIG. 10.8. Electrical conductivity of Apollo 11 and 12 lunar materials as a function of temperature in comparison with that of typical terrestrial rocks (NAGATA, 1971). 1. No. 10024–22: Apollo 11 crystalline rock; 2. No. 12053–47 (I): Apollo 12 crystalline rock; 3. No. 12053–47 (II): Another specimen cut from the same sample as above; 4. No. 10048–55: Apollo 11 microbreccia; 5. Terrestrial olivine; 6. Terrestrial peridotite.

Figure 10.8 shows how σ changes with the reciprocal of absolute temperature T for a number of lunar rock samples and terrestrial olivine and peridotite. As can be seen in the figure, the conductivities of the Apollo samples are outstandingly higher than those of terrestrial rock samples. Accordingly, a temperature distribution proposed by ENGLAND *et al.* (1968) would lead to an electrical conductivity higher than that obtained from the analyses of electromagnetic response by a factor of 1,000 or more in the deep interior of the moon.

In order to solve such a contradiction, we have to assume either that the substance constituting the moon's deep interior is different from that near the surface or that the temperature there is substantially lower than that supposed by ENGLAND *et al.* (1968).

It is known that metallic iron is abundant in lunar rocks, so that it may be that the electrical conductivity in the moon takes on a fairly high value. This may also be the case for meteorites which include other free metals along with iron.

10.9 Temperature and Electrical Conductivity

The electrical conductivity, denoted by σ, of silicate minerals constituting the earth is also known to be highly dependent on temperature and similarly to moon rocks, it can be expressed as a function of the absolute temperature T as

$$\sigma = \sum_i \sigma_i \exp\left(-A_i/kT\right)$$

where A_i and σ_i are the activation energy and a constant depending on conduction mechanism. k is the Boltzmann constant. Activation energy indicates the energy required for generation and transport of charge carriers within mineral structure. In the earth three kinds of conduction mechanisms are considered in general: impurity, intrinsic, and ionic conduction. Impurity conduction is dominant at low temperatures and in this case the activation energy necessary for transporting charge carriers is the most important, while the other two are supposed to be likely mechanisms at relatively high temperatures. However, it is unknown at present which one of the above two mechanisms is actually dominant in the earth's mantle (DUBA, 1976).

There have been many reports on measurements of the electrical conductivity of olivine for various temperature ranges, since olivine is the

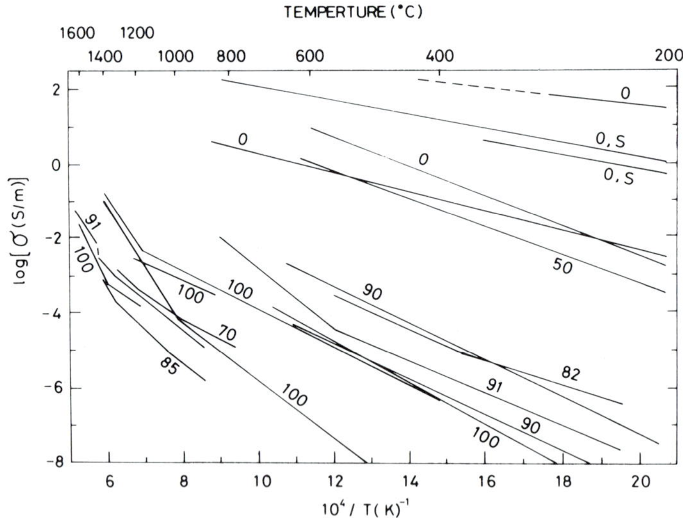

FIG. 10.9. Electrical conductivity of olivine as a function of temperature. Numerals indicate the forsterite component in percent for the samples used in conductivity experiments. *S* denotes the spinel phase (DUBA, 1976).

main constituent of the earth's mantle material (e.g. NORITOMI, 1961; BRADLEY *et al.*, 1964; AKIMOTO and FUJISAWA, 1965; HAMILTON, 1965; MIZUTANI and KANAMORI, 1967; SCHOBER, 1971; KOBAYASHI and MARUYAMA, 1971; DUBA, 1972; DUBA *et al.*, 1974). DUBA (1976) compiled some of the reported data as shown in Fig. 10.9. It is well recognized that the temperature dependence is described by the above expression, although activation energy is different from one sample to another. An incredibly large discrepancy between different data is evident, partly because of sample differences, ranging from pure forsterite, as shown by (Fo, 100), to pure fayalite, (Fo, 0). The tendency for the conductivity to become higher as the forsterite component decreases can be recognized.

The ratio of fayalite to forsterite for the earth's mantle is supposed to be 1:9 or so and some samples of this composition are also included in Fig. 10.9. However, even samples of this composition show a considerable scatter in measured conductivity values: as much as two or more orders of magnitude. The main reason would be different experimental conditions, particularly in oxygen fugacity. In fact, DUBA and NICHOLLS (1973) clearly demonstrated that differences in oxygen fugacity caused discrepancies in conductivity values by as much as three orders of magnitude for the same sample. DUBA *et al.* (1973) also pointed out the importance of the oxidation state of iron as a factor controlling the electrical conductivity of olivine.

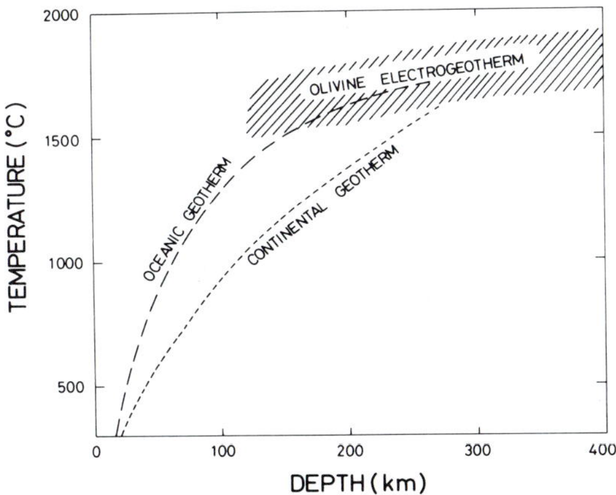

FIG. 10.10. Temperature distribution in the upper mantle. The olivine electrogeotherm represents the temperature range at the respective depth as inferred from the temperature dependence of olivine with a 90% forsterite component and the electrical conductivity distribution in the earth. The broken and dotted lines indicate Ringwood's oceanic and continental geotherms (DUBA, 1976).

The temperature dependence of electrical conductivity enables us to infer the temperature distribution within the earth and the moon from the global electrical conductivity distribution. DUBA *et al.* (1974) and DUBA (1976) used the most reliable conductivity data experimentally determined under oxygen fugacity control and derived the temperature distribution within the earth as shown in Fig. 10.10. It is not meaningful to use this method for inferring the temperature distribution at shallower depth, because the global conductivity distribution at shallow depth cannot be accurately deduced. At shallow depths there are many conductivity anomalies over the world; most of them are due to other factors such as conducting water and partial melting of mantle material (see Chapter 12). Similar studies have also been undertaken for the moon (e.g. DYAL and PARKIN, 1973; TOLLAND, 1974; VANYAN, 1980).

10.10 Phase Transformation and Electrical Conductivity

It has been well known that the olivine—modified spinel transformation in the (Mg, Fe)$_2$SiO$_4$ system occurs at the pressure corresponding to about 400 km depth in the earth (e.g. AKIMOTO, 1972). Another transformation is also supposed to occur at a depth of 650 km or so, transformation from spinel or modified spinel to a denser strontium plumbate structure (YAGI and AKIMOTO, 1973).

Although the electrical conductivity depends little on pressure itself (e.g. DUBA *et al.*, 1974), the phase transformation is likely to cause a sharp increase in electrical conductivity as demonstrated by AKIMOTO and FUJISAWA (1965) and YAGI and AKIMOTO (1973). In the case of Fe_2SiO_4, the conductivity increase amounted to about two orders of magnitude in association with the olivine—modified spinel transformation (AKIMOTO and FUJISAWA, 1965). In the case of Mn_2GeO_4 for which YAGI and AKIMOTO (1973) conducted experiments, they observed an abrupt change amounting to half and one third of one order of magnitude for the olivine—modified spinel transformation and the modified spinel—strontium plumbate transformation, respectively.

On the other hand, DUBA (1976) cast doubt on the amount of the conductivity change, on the grounds that the oxidation condition during the experiments is uncertain. In any case, an abrupt change in conductivity is likely to occur in association with phase transformation and the marked increase in conductivity at a depth of 400 $\sim$ 600 km in the earth (see Fig. 10.1) most likely corresponds to one of the above two kinds of transformation or both.

CHAPTER 11

CHAPTER 11

ELECTROMAGNETIC INDUCTION IN A PLANE CONDUCTOR

In contrast to global induction problems in which electromagnetic induction in a spherical conductor is important, problems of local induction have been solved on the assumption that the earth is a plane conductor. The theories that will be described in this chapter have contributed significantly to an understanding of various kinds of ocean effects and also to interpretation of short-period geomagnetic variation anomalies in terms of electrical conductivity anomalies in the crust and the upper mantle.

11.1 Price's Theory

11.1.1 Fundamental equations and their solutions

The following description is based on a basic theory of electromagnetic induction in a semi-infinite conductor which was put forward by PRICE (1950).

Suppose for simplicity that the electrical conductivity, denoted by σ, and the permeability, μ, are both constant. Then, as was shown in Section 9.1, the electric or magnetic field vector, denoted by $\boldsymbol{F}$, inside the conductor satisfies

$$\nabla^2 \boldsymbol{F} = 4\pi\sigma\mu\partial \boldsymbol{F}/\partial t \tag{11.1}$$

while outside the conductor $\boldsymbol{F}$ satisfies

$$\nabla^2 \boldsymbol{F} = 0. \tag{11.2}$$

Here we denote the surface of the semi-infinite conductor by $z = 0$ in Cartesian coordinates and express the electric field as

$$\boldsymbol{E} = Z(z, t)\boldsymbol{F}(x, y). \tag{11.3}$$

Substituting (11.3) into (9.6), we derive

$$Z(\partial F_x/\partial x + \partial F_y/\partial y) + F_z\partial Z/\partial z = 0 \tag{11.4}$$

(11.4) is satisfied if

$$F_z = 0, \quad \partial F_x/\partial x + \partial F_y/\partial y = 0 \tag{11.5}$$

or

$$(\partial F_x/\partial x + \partial F_y/\partial y)/F_z = -(\partial Z/\partial z)/Z = -\alpha \tag{11.6}$$

where α is a constant.

Substituting (11.3) into (11.1) and (11.2), we obtain

$$\partial^2 \boldsymbol{F}/\partial x^2 + \partial^2 \boldsymbol{F}/\partial y^2 = (1/Z)(4\pi\sigma\mu\partial Z/\partial t - \partial^2 Z/\partial z^2)\boldsymbol{F} \qquad (z<0) \tag{11.7}$$

$$\partial^2 \boldsymbol{F}/\partial x^2 + \partial^2 \boldsymbol{F}/\partial y^2 = -(1/Z)(\partial^2 Z/\partial z^2)\boldsymbol{F} \qquad (z>0) \tag{11.8}$$

where we assumed that the conductor occupies the half space $z<0$. Since the right-hand side is independent of x and y, we can write it as $-\lambda^2$ and then

$$\partial^2 \boldsymbol{F}/\partial x^2 + \partial^2 \boldsymbol{F}/\partial y^2 + \lambda^2 \boldsymbol{F} = 0 \tag{11.9}$$

$$\partial^2 Z/\partial z^2 - \lambda^2 Z - 4\pi\sigma\mu\partial Z/\partial t = 0 \qquad (z<0) \tag{11.10}$$

$$\partial^2 Z/\partial z^2 - \lambda^2 Z = 0 \qquad (z>0). \tag{11.11}$$

The case of (11.5): the first solution

In this case $\boldsymbol{F}$ can be written as

$$\boldsymbol{F} = \begin{cases} \partial P/\partial y \\ -\partial P/\partial x \\ 0 \end{cases} \tag{11.12}$$

According to (11.9), P must satisfy

$$\partial^2 P/\partial x^2 + \partial^2 P/\partial y^2 + \lambda^2 P = 0. \tag{11.13}$$

Outside the conductor $(z>0)$, we obtain from (11.11)

$$Z = A^*(t)\mathrm{e}^{\lambda z} + B^*(t)\mathrm{e}^{-\lambda z} \tag{11.14}$$

and then

$$\boldsymbol{E} = [A^*(t)\mathrm{e}^{\lambda z} + B^*(t)\mathrm{e}^{-\lambda z}]\begin{cases} \partial P/\partial y \\ -\partial P/\partial x \\ 0 \end{cases} \qquad (z>0). \tag{11.15}$$

Therefore, in this case the electric current flows parallel to the conductor surface $(z = 0)$.

The continuity condition for the tangential component of $\boldsymbol{E}$ at $z = 0$ yields

$$Z(-0, t) = A^*(t) + B^*(t) \tag{11.16}$$

From (9.2) we obtain

$$\mu\partial \boldsymbol{H}/\partial t = -\begin{cases}(\partial Z/\partial z)(\partial P/\partial x)\\(\partial Z/\partial z)(\partial P/\partial y)\\\lambda^2 ZP\end{cases} \qquad (z<0) \tag{11.17}$$

and

$$\partial \boldsymbol{H}/\partial t = -\lambda \operatorname{grad}\left[\{A^*(t)\mathrm{e}^{\lambda z} - B^*(t)\mathrm{e}^{-\lambda z}\}P\right] \qquad (z>0). \tag{11.18}$$

The continuity condition for the tangential component of $\boldsymbol{H}$ yields

$$Z'(-0,\ t) = \lambda\mu\{A^*(t) - B^*(t)\}. \tag{11.19}$$

In the case of electromagnetic induction by a magnetic field variation of external origin, we can impose an additional condition on $Z(z, t)$;

$$Z(z,\ t)\to 0 \quad (z\to -\infty). \tag{11.20}$$

The magnetic field outside the conductor can be expressed in terms of the magnetic potential W which satisfies (9.25). In the present case, W is written as

$$W = -\{A(t)\mathrm{e}^{\lambda z} + B(t)\mathrm{e}^{-\lambda z}\}P(x, y) \qquad (z>0) \tag{11.21}$$

(9.11), (11.21) and (11.18) yield

$$\lambda A^*(t) = -\partial A(t)/\partial t, \qquad \lambda B^*(t) = \partial B(t)/\partial t. \tag{11.22}$$

If a solution of (11.10) is obtained under the conditions (11.16), (11.19) and (11.20), we can determine the magnetic field induced by a given external field.

The case of (11.6): the second solution

In this case, we have

$$Z = C(t)\ \mathrm{e}^{\alpha z}. \tag{11.23}$$

For $z<0$ (11.10) and (11.23) yield

$$C(t) = C_0\ \mathrm{e}^{-\beta t} \tag{11.24}$$

where C_0 is a constant and

$$\beta = (\lambda^2 - \alpha^2)/(4\pi\sigma\mu). \tag{11.25}$$

From (11.6) and (11.9), we derive the following relations:

$$\begin{aligned}F_x &= (\alpha/\lambda^2)(\partial F_z/\partial x) + \partial P_1/\partial y\\ F_y &= (\alpha/\lambda^2)(\partial F_z/\partial y) - \partial P_1/\partial x\end{aligned} \tag{11.26}$$

where P_1 satisfies (11.13). The terms containing P_1 have already formed the first solution and hence they can be excluded. Since F_z satisfies (11.13), we can write $F_z = \lambda P$. The electric field for $z<0$ is now obtained as

$$E = C_0\, e^{-\beta t+\alpha z}\begin{cases}(\alpha/\lambda)\partial P/\partial x\\ (\alpha/\lambda)\partial P/\partial y \qquad (z<0).\\ \lambda P\end{cases} \tag{11.27}$$

For $z>0$, α is obtained from (11.11) and (11.23) as $\alpha = \pm\lambda$. Then the electric field is given by

$$E = \operatorname{grad}[A_1(t)e^{\lambda z} + B_1(t)e^{-\lambda z}]P(x, y) \tag{11.28}$$

(11.28) and (9.2) yield

$$\mu\partial H/\partial t = -C_0 e^{-\beta t+\alpha z}(\lambda - \alpha^2/\lambda)\begin{cases}\partial P/\partial y\\ -\partial P/\partial x \qquad (z<0)\\ 0\end{cases} \tag{11.29}$$

$$\partial H/\partial t = 0 \qquad (z>0).$$

In this case, the magnetic field is confined within the conductor. Therefore, the second solution has no bearing on problems of electromagnetic induction by magnetic field variations of external origin.

11.1.2 Indeterminate problem

If $\lambda=0$ in the previous section, the magnetic potential for the external field is written as

$$W_0 = -(A_0 z + B_0)P_0(x, y) \tag{11.30}$$

where P_0 satisfies

$$\partial^2 P_0/\partial x^2 + \partial^2 P_0/\partial y^2 = 0. \tag{11.31}$$

A uniform inducing field is included in this case.

Substituting $\lambda = 0$ in (11.10), we derive a solution satisfying the condition (11.20) as

$$Z = C\, e^{\gamma z} \tag{11.32}$$

where

$$\gamma = (4\pi\sigma\mu p)^{1/2}, \qquad p = \partial/\partial t. \tag{11.33}$$

Then the electric and magnetic fields are obtained as

$$E = C\, e^{\gamma z}\begin{cases}\partial P_0/\partial y\\ -\partial P_0/\partial x \qquad (z<0)\\ 0\end{cases} \tag{11.34}$$

$$E=(az+b)\begin{cases}\partial P_0/\partial y\\ -\partial P_0/\partial x \quad (z>0)\\ 0\end{cases} \tag{11.35}$$

$$\mu pH=-C\gamma\, e^{\gamma z}\begin{cases}\partial P_0/\partial x\\ \partial P_0/\partial y \quad (z<0)\\ 0\end{cases} \tag{11.36}$$

$$H=\begin{cases}(Az+B)\partial P_0/\partial x\\ (Az+B)\partial P_0/\partial y \qquad (z>0).\\ AP_0\end{cases} \tag{11.37}$$

The continuity conditions at $z = 0$ for the normal component of $\boldsymbol{B}$ and the tangential component of $\boldsymbol{H}$ yield

$$\begin{aligned} A&=0\\ B &= -(C\gamma)/(\mu p). \end{aligned} \tag{11.38}$$

Expressing the induced field as

$$W_1 = -(A_1 z + B_1)P_0(x, y). \tag{11.39}$$

A and B are written as

$$A = A_0 + A_1, \qquad B = B_0 + B_1. \tag{11.40}$$

Then the induced field is obtained as

$$\begin{aligned} A_1&=-A_0\\ B_1 &= -(C\gamma)/(\mu p) - B_0. \end{aligned} \tag{11.41}$$

(11.41) implies that the vertical component of the inducing field is cancelled out by the induced field, while the relation between the inducing and induced tangential components remains indeterminate because C is unknown.

Equivalently we may treat the problem of electromagnetic induction in a semi-infinite conductor, due to an inducing field having an infinite wavelength, by considering the case a (radius) $\to \infty$ in the electromagnetic induction theory for a spherical conductor. If $a \to \infty$ in (9.32), $i_n/e_n \to n/(n + 1)$ and then from (9.27) $H_r = 0$ at $\rho = 1$ (surface), while the ratio of the tangential component of the induced field to that of the inducing field is $n/(n + 1)$. Since n can be any positive integer in the limit $a \to \infty$, the ratio is indeterminate in the range $1/2 \sim 1$.

It is interesting to note that the ratio of the electric to magnetic fields can be determined even in this case. From (11.34) and (11.36) we can eliminate the unknown C and derive

$$E_y/H_x = -E_x/H_y = \mu p/\gamma. \qquad (11.42)$$

The relation (11.42) is a basic equation in magnetotellurics (see Section 12.3).

11.2 Relation between External and Internal Fields

In this section we consider electromagnetic induction in a laterally uniform conductor by an external magnetic field variation having a finite scale length. Then we examine the relation between external and internal fields. In the following, the Cartesian coordinate system is used with z positive downward.

Let us first consider the case of a uniform conductor. Within the conductor the vector potential $\boldsymbol{A}$ satisfies (9.17). Its typical solution is given by

$$\boldsymbol{A}_{mn} = -C_{mn} e^{-\sqrt{k^2+\gamma^2}z} (\boldsymbol{n} \times \text{grad}\, V_{mn}) \qquad (11.43)$$

where

$$V_{mn} = e^{i(mx+ny)}, \quad k^2 = m^2 + n^2 \qquad (11.44)$$

and γ is given by (11.33). $\boldsymbol{n}$ denotes a unit vector in the z direction. The magnetic field in the conductor is derived from (11.43) as

$$\mu \boldsymbol{H} = \begin{cases} -C_{mn}\sqrt{k^2+\gamma^2}\, e^{-\sqrt{k^2+\gamma^2}z}\partial V_{mn}/\partial x \\ -C_{mn}\sqrt{k^2+\gamma^2}\, e^{-\sqrt{k^2+\gamma^2}z}\partial V_{mn}/\partial y \\ C_{mn}k^2 e^{-\sqrt{k^2+\gamma^2}z} V_{mn}. \end{cases} \qquad (11.45)$$

Outside the conductor, a typical magnetic potential is given by

$$W_{mn} = (e_{mn} e^{-kz} + i_{mn} e^{kz}) V_{mn} \qquad (11.46)$$

where e_{mn} and i_{mn} are external and internal coefficients of magnetic potential, respectively. Then the magnetic field outside the conductor is obtained as

$$\boldsymbol{H} = \begin{cases} -(e_{mn}e^{-kz} + i_{mn}e^{kz})\partial V_{mn}/\partial x \\ -(e_{mn}e^{-kz} + i_{mn}e^{kz})\partial V_{mn}/\partial y \\ k(e_{mn}e^{-kz} - i_{mn}e^{kz}) V_{mn}. \end{cases} \qquad (11.47)$$

From the boundary conditions at the surface ($z=D$), we derive

$$\left.\begin{aligned} \mu(ee^{-kD} + ie^{kD}) &= C\sqrt{k^2+\gamma^2}\, e^{-\sqrt{k^2+\gamma^2}D} \\ ee^{-kD} - ie^{kD} &= Cke^{-\sqrt{k^2+\gamma^2}D} \end{aligned}\right\} \qquad (11.48)$$

where subscripts m and n are omitted. The relation between external and internal fields is thus given by

$$i = e^{-2kD}\frac{\sqrt{k^2+\gamma^2}-\mu k}{\sqrt{k^2+\gamma^2}+\mu k}e. \tag{11.49}$$

We can treat the case of a laterally uniform conductor by dividing the plane conductor into N layers, each layer having a uniform conductivity, as is shown in Fig. 11.1. A typical vector potential in the r'th layer is expressed as

$$\boldsymbol{A}_r = -(E_r e^{-k_r z} + I_r e^{k_r z})(\boldsymbol{n} \times \operatorname{grad} V) \tag{11.50}$$

where subscripts m and n are omitted and a periodic variation with angular frequency ω is assumed. Then the magnetic field is obtained as

$$\mu\boldsymbol{H}_r = \begin{cases} -k_r(E_r e^{-k_r z} - I_r e^{k_r z})\dfrac{\partial V}{\partial x} \\ -k_r(E_r e^{-k_r z} - I_r e^{k_r z})\dfrac{\partial V}{\partial y} \\ \lambda^2(E_r e^{-k_r z} + I_r e^{k_r z})V. \end{cases} \tag{11.51}$$

The boundary conditions at $z = D_{r+1}$ (($r+1$)'th boundary) yield

$$\begin{aligned} k_{r+1}(E_{r+1}e^{-k_{r+1}D_{r+1}} - I_{r+1}e^{k_{r+1}D_{r+1}}) &= k_r(E_r e^{-k_r D_{r+1}} - I_r e^{k_r D_{r+1}}) \\ E_{r+1}e^{-k_{r+1}D_{r+1}} + I_{r+1}e^{k_{r+1}D_{r+1}} &= E_r e^{-k_r D_{r+1}} + I_r e^{k_r D_{r+1}} \end{aligned} \tag{11.52}$$

where

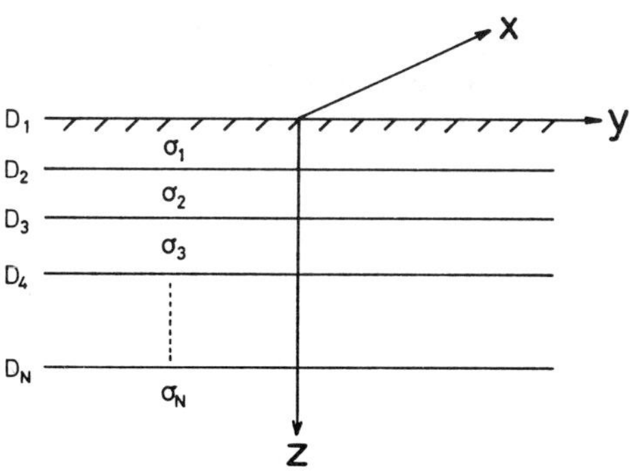

FIG. 11.1 A layered earth model.

$$k_r^2 = m^2 + n^2 + 4\pi\sigma_r\mu\omega i. \tag{11.53}$$

μ has been assumed to be constant (equal to μ_0: the value in a vacuum). σ_r is the electrical conductivity in the r'th layer. From (11.52) we derive

$$k_r G_r(D_{r+1}) = k_{r+1} G_{r+1}\,(D_{r+1}) \tag{11.54}$$

where

$$G_r(z) = \frac{E_r e^{-k_r z} - I_r e^{k_r z}}{E_r e^{-k_r z} + I_r e^{k_r z}}. \tag{11.55}$$

Thus we obtain a reccurrence relation between successive layers:

$$G_r(D_r) = \frac{C_{r+1} + k_r \tanh(k_r d_r)}{k_r + C_{r+1} \tanh(k_r d_r)} \tag{11.56}$$

where

$$\begin{aligned} d_r &= D_{r+1} - D_r \\ C_{r+1} &= k_{r+1} G_{r+1}(D_{r+1}). \end{aligned} \tag{11.57}$$

In the bottom (N'th) layer, $I_N = 0$, and hence $G_N(z) = 1$. Starting with G_N $(= 1)$, we can obtain G_r successively up to G_1. Suppose the E_0 and I_0 denote the external and internal fields outside the conductor. Then from the boundary conditions at $z = D_1$, we obtain

$$I_0 = \frac{k_1 G_1(D_1) - \lambda}{k_1 G_1(D_1) + \lambda} e^{-2\lambda D_1} E_0 \tag{11.58}$$

where $\lambda^2 = m^2 + n^2$. (11.58) gives the relation between the external and internal fields for a layered earth model.

11.3 Non-Uniform Semi-Infinite Conductor

11.3.1 The two-dimensional case

A typical two-dimensional structure with a horizontal plane boundary is the conducting half space, divided by a vertical plane into two regions having different conductivities. Such a model is shown in Fig. 11.2 with the coordinate system to be used below.

In the model shown in Fig. 11.2, all quantities are independent of x. Then any solution of (9.1) and (9.2) is separated into two cases

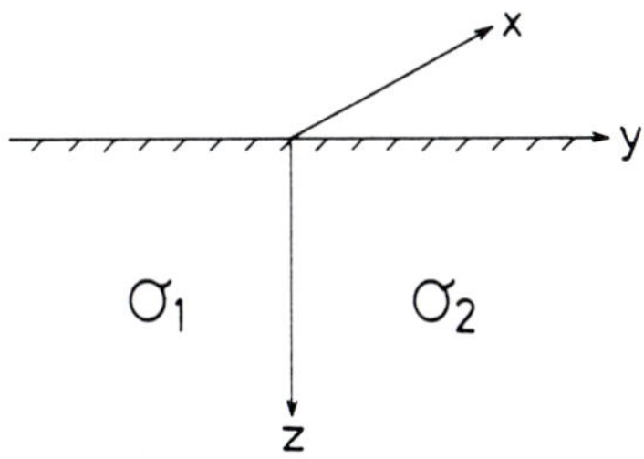

FIG. 11.2. A typical two-dimensional model. The x-z plane at $y = 0$ is taken as a conductivity boundary.

$$\begin{aligned} E_y &= \frac{\omega}{\eta^2} \frac{\partial H_x}{\partial z} \\ E_z &= -\frac{\omega}{\eta^2} \frac{\partial H_x}{\partial y} \\ \frac{\partial E_z}{\partial y} - \frac{\partial E_y}{\partial z} &= -i\omega H_x \end{aligned} \tag{11.59}$$

and

$$\begin{aligned} H_y &= \frac{i}{\omega} \frac{\partial E_x}{\partial z} \\ H_z &= -\frac{i}{\omega} \frac{\partial E_x}{\partial y} \\ \frac{\partial H_z}{\partial y} - \frac{\partial H_y}{\partial z} &= \frac{\eta^2}{\omega} E_x \end{aligned} \tag{11.60}$$

where a slow and periodic variation is assumed and thus $\eta^2 = 4\pi\sigma\omega$. Equations (11.59) and (11.60) have been called the H-polarization and E-polarization cases, respectively.

WEAVER (1963) treated these two cases analytically by using the Fourier transformation method, although his treatment is based on rather approximate boundary conditions. In the meantime, JONES and PRICE (1970) discussed the same problem, taking into account more appropriate boundary conditions at infinity. The following is based on JONES and PRICE (1970).

In the H-polarization case, we derive

$$\frac{\partial^2 H_x}{\partial y^2} + \frac{\partial^2 H_x}{\partial z^2} = i\eta^2 H_x. \tag{11.61}$$

Since the effect of the conductivity discontinuity can be neglected at distances far enough from the discontinuity, (11.61) is reduced to

$$\frac{\partial^2 H_x}{\partial z^2} = i\eta^2 H_x \tag{11.62}$$

as $y \to +\infty$ or $-\infty$. Since H_x vanishes as $z \to +\infty$, a solution of (11.62) is obtained as

$$H_x = H_0 \exp\left\{-\frac{1}{\sqrt{2}}(1+i)\eta z\right\} \tag{11.63}$$

where H_0 is constant because H_x is independent of y and z in the non-conducting region and hence $H_x = H_0$ at $z = 0$.

In the E-polarization case, an equation similar to (11.61) is derived:

$$\frac{\partial^2 E_x}{\partial y^2} + \frac{\partial^2 E_x}{\partial z^2} = i\eta^2 E_x. \tag{11.64}$$

When $y \to +\infty$ or $-\infty$, (11.64) is reduced to

$$\frac{\partial^2 E_x}{\partial z^2} = i\eta^2 E_x. \tag{11.65}$$

Its solution within the conductor is

$$E_x = E_0 \exp\left\{-\frac{1}{\sqrt{2}}(1+i)\eta z\right\} \quad (z > 0). \tag{11.66}$$

In this case, E_0 at $y \to +\infty$ may not be the same as E_0 at $y \to -\infty$. In the conductor ($z > 0$), (11.66) and the first equation of (11.60) yield

$$H_y = \frac{\eta}{\sqrt{2i\omega}}(1+i)E_0 \exp\left\{-\frac{1}{\sqrt{2}}(1+i)\eta z\right\}. \tag{11.67}$$

For $z < 0$, the last equation of (11.60) implies $\partial H_y/\partial z \to 0$ as $y \to \pm\infty$. Hence $H_y = H_0$ (const). Since H_y must be continuous at $z = 0$,

$$H_0 = \frac{1}{\sqrt{2}}(1-i)\frac{\eta}{\omega}E_0. \tag{11.68}$$

(11.68) and the first equation of (11.60) yield

$$\frac{\partial E_x}{\partial z} = -\frac{1}{\sqrt{2}}(1+i)\eta E_0. \tag{11.69}$$

A solution of (11.69) satisfying the boundary condition at $z = 0$ is derived as

$$E_x = E_0\left\{1 - \frac{1}{\sqrt{2}}(1 + i)\eta z\right\} \quad (z < 0). \tag{11.70}$$

Another boundary condition is required at $z = -h$. It could reasonably be assumed that the perturbation due to a vertical conductivity discontinuity is of local nature and its effect does not extend beyond a certain distance, say, h. Now we can determine $H_y(= H_0)$ at $z = -h$ and hence E_x at $z = -h$ from (11.60).

In the above discussion only values at boundaries could be determined and a numerical method must be applied to solve the H- and E-polarization cases. In fact, JONES and PRICE (1970) derived finite difference quations from (11.61) and (11.64) and used the Gauss-Seidel iteration method. WEAVER and THOMSON (1973) treated the same problem in a different way. Instead of applying a uniform inducing field, they first considered a periodic line current at a finite height, say h, above the discontinuity and obtained an analytical solution. In the limit of $h \to \infty$, such a line current gives rise to a uniform magnetic field. They compared the results for $h \to \infty$ with Jones and Price's numerical estimation and found good agreement for the electric field, although some discrepancies were noticed for the magnetic field.

The method formulated by JONES and PRICE (1970) has been used extensively to examine the electromagnetic response of various two-dimensional structures. JONES and PASCOE (1971) and PASCOE and JONES (1972) extended the above method, in which equal grid intervals had been used in numerical calculations, to the case of variable grid intervals. The new method turned out to be much more effective; it enables one to examine the electromagnetic response for any two-dimensional structures without expanding significantly the number of grid points. Meanwhile, however, WILLIAMSON *et al.* (1974) and also BREWITT-TAYLOR and WEAVER (1976) pointed out some defects in their finite difference formulation, particularly for the H-polarization case.

In the case of a perfect conductor, the use of a function, ψ, for magnetic lines of force would be more effective in view of its constant value at the conductor surface. In a two-dimensional problem, such as the one shown in Fig. 11.3, ψ satisfies, for example, outside the conductor

$$\partial^2\psi/\partial x^2 + \partial^2\psi/\partial z^2 = 0. \tag{11.71}$$

The x and z components of the magnetic field are derived from ψ as

$$H_x = -\partial\psi/\partial z, \quad H_z = \partial\psi/\partial x. \tag{11.72}$$

RIKITAKE (1969) examined the electromagnetic response of a perfect conductor with a depression of semi-triangular shape, as shown in Fig. 11.3, using the relaxation method.

RIKITAKE and HONKURA (1973) and RIKITAKE (1975) extended the method to the case of finite conductivity. Within the conductor ψ satisfies

$$\partial\psi/\partial t = (4\pi\sigma)^{-1}\nabla^2\psi. \tag{11.73}$$

At conductivity boundaries, the continuity condition for $\boldsymbol{H}$ and $\boldsymbol{B}$ are derived and transferred to conditions in ψ. Figure 11.4 shows one example of numerical calculation; magnetic lines of force which were initially uniform are perturbed by the presence of a two-dimensional conductor with a square cross-section. In this figure the electromagnetic response is shown for various values of parameter Ω defined by

$$\Omega = 4\pi\sigma\omega D^2 \tag{11.74}$$

where D indicates a characteristic length which, in the present case, is represented by the grid interval.

Another numerical technique called the finite element method has also been applied to problems of electromagnetic induction in two-dimensional structures and some parameters describing electromagnetic response, particularly for magnetotellurics, have been evaluated numerically (e.g. SILVESTER and HASLAM, 1972; RODI, 1976; KAIKKONEN, 1977).

On the other hand, an analytical approach to two-dimensional problems has been attempted for rather simple models. RIKITAKE (1965) examined the electromagnetic response of a semi-infinite conductor with an undulatory surface. In this case, the conductor is assumed to be perfectly conducting so

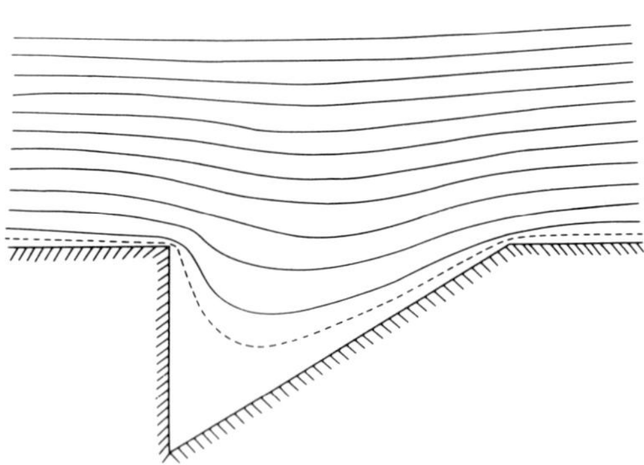

FIG. 11.3. Magnetic lines of force over a half-equilateral-triangle depression in a perfectly conducting medium (RIKITAKE, 1969).

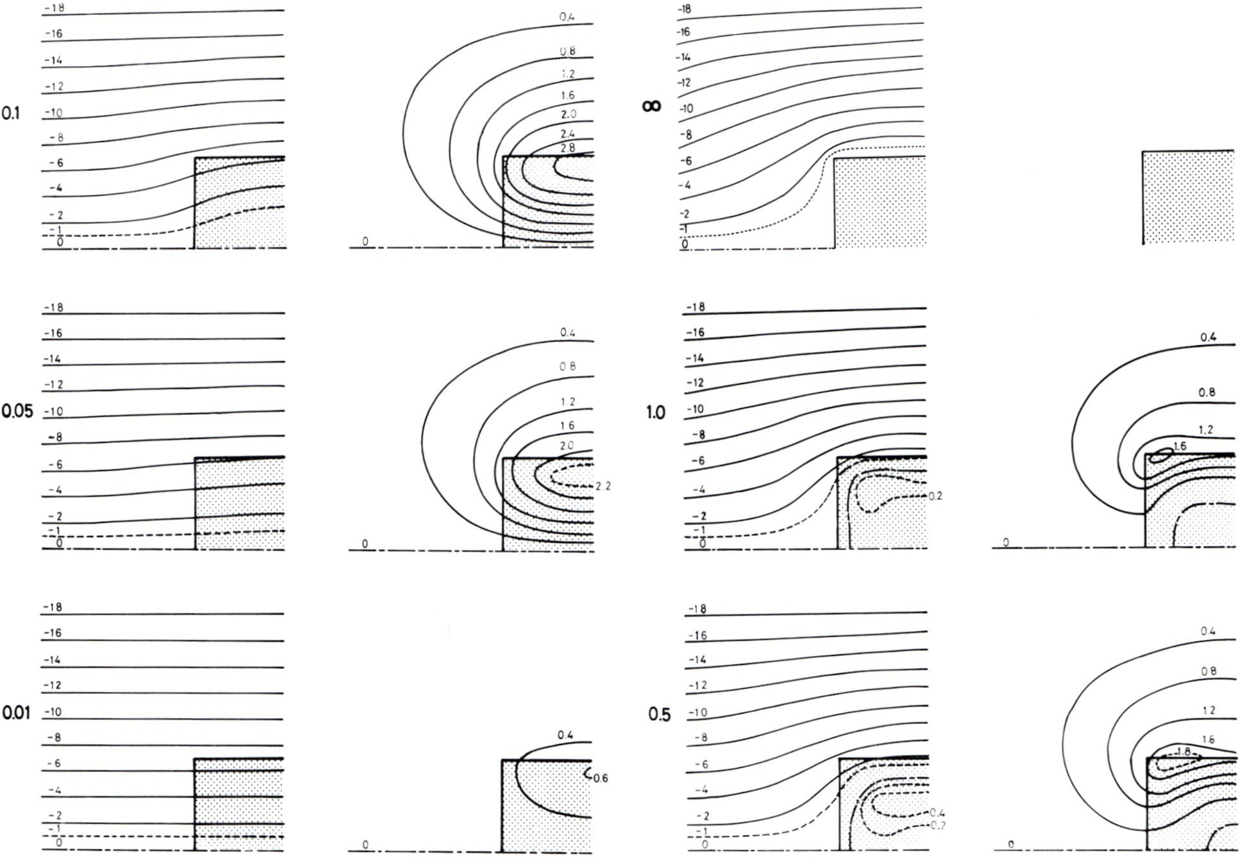

FIG. 11.4. Magnetic lines of force perturbed by a two-dimensional conductor with square cross-section. The real and imaginary parts are shown for various values of parameter Ω.

that the normal component of the magnetic field must vanish at the conductor surface. Thus the magnetic potential above the conductor can be determined. YUKUTAKE'S (1967) work, in which the case of a slightly inclined surface was investigated, is based on the reflection of plane waves. TREUMANN (1970a, b) treated the E-polarization case for a two-dimensional sheet-like conductor by means of an integral equation. Some other analytical approaches to two-dimensional problems are described in a review article by HOBBS (1975b).

Two-dimensional problems in electromagnetic induction have also been studied extensively in the U.S.S.R. as presented in review articles by BERDICHEVSKIY and DMITRIEV (1976a, b). Problems of the E-polarization case have usually been solved by using the integral equation method, while for the H-polarization case, a technique called the projection method has mostly been used.

11.3.2 The three-dimensional case

RAICHE (1974) and WEIDELT (1975a) put forward a method to solve a three-dimensional problem in electromagnetic induction by means of integral equations. An anomalous body is assumed to be embedded in a uniform or stratified structure. In general, the electric field $\boldsymbol{E}$ is governed by

$$\text{curl curl } \boldsymbol{E}(\boldsymbol{r})+k^2(\boldsymbol{r})\boldsymbol{E}(\boldsymbol{r})=-4\pi\omega i\, \boldsymbol{j}_e(\boldsymbol{r}) \tag{11.75}$$

where

$$k^2(\boldsymbol{r})=4\pi\omega i\sigma(\boldsymbol{r}) \tag{11.76}$$

and $\boldsymbol{J}_e$ represents the external electric current giving rise to the external magnetic field. The exp $(i\omega t)$ form is assumed, as usual, for the time variation. Now we separate $\sigma(\boldsymbol{r})$ into the normal and anomalous parts as

$$\sigma(\boldsymbol{r})=\sigma_n(\boldsymbol{r})+\sigma_a(\boldsymbol{r}). \tag{11.77}$$

Similarly,

$$\begin{aligned} k^2(\boldsymbol{r})&=k_n^2(\boldsymbol{r})+k_a^2(\boldsymbol{r}) \\ \boldsymbol{E}(\boldsymbol{r})&=\boldsymbol{E}_n(\boldsymbol{r})+\boldsymbol{E}_a(\boldsymbol{r}). \end{aligned} \tag{11.78}$$

The normal electric field $\boldsymbol{E}_n$ satisfies

$$\text{curl curl } \boldsymbol{E}_n(\boldsymbol{r}) + k_n^2(\boldsymbol{r})\boldsymbol{E}_n(\boldsymbol{r}) = -4\pi\omega i\boldsymbol{j}_e(\boldsymbol{r}). \tag{11.79}$$

Given $\boldsymbol{j}_e(\boldsymbol{r})$, (11.79) can be solved in a way similar to the case of a layered-earth model as described in Section 11.2. On the other hand, $\boldsymbol{E}_a$ satisfies

$$\text{curl curl } \boldsymbol{E}_a(\boldsymbol{r})+k_n^2(\boldsymbol{r})\boldsymbol{E}_a(\boldsymbol{r}) =-k_a^2(\boldsymbol{r})\boldsymbol{E}(\boldsymbol{r}). \tag{11.80}$$

Provided that $\boldsymbol{G}_i(\boldsymbol{r}_0|\boldsymbol{r})$ represents the electric field at $\boldsymbol{r}$ due to an oscillating electric dipole of unit moment, situated at $\boldsymbol{r}_0$, with its direction along the i axis, $\boldsymbol{G}_i(\boldsymbol{r}_0|\boldsymbol{r})$ satisfies in the normal (stratified) structure

$$\operatorname{curl}\operatorname{curl}\boldsymbol{G}_i(\boldsymbol{r}_0|\boldsymbol{r})+k_n^2(\boldsymbol{r})\boldsymbol{G}_i(\boldsymbol{r}_0|\boldsymbol{r})=\boldsymbol{n}_i\delta(\boldsymbol{r}-\boldsymbol{r}_0) \tag{11.81}$$

where $\boldsymbol{n}_i$ denotes the unit vector along the i axis and $\delta(\boldsymbol{r}-\boldsymbol{r}_0)$ the δ-function. Multiplying (11.81) by $\boldsymbol{E}_a(\boldsymbol{r})$ and (11.80) by $\boldsymbol{G}_i(\boldsymbol{r}_0|\boldsymbol{r})$, respectively, and integrating the difference with respect to $\boldsymbol{r}$, we obtain

$$E_{ai}(\boldsymbol{r}_0)=-\int k_a^2(\boldsymbol{r})\boldsymbol{G}_i(\boldsymbol{r}_0|\boldsymbol{r})\boldsymbol{E}(\boldsymbol{r})\mathrm{d}V \quad (i=1,2,3) \tag{11.82}$$

In the derivation of (11.82), we applied Green's vector theorem:

$$\begin{aligned}&\int\{\boldsymbol{a}\cdot\operatorname{curl}\operatorname{curl}\boldsymbol{b}-\boldsymbol{b}\cdot\operatorname{curl}\operatorname{curl}\boldsymbol{a}\}\,\mathrm{d}V\\&\quad=\int\{(\boldsymbol{n}\times\boldsymbol{b})\cdot\operatorname{curl}\boldsymbol{a}-(\boldsymbol{n}\times\boldsymbol{a})\cdot\operatorname{curl}\boldsymbol{b}\}\,\mathrm{d}S.\end{aligned} \tag{11.83}$$

The condition that $\boldsymbol{E}_a\to 0$ and $\boldsymbol{G}_i\to 0$ at infinity is also taken into account.

If $\boldsymbol{G}_i(\boldsymbol{r}_0|\boldsymbol{r})$ $(i=1,2,3)$ can be determined for the normal structure, $E_{ai}(\boldsymbol{r}_0)$ should be derived from (11.82) in which integration is required only in the anomalous region. The combination of $\boldsymbol{G}_1$, $\boldsymbol{G}_2$ and $\boldsymbol{G}_3$ yields the Green's tensor. WEIDELT (1975a) derived elements of the Green's tensor for a stratified earth model and showed some results of computations for typical three-dimensional structures.

As in the two-dimensional case, numerical techniques have also been applied to three-dimensional problems in electromagnetic induction. Jones and his co-workers extended their finite difference technique to the three-dimensional case and examined perturbation by three-dimensional bodies with special reference to induction anomalies due to sea effects such as current channeling and the island effect (JONES and PASCOE, 1972; LINES and JONES, 1973; JONES and LOKKEN, 1975; RAMASWAMY *et al.*, 1976). REDDY *et al.* (1977) applied the finite element method to numerical computations for a simple three-dimensional structure.

11.4 Thin Plane Sheet

Suppose that a conducting plane sheet lies at $z=0$ in Cartesian coordinates (x,y,z), with z positive downward. It is obvious that

$$W_{i+}=-W_{i-} \tag{11.84}$$

and the current function defined by (9.68) becomes

$$\Psi=(1/2\pi)W_{i+} \tag{11.85}$$

(9.66) and (11.84) yield

$$-\rho(\partial^2 W_{i+}/\partial z^2) + \text{grad}\,\rho\cdot\text{grad}\,W_{i+} = -2\pi\partial(\partial W_e/\partial z + \partial W_i/\partial z)/\partial t \tag{11.86}$$

and also

$$\rho(\partial^2\Psi/\partial x^2 + \partial^2\Psi/\partial y^2) + (\partial\rho/\partial x)(\partial\Psi/\partial x) + (\partial\rho/\partial y)(\partial\Psi/\partial y) = \partial H_z/\partial t. \tag{11.87}$$

11.4.1 The uniform case

In this case, (11.86) is reduced to

$$\rho(\partial^2 W_i/\partial z^2)_+ = 2\pi p(\partial W_e/\partial z + \partial W_i/\partial z)_+ \tag{11.88}$$

where $\partial/\partial t$ is denoted by p, which can be replaced with $i\omega$ for a periodic variation exp $(i\omega t)$. As a typical inducing potential, we assume

$$W_e = A_0 \mathrm{e}^{-\lambda z}\sin\lambda x \tag{11.89}$$

where A_0 is a constant. Then the induced potential is derived as

$$W_i = \begin{cases} -B\mathrm{e}^{-\lambda z}\sin\lambda x & z>0 \\ B\mathrm{e}^{\lambda z}\sin\lambda x & z<0. \end{cases} \tag{11.90}$$

Substituting (11.89) and (11.90) into (11.88), we obtain

$$B = 2\pi\omega i(\rho\lambda + 2\pi\omega i)^{-1}A_0. \tag{11.91}$$

Thus the current function is obtained as

$$\Psi = -\frac{A_0}{2\pi}(U + iV)\sin\lambda x \tag{11.92}$$

where

$$U = \frac{1}{1+\alpha^2}, \quad V = \frac{\alpha}{1+\alpha^2}, \quad \alpha = \frac{\rho\lambda}{2\pi\omega}. \tag{11.93}$$

11.4.2 The non-uniform case

RIKITAKE (1967b) examined electromagnetic induction in a non-uniform thin sheet for an inducing field expressed as

$$W_e = A\mathrm{e}^{-qz}\cos qy. \tag{11.94}$$

The resistivity of the sheet, ρ, is assumed to be a function of x only.

In this case, the potential for the induced field is given as

$$W_i = \cos qy \sum_{m=0}^{\infty} \mathrm{e}^{\sqrt{m^2p^2+q^2}z}(i_{mc}\cos mpx + i_{ms}\sin mpx). \tag{11.95}$$

The current function is expressed as

$$\Psi = -\cos qy \sum_{m=0}^{\infty} (K_{mc}\cos mpx + K_{ms}\sin mpx) \tag{11.96}$$

where

$$i_{mc} = 2\pi K_{mc}, \quad i_{ms} = 2\pi K_{ms}. \tag{11.97}$$

Substituting (11.96) into (11.87), we have

$$\begin{aligned}&\sum_{m=0}^{\infty} [\rho(m^2p^2 + q^2)\cos mpx + (\mathrm{d}\rho/\mathrm{d}x)mp\sin mpx]K_{mc}\\ &\quad + \sum_{m=0}^{\infty} [\rho(m^2p^2 + q^2)\sin mpx - (\mathrm{d}\rho/\mathrm{d}x)mp\cos mpx]K_{ms}\\ &= (\partial/\partial t)\left[Aq - 2\pi \sum_{m=0}^{\infty} \sqrt{m^2p^2 + q^2}(K_{mc}\cos mpx + K_{ms}\sin mpx)\right].\end{aligned} \tag{11.98}$$

Multiplying (11.98) by cos Mpx and integrating with respect to x, from $-\pi/p$ to π/p, we obtain

$$\begin{aligned}&\sum_{m=0}^{\infty}\left[(m^2p^2 + q^2)\int_{-\pi/p}^{\pi/p} \rho(x)\cos mpx\cos Mpx\,\mathrm{d}x\right.\\ &\quad\left. + mp\int_{-\pi/p}^{\pi/p}(\mathrm{d}\rho/\mathrm{d}x)\sin mpx\cos Mpx\,\mathrm{d}x\right]K_{mc}\\ &\quad + \sum_{m=0}^{\infty}\left[(m^2p^2 + q^2)\int_{-\pi/p}^{\pi/p}\rho(x)\sin mpx\cos Mpx\,\mathrm{d}x\right.\\ &\quad\left. - mp\int_{-\pi/p}^{\pi/p}(\mathrm{d}\rho/\mathrm{d}x)\cos mpx\cos Mpx\,\mathrm{d}x\right]K_{ms}\\ &= \begin{cases}(2\pi/p)(\mathrm{d}/\mathrm{d}t)[Aq - 2\pi qK_{0c}] & M = 0\\ -(\pi/p)2\pi\sqrt{M^2p^2 + q^2}\;\mathrm{d}K_{Mc}/\mathrm{d}t & M > 0.\end{cases}\end{aligned} \tag{11.99}$$

Introducing a dimensionless resistivity ρ_1 defined by ρ/ρ_0,

$$\begin{aligned}&\sum_{m=0}^{\infty} [\{A_1(m,M) + A_3(m,M)\}K_{mc} + \{B_1(m,M) - B_3(m,M)\}K_{ms}]\\ &= \begin{cases}(2\pi/\rho_0p)(q/p)(\mathrm{d}/\mathrm{d}t)(A - 2\pi K_{0c}) & M = 0\\ -(2\pi^2/\rho_0p)\sqrt{M^2 + q^2/p^2}\;\mathrm{d}K_{Mc}/\mathrm{d}t & M > 0\end{cases}\end{aligned} \tag{11.100}$$

where

$$\begin{aligned}
A_1(m, M) &= (m^2 + q^2/p^2) \int_{-\pi}^{\pi} \rho_1(s) \cos ms \cos Ms \, ds \\
A_3(m, M) &= m \int_{-\pi}^{\pi} (d\rho_1/ds) \sin ms \cos Ms \, ds \\
B_1(m, M) &= (m^2 + q^2/p^2) \int_{-\pi}^{\pi} \rho_1(s) \sin ms \cos Ms \, ds \\
B_3(m, M) &= m \int_{-\pi}^{\pi} (d\rho_1/ds) \cos ms \cos Ms \, ds .
\end{aligned} \tag{11.101}$$

Similar operation with respect to $\sin Mpx$ yields

$$\sum_{m=0}^{\infty} [\{A_2(m, M) + A_4(m, M)\} K_{mc} + \{B_2(m, M) - B_4(m, M)\} K_{ms}]$$

$$= \begin{cases} 0 & M = 0 \\ -(2\pi^2/\rho_0 p)\sqrt{M^2 + q^2/p^2} \, dK_{Ms}/dt & M > 0 \end{cases} \tag{11.102}$$

where

$$\begin{aligned}
A_2(m, M) &= (m^2 + q^2/p^2) \int_{-\pi}^{\pi} \rho_1(s) \cos ms \sin Ms \, ds \\
A_4(m, M) &= m \int_{-\pi}^{\pi} (d\rho_1/ds) \sin ms \sin Ms \, ds \\
B_2(m, M) &= (m^2 + q^2/p^2) \int_{-\pi}^{\pi} \rho_1(s) \sin ms \sin Ms \, ds \\
B_4(m, M) &= m \int_{-\pi}^{\pi} (d\rho_1/ds) \cos ms \sin Ms \, ds .
\end{aligned} \tag{11.103}$$

On the basis of the above method, RIKITAKE (1968c) examined the electromagnetic response for a resistivity distribution given by

$$\rho = \rho_0(1 - \varepsilon \cos rx) \qquad \varepsilon < 1. \tag{11.104}$$

The response of this thin-sheet turned out to be highly dependent on the wavelength of the external magnetic field. A resonance-like response was found when the wavelength of the external field approaches that of the resistivity distribution.

The case in which ρ is a function of both x and y was treated numerically by SASAI (1968) and HONKURA (1971), with the assumption that the local area under consideration is surrounded by a uniform structure. In such a case, we can separate Ψ into normal (Ψ_0) and anomalous (ψ) parts as

$$\Psi = \Psi_0 + \psi. \tag{11.105}$$

Ψ_0 satisfies

$$\rho_0(\partial^2\Psi_0/\partial x^2 + \partial^2\Psi_0/\partial y^2) = \partial(H_{ze} + H_{zio})/\partial t \tag{11.106}$$

where ρ_0 is the resistivity of a uniform sheet outside the anomalous area. Substituting (11.105) into (11.87) and using (11.106), we obtain

$$\begin{aligned}\rho(\partial^2\psi/\partial x^2 + \partial^2\psi/\partial y^2) &+ \text{grad}\,\rho\cdot\text{grad}\,\psi \\ &= -(\rho-\rho_0)(\partial^2\Psi_0/\partial x^2 + \partial^2\Psi_0/\partial y^2) \\ &\quad - \text{grad}\,(\rho-\rho_0)\cdot\text{grad}\,\Psi_0 + \partial(H_{zi} - H_{zio})/\partial t.\end{aligned} \tag{11.107}$$

The last term of the right-hand side of (11.107) is usually called the self-induction term. It is negligible unless the size of the anomalous body is large compared to the characteristic length for the period of the inducing magnetic field. Provided that the self-induction term is neglected, the perturbation of the electric current flowing in the anomalous area is reduced to the distortion of a direct electric current, although electromagnetic induction controls the overall current.

HONKURA (1971) solved (11.107) numerically for the case of Miyake-jima Island. Since the island is a small circular one, less than 10 km in diameter, a 40 km × 40 km square was taken as the anomalous area, including the island. Figure 11.5 shows the pattern of electric current flow in the case of an inducing magnetic field having a north-south polarization. In this calculation, however, the self-induction effect is ignored, in view of the narrow spatial coverage of the island effect.

The effect of self-induction was examined by RIKITAKE (1970) for an anomaly of circular shape. First, the current function $\Psi^{(0)}$, which includes the distortion effect, and the magnetic field $H_{z0}^{(0)}$ at $z = 0$ due to $\Psi^{(0)}$ can be derived from the theory of ASHOUR and CHAPMAN (1965) for the DC current.

As a first approximation, we express Ψ as

$$\Psi = \Psi^{(0)} + \Psi^{(1)}. \tag{11.108}$$

Then $\Psi^{(1)}$ must satisfy

$$\rho\nabla^2\Psi^{(1)} = \partial H_{z0}^{(0)}/\partial t. \tag{11.109}$$

$H_{z0}^{(1)}$ at $z = 0$ due to $\Psi^{(1)}$ can easily be calculated. The second approximation is evaluated similarly. By continuing this procedure successively, convergence will be attained eventually.

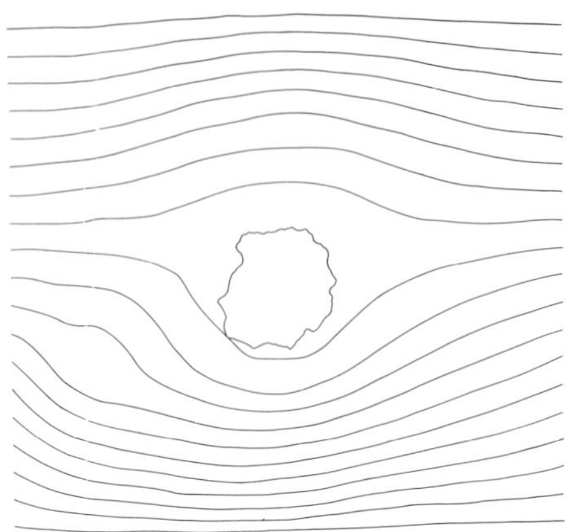

FIG. 11.5. The pattern of electric current flow around a small island, Miyake-jima Island, for a varying magnetic field having a north-south polarization (HONKURA, 1971).

It was then confirmed that the self-induction term is really negligible for a small island, such as Miyake-jima Island, as far as a period of 30 min or so is concerned. However, it must be kept in mind that for a large anomalous body the self-induction term cannot be neglected and, moreover, the above method may not be applicable because of the divergent behavior of the series.

11.5 Thin Sheet Underlain by a Conductor

11.5.1 Uniform sheet

Suppose that a uniform sheet is underlain by a laterally uniform conductor as is shown in Fig. 11.6, where the conductivity, σ, is a function of z only. As an inducing field, we assume (11.89).

In the non-conducting regions, the scalar potential satisfies the Laplace equation and its typical solution for $z<0$ is given as

$$W_- = (A_0\,\mathrm{e}^{-\lambda z} + C\,\mathrm{e}^{\lambda z})\sin\lambda x. \tag{11.110}$$

Then the magtic field is derived as

$$\boldsymbol{H} = \begin{cases} -\lambda(A_0\,\mathrm{e}^{-\lambda z} + C\,\mathrm{e}^{\lambda z})\cos\lambda x \\ 0 \\ \lambda(A_0\,\mathrm{e}^{-\lambda z} - C\,\mathrm{e}^{\lambda z})\sin\lambda x \end{cases} \tag{11.111}$$

For $0<z<D_1$, the potential can be written as

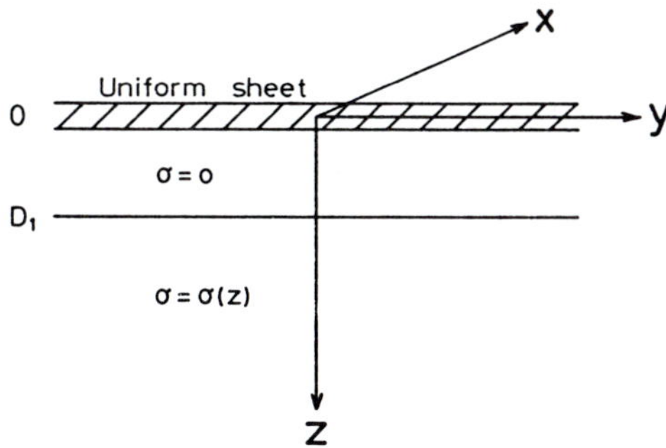

FIG. 11.6. A model of a thin sheet underlain by a conductor.

$$W_+ = (R\,\mathrm{e}^{-\lambda z} + S\,\mathrm{e}^{\lambda z}) \sin \lambda x. \tag{11.112}$$

The magnetic field is obtained as

$$\boldsymbol{H} = \begin{cases} -\lambda(R\,\mathrm{e}^{-\lambda z} + S\,\mathrm{e}^{\lambda z}) \cos \lambda x \\ 0 \\ \lambda(R\,\mathrm{e}^{-\lambda z} - S\,\mathrm{e}^{\lambda z}) \sin \lambda x. \end{cases} \tag{11.113}$$

In the conducting region, a typical solution of (9.17) is expressed as

$$\boldsymbol{A} = \boldsymbol{n} \times \operatorname{grad} u(x, z) \tag{11.114}$$

where $\boldsymbol{n}$ denotes a unit vector in the z direction. For a laterally uniform structure $u(x, z)$ can be written as

$$u(x, z) = v(z) \sin \lambda x. \tag{11.115}$$

$v(z)$ must satisfy

$$\frac{\mathrm{d}^2 v(z)}{\mathrm{d}z^2} - (\lambda^2 + k^2) v(z) = 0 \tag{11.116}$$

where

$$k^2 = 4\pi\sigma\omega i. \tag{11.117}$$

The magnetic field in the semi-infinite conductor is obtained as

$$\boldsymbol{H} = \begin{cases} -\lambda \dfrac{\mathrm{d}v(z)}{\mathrm{d}z} \cos \lambda x \\ 0 \\ -\lambda^2 v(z) \sin \lambda x. \end{cases} \tag{11.118}$$

In the sheet, the current function is defined and expressed as

$$\Psi = K \sin \lambda x \tag{11.119}$$

where K is a constant to be determined. The boundary conditions at $z=D_1$ yield

$$\begin{aligned} Re^{-\lambda D_1} + Se^{\lambda D_1} &= \frac{\mathrm{d}v(D_1)}{\mathrm{d}z} \\ Re^{-\lambda D_1} - Se^{\lambda D_1} &= -\lambda v(D_1). \end{aligned} \tag{11.120}$$

From (9.68), (11.110) and (11.112), and from the continuity condition of H_z at $z = 0$,

$$\begin{aligned} R+S &= A_0+C+4\pi K \\ R-S &= A_0-C. \end{aligned} \tag{11.121}$$

From (11.87), (11.111) and (11.119),

$$\rho\lambda K = -i\omega(A_0 - C). \tag{11.122}$$

(11.120), (11.121) and (11.122) yield

$$K = -\frac{\omega i(1-Q)}{\rho\lambda + 2\pi\omega i(1-Q)} A_0 \tag{11.123}$$

where

$$Q = \frac{\dfrac{\mathrm{d}v(D_1)}{\mathrm{d}z} + \lambda v(D_1)}{\dfrac{\mathrm{d}v(D_1)}{\mathrm{d}z} - \lambda v(D_1)} \mathrm{e}^{-2\lambda D_1}. \tag{11.124}$$

The current function represented by K in (11.123) contains, through the parameter Q, information on the conductivity distribution below the sheet. Estimating the current function from the observed island effect, HONKURA (1973) and HONKURA *et al.* (1974, 1981) inferred the conductivity structure in the upper mantle (see Subsection 12.2.2).

11.5.2 Non-uniform sheet

With application to the coastal effect in mind, BAILEY (1977) investigated the inductive response for the H-polarization case at the edge of a perfectly conducting sheet. The electrical conductivity of the half space representing the crust and mantle is assumed to be uniform. His method is based on the Fourier transformation of (11.61). The solution is expressed as

$$H_y(x, z) = \int_{-\infty}^{\infty} f(k, z) \exp(\sqrt{k^2 + i\eta^2} z) \exp(ikx) \, dk \qquad (11.125)$$

where $f(k, z)$ is an unknown function which must be determined from the boundary conditions. Bailey applied the Wiener-Hopf technique to the present problem and obtained an exact analytical solution. On the other hand, FISHER *et al.* (1978) treated the same problem somewhat differently for both the H- and E-polarization cases.

As an extension of Bailey's work, NICOLL and WEAVER (1977) added a perfectly conducting mantle layer and examined the edge effect, also using the Wiener-Hopf technique. DAWSON and WEAVER (1979) treated the H-polarization case for a structure, characterized by a pair of half-sheets of different conductivities, which is underlain by a uniform half-space. This model is more realistic than the model of a perfectly conducting sheet. GREEN and WEAVER (1978) investigated the H-and E-polarization cases for a model in which the integrated conductivity of the sheet is variable, although the underlying half-space is assumed to be uniform.

A more general three-dimensional case was solved by VASSEUR and WEIDELT (1977), with special reference to the channeling effect in the Pyrenean region, where a notable anomaly in the geomagnetic variation has been found (BABOUR *et al.*, 1976). Their model is characterized by an anomalous sheet of local extent which is embedded in a uniform sheet, and the sheet, as a whole, is underlain by a layered half-space. The technique used to solve the problem is similar to the one applied to the three-dimensional case for a half space (WEIDELT, 1975a). Whenever the position vector $\boldsymbol{r}$ indicates the inside of the sheet, (11.79) and (11.80) are replaced with equations for thin sheets. Then the anomalous electric field is derived as a solution of an integral equation similar to (11.82).

11.6 Thin Circular Disk

ASHOUR (1965b) solved a problem on electromagnetic induction within a circular thin disk using an inducing magnetic field represented by a varying magnetic dipole above a conductor. Let us take the cylindrical coordinates (r, ϕ, z) with the origin situated at the center of the disk. We can then express the inducing potential due to the varying dipole as

$$W_e = M(z - b)\{(z - b)^2 + r^2\}^{-3/2} \qquad (11.126)$$

where M denotes the moment of the dipole which is assumed to lie at a point $r = 0$, $z = b$. The induced potential can be expressed, from symmetry conditions, as

$$W_i = \frac{M}{a}\int_0^\infty f(ka)J_0(kr)\mathrm{e}^{-kz}\mathrm{d}k \qquad (11.127)$$

for $z>0$, where a denotes the radius of the disk and J_0 is the Bessel function. $f(ka)$ is an unknown function to be determined. Since $W_i(r,\phi,z) = -W_i(r,\phi,-z)$ and the potential must be continuous outside the disk, $W_i = 0$ at $z = 0$ for $r > a$.

Using new parameters defined by $u = ak$ and $x = r/a$, we obtain

$$\int_0^\infty f(u)J_0(ux)\,\mathrm{d}u = 0 \qquad (x>1). \qquad (11.128)$$

Within the disk, (11.88) yields

$$\int_0^\infty (u^2+\alpha u)f(u)J_0(ux)\,\mathrm{d}u = \alpha\{(\beta^2+x^2)^{-3/2} - 3\beta^2(\beta^2+x^2)^{-3/2}\} \quad (x<1) \qquad (11.129)$$

where $\alpha = 2\pi ap/\rho$ and $\beta = b/a$. Multiplying (11.129) by x and integrating with respect to x form 0 to x, and differentiating (11.128) with respect to x, we obtain

$$\begin{aligned} &\int_0^\infty (u+\alpha)f(u)J_1(ux)\,\mathrm{d}u = -\alpha x(\beta^2+x^2)^{-3/2} \quad (x<1) \\ &\int_0^\infty uf(u)J_1(ux)\,\mathrm{d}u = 0. \quad (x>1) \end{aligned} \qquad (11.130)$$

Since the electric current is given by

$$\boldsymbol{i}_s = -\boldsymbol{n}\times\mathrm{grad}\,\Psi, \quad \Psi = W_{i+}/2\pi \qquad (11.131)$$

$\boldsymbol{i}_s$ at $r=ax$ is derived as

$$\boldsymbol{i}_s(x) = \begin{cases} 0 \\ -\dfrac{1}{2\pi a}\dfrac{\partial W_i(0,x)}{\partial x} \\ 0. \end{cases} \qquad (11.132)$$

Denoting the ϕ component of $\boldsymbol{i}_s$ by I,

$$I(x) = \frac{M}{2\pi a^3}\int_0^\infty uf(u)J_1(ux)\,\mathrm{d}u. \qquad (11.133)$$

ASHOUR (1965b) further showed that $I(x)$ in (11.133) satisfies the Fredholm integral equation:

$$I(x) = \frac{2\gamma M x}{a^3(\beta^2 + x^2)^{3/2}} + \frac{\gamma}{\pi x}\int_0^1 I(x')M(x,x')\,\mathrm{d}x' \tag{11.134}$$

where

$$\begin{aligned}\gamma &= -\beta/4\pi \\ M(x,x') &= 4\pi^2 x x' \int_0^\infty J_1(ux)J_1(ux')\,\mathrm{d}u.\end{aligned} \tag{11.135}$$

ASHOUR (1965b) examined various cases and showed, as one of the examples, that an analytical solution for electromagnetic induction in a perfectly conducting disk by a uniform inducing field is given by

$$I(x) = -\frac{H}{\pi^2}x(1-x^2)^{-1/2} \tag{11.136}$$

where H denotes the intensity of the inducing magentic field. (11.136) clearly shows that the induced electric current tends to flow along the periphery.

11.7 Inverse Problem

A one-dimensional inverse problem was solved by WEIDELT (1972) for a layered earth model. As a response function, he chose $c(\omega)$ which is defined by

$$c(\omega) = -E(0,\omega)\Big/\frac{\mathrm{d}E(0,\omega)}{\mathrm{d}z} = -E(0,\omega)/4\pi\omega i H(0,\omega). \tag{11.137}$$

The function $c(\omega)$ has often been used in magnetotellurics since the apparent resistivity $\rho_a(\omega)$ can be calculated from $|c(\omega)|^2$ for a layered earth model.

One-dimensional inversion was also treated by OLDENBURG (1979) with special reference to practical applications to magnetotellurics. His formulation is very similar to that of PARKER (1971), for a spherical earth, and will be described briefly in the following.

In one-dimensional structures we can express $\boldsymbol{E}$ and $\boldsymbol{H}$ as $\boldsymbol{E} = (0, E_y(z), 0)$ and $\boldsymbol{H} = (H_x(z), 0, 0)$, respectively, where variations are assumed to be expressed by exp $(i\omega t)$. The Maxwell equations yield

$$\frac{\mathrm{d}^2 E_y}{\mathrm{d}z^2} - 4\pi i\omega\sigma(z)E_y = 0. \tag{11.138}$$

Defining $R(z, \omega)$ by

$$R(z, \omega) = \frac{H(z, \omega)}{E(z, \omega)} \tag{11.139}$$

we select $R(0, \omega)$ as a response function. From (9.2),

$$H(z, \omega) = \frac{1}{i\omega} \frac{\mathrm{d}E(z, \omega)}{\mathrm{d}z}. \tag{11.140}$$

Thus (11.138) is transformed to a differential equation of $R(z, \omega)$:

$$\frac{\mathrm{d}R}{\mathrm{d}z} + i\omega R^2 - 4\pi\sigma = 0. \tag{11.141}$$

Now let us consider preturbations $\delta\sigma$ and δR; that is $\sigma' = \sigma + \delta\sigma$ and $R' = R + \delta R$. Substituting these into (11.141) and neglecting higher order terms in perturbation, we obtain

$$\frac{\mathrm{d}}{\mathrm{d}z}(\delta R) + 2\omega i R \delta R - 4\pi\delta\sigma = 0. \tag{11.142}$$

Hence $\delta R(0, \omega)$ is derived as

$$\delta R(0, \omega) = -4\pi \int_0^\infty \left\{\frac{E(z, \omega)}{E(0, \omega)}\right\}^2 \delta\sigma(z)\, \mathrm{d}z. \tag{11.143}$$

Suppose that $R(0, \omega_j)$ is the observed response function for ω_j ($j=1,\cdots,N$) and $R' \cdot (0, \omega_j)$ is the response function calculated for a starting model $\sigma(z)$. Imposing the condition that $\int(\delta\sigma/\sigma)^2 \mathrm{d}z$ must be minimized, we can solve $\delta\sigma(z)$ from N equations in (11.143) and a new model $\sigma'(z) = \sigma(z) + \delta\sigma(z)$ is constructed. The same procedure, if applied successively, will yield a model which gives rise to a response function very close to that observed.

No general method has been put forward for the two-dimensional inversion problem. However, WEIDELT (1975b) found solutions for two special cases; for an undulatory boundary between an insulator and a perfect conductor and for a non-uniform thin sheet conductor. In both cases, the anomalous area is surrounded by the normal structure.

CHAPTER 12

CHAPTER 12

ELECTRICAL CONDUCTIVITY ANOMALIES

Since the early 1950's it has been noticed that the vertical component of geomagnetic variations having periods shorter than a few hours differs considerably in its behavior from one site to another within a distance of some tens of kilometers. This kind of anomaly in the geomagnetic variation can be accounted for neither by spatial dependence of external fields nor by internal fields arising through electromagnetic induction in a laterally uniform earth. Anomalies have thus been interpreted as indicating lateral heterogeneity in the conductivity distribution; denoted by the term 'conductivity anomaly' (CA).

Investigation of conductivity anomalies of local nature usually requires temporary observations at some sites in addition to magnetic observatories which exist already. Intensive observations of short-period geomagnetic variations have been undertaken in Japan and Germany (e.g. RIKITAKE, 1966a) and systematic studies of conductivity anomalies were initiated. CA studies have been so fashionable since then that many CA's have been discovered over the world, and have been reviewed from time to time (RIKITAKE, 1966a, 1971; HUTTON, 1976a; HONKURA, 1978a).

12.1 Anomalies of Short-Period Geomagnetic Variation

A geomagnetic variation of external origin would be regarded as being approximately spatially uniform within a certain area of local extent. Nevertheless, it is not rare to observe a vertical field which is quite different in character from one at a nearby site. Figure 12.1 shows one typical example of such anomalies in the vertical component. The spatial distribution of the vertical component is fairly irregular, while the horizontal component is characterized by a rather smooth spatial dependence.

On some occasions, anomalies have been observed in the horizontal component as well (e.g. NISHIDA, 1976). Anomalies in the horizontal com-

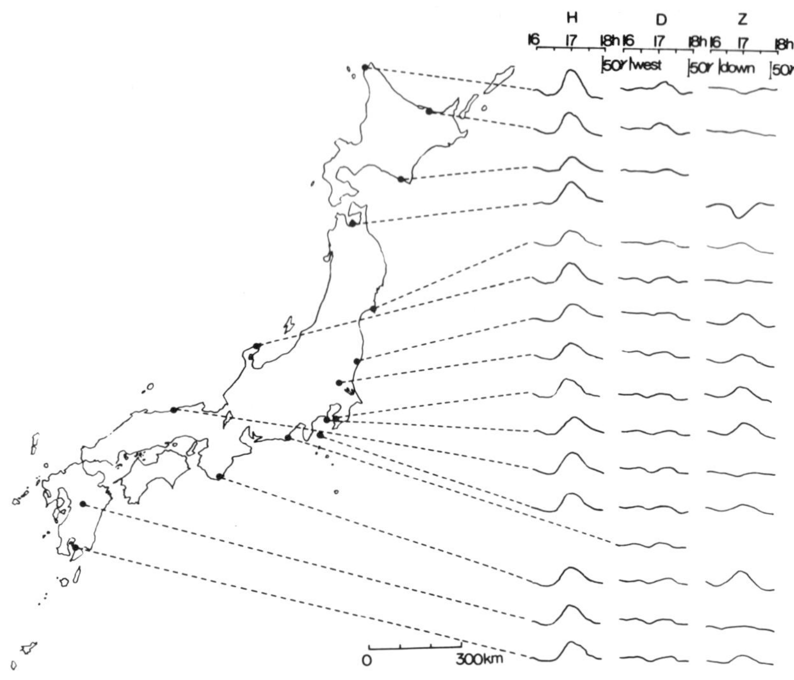

FIG. 12.1. An example of anomalous Z variations for a bay-type event as obtained from simultaneous observations at stations in Japan.

ponent are in most cases represented by an enhancement in amplitude compared with nearby sites, usually resulting from local concentration of induced electric currents in highly conducting zones.

In order to judge whether or not a geomagnetic field observed at a site is anomalous, or to what extent it is anomalous, the general distribution of a source (external) field over the entire earth must be understood, at least to some extent, since source field features are highly regionally dependent and also differ from one type of variation to another (see Chapter 8).

For instance, it is not unusual in substorm events at high-latitudes that the amplitude of a vertical field variation is as large as that of the horizontal field variation. This phenomenon can reasonably be interpreted in terms of strong source currents called the westward and eastward electrojets flowing in the ionosphere at high-latitudes (see Section 8.4). If the amplitude of a vertical field variation is very small at high-latitudes, in contrast to the general feature mentioned above, such a vertical field variation should be considered anomalous. One representative anomaly of this sort is the Mould Bay anomaly found in the Canadian Arctic (WHITHAM, 1963; NIBLETT and

WHITHAM, 1970; DELAURIER *et al.*, 1974).

This kind of anomaly is characterized by a reduction in the amplitude of high-frequency (short-period) components and may be called the *high-cut* ΔZ type (UYEDA and RIKITAKE, 1970). It can be interpreted as indicating the existence of a horizontally extensive conductor near the earth's surface. As described in the preceding chapter, the theory of electromagnetic induction in a plane conductor indicates that high-frequency (short-period) components of an external Z variation, for which the conductor tends to respond as a nearly perfect conductor, are partly cancelled by those of an internal (induced) variation. In the analysis of this kind of anomaly, $\Delta Z/\Delta H$ is usually used as a response function.

A vertical field of large amplitude for short-period geomagnetic variations such as bay-type events is often observed in mid-latitudes, where as a general feature, the vertical field amplitude should be much smaller than the horizontal. Usually such an anomalous vertical field is highly dependent on the polarization of the horizontal field. This kind of anomaly may be called the *directional variation* type (UYEDA and RIKITAKE, 1970), and most of the reported anomalies of short-period geomagnetic variation are classified into this type.

This type of anomaly is due to lateral inhomogeneity of the conductivity distribution. A typical mechanism for a directional variation type anomaly is shown in terms of magnetic lines of force in Fig. 12.2, along with a schematic illustration of high-cut ΔZ type anomaly. Another typical cause is local concentration of induced currents in a highly conducting zone. In this case, anomalies in the horizontal component often occur also. This phenomenon is sometimes referred to as the channeling effect.

Good correlation between the vertical and horizontal fields in directional variation type anomalies enables one to derive an empirical relation:

$$\Delta Z = A\Delta H + B\Delta D \tag{12.1}$$

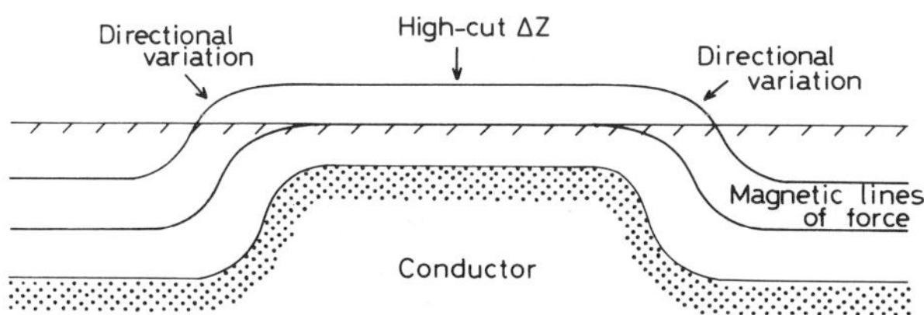

FIG. 12.2. Schematic representation of anomalies of directional variation and high-cut ΔZ types. (UYEDA and RIKITAKE, 1970)

where ΔZ, ΔH, and ΔD are variations of the downward, northward, and eastward components, respectively, with respect to geomagnetic coordinates. A and B are frequency- or period-dependent coefficients. In most cases, (12.1) is treated in the frequency domain, so that A and B are complex functions of frequency and are usually called transfer functions. ΔH is sometimes replaced by ΔX, and ΔD by ΔY, where ΔX and ΔY are geographically northward and eastward components, respectively.

Transfer functions A and B characterize the polarization dependence of the electromagnetic response of an anomalous conductor in the earth; that is, dependence of anomalous response on the direction in which an inducing magnetic field varies. For instance, in a purely two-dimensional case, anomalous Z variations will not be observed on the earth's surface for the H-polarization case, whereas for the E-polarization case a striking anomaly is expected (see Section 11.3).

A representation by an arrow called the induction arrow is often used to illustrate a directional variation type anomaly. The in-phase and out-of-phase (quadrature) arrows are determined from the in-phase and out-of-phase parts of the transfer functions, respectively. In general, the in-phase part is much larger than the out-of-phase part, and hence the in-phase arrow is used more often. The length of the in-phase arrow is determined by $(Au^2 + Bu^2)^{1/2}$ where subscript u denotes the in-phase part. The direction of the arrow is defined by the angle $\tan^{-1}(Bu/Au)$ measured clockwise from south. The arrow usually tends to point towards an anomalous conductor giving rise to a geomagnetic variation anomaly.

It is also possible to interpret (12.1) as indicating a plane in which the vector of magnetic variation is confined. Such a plane has sometimes been called the Rikitake and Yokoyama plane (Rikitake and Yokoyama, 1955b). This interpretation is well understood in terms of magnetic lines of force intersecting the earth's surface, as shown in Fig. 12.2. Another conventionally used arrow, called the Parkinson vector, can be derived by projecting a vector which is normal to the Rikitake and Yokoyama plane onto the earth's surface (Parkinson, 1959, 1962), as is shown in Fig. 12.3. The induction arrow is nearly identical to the Parkinson vector; only the length is slightly different. Another arrow representation called the Wiese vector is sometimes used. It is essentially the same as the Parkinson vector except for a 180° difference in direction (Wiese, 1962a, b).

In discussing anomalies of the horizontal components as well as the vertical, the use of a more general form

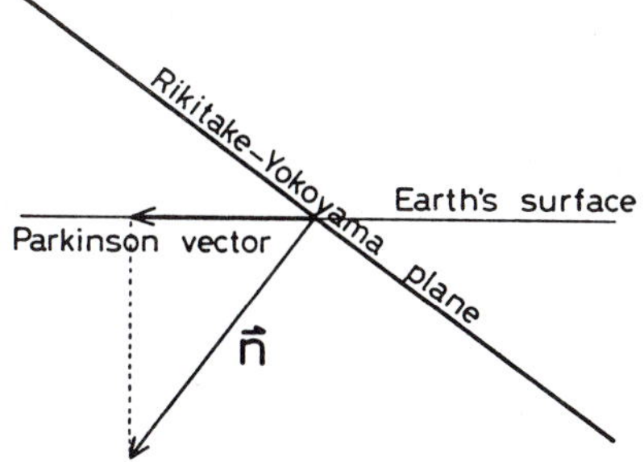

FIG. 12.3. Relation between the preferred plane and Parkinson vector.

$$\begin{pmatrix} \Delta H_a \\ \Delta D_a \\ \Delta Z_a \end{pmatrix} = \begin{pmatrix} h_H & h_D & h_Z \\ d_H & d_D & d_Z \\ z_H & z_D & z_Z \end{pmatrix} \begin{pmatrix} \Delta H_n \\ \Delta D_n \\ \Delta Z_n \end{pmatrix} \tag{12.2}$$

is more suitable to express anomalous fields ΔH_a, ΔD_a, and ΔZ_a when normal fields ΔH_n, ΔD_n, and ΔZ_n are inferred (e.g. SCHMUCKER, 1970). If the spatial extent of an anomalous area is limited, the observed field at a site far from the anomalous area would be considered as the normal one. Then anomalous fields would be derived by subtracting the normal from the observed fields at sites under investigation.

In mid-latitudes, ΔZ_n is negligibly small for substorm-type variations and (12.2) is reduced to

$$\begin{pmatrix} \Delta H_a \\ \Delta D_a \\ \Delta Z_a \end{pmatrix} = \begin{pmatrix} h_H & h_D \\ d_H & d_D \\ z_H & z_D \end{pmatrix} \begin{pmatrix} \Delta H_n \\ \Delta D_n \end{pmatrix}. \tag{12.3}$$

The total (anomalous+normal) fields are expressed as

$$\begin{pmatrix} \Delta H \\ \Delta D \\ \Delta Z \end{pmatrix} = \begin{pmatrix} h_H+1 & h_D \\ d_H & d_D+1 \\ z_H & z_D \end{pmatrix} \begin{pmatrix} \Delta H_n \\ \Delta D_n \end{pmatrix} \tag{12.4}$$

On the other hand, (12.1) is written as

$$(A,\ B,\ -1) \begin{pmatrix} \Delta H \\ \Delta D \\ \Delta Z \end{pmatrix} = 0. \tag{12.5}$$

Substitution of (12.4) into (12.5) yields

$$(A,\ B,\ -1)\begin{pmatrix} h_H+1 & h_D \\ d_H & d_D+1 \\ z_H & z_D \end{pmatrix} = (0,0) \tag{12.6}$$

because ΔH_n and ΔD_n are implicitly assumed to vary independently. Thus we obtain

$$\begin{pmatrix} h_H+1 & d_H \\ h_D & d_D+1 \end{pmatrix} \begin{pmatrix} A \\ B \end{pmatrix} = \begin{pmatrix} z_H \\ z_D \end{pmatrix}. \tag{12.7}$$

If no anomalies exist in the horizontal components, $h_H = h_D = d_H = d_D = 0$, and hence $A = z_H$, $B = z_D$. A more detailed discussion of the relationship between (12.1) and (12.2) was made by LILLEY and BENNETT (1973) and PARKINSON and HOBBS (1979).

It is certainly preferrable to carry out simultaneous observations over a two-dimensional array of sites covering an anomalous area. In this case, a more direct method of analysis can be used to delineate a zone of anomalous geomagnetic variations. First, Fourier amplitudes and phases of three components are calculated for events recorded simultaneously at the array of sites. Then contours of these estimates are drawn on a map covering the array. A typical example of a Fourier amplitude map is shown in Fig. 12.4 (REITZEL *et al.*, 1970). In the map, an anomaly can easily be identified, not only for the vertical component but also for the horizontal components. As another merit

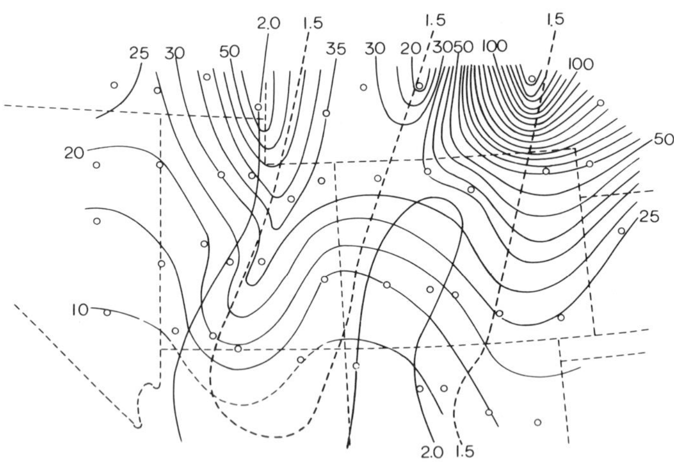

FIG. 12.4. Fourier amplitude map for the Z component for a substorm event in the southwestern United States. The contours are labeled in units of 0.1 nT. Contours denoted by numerals 2.0 and 1.5 indicate heat flow. (REITZEL *et al.*, 1970)

of this method, a regional variation of the source field can also be inferred.

12.2 Effects of Sea-Water and Sediments

Short-period geomagnetic variations are highly sensitive to surface conductors such as seas and thick sedimentary layers. The electrical conductivity of sea-water is as high as $3 \sim 4$ S/m, depending on its salinity and temperature. In the case of surface sediments, the electrical conductivity is controlled mostly by saline water contained in pores and usually amounts to about 0.1 S/m. As will be described later, conductivity anomalies in the lower crust and upper mantle are characterized by a conductivity of 0.1 S/m or so, which is comparable to that of sedimentary layers and much lower than that of sea water. Consequently, the effect of surface conductors on short-period geomagnetic variations cannot be neglected for studies of deep conductivity anomalies.

In view of the circumstance that most tectonically active regions such as spreading and subduction zones are located in ocean areas, it is of great importance to have a clear understanding of sea-water effects. The following subsections will be devoted to a description of sea-water effects such as the coast, island, peninsula, and channeling effects, with some typical examples.

12.2.1 Coast effect

A coast forms a clear conductivity boundary separating a highly conducting sea from a poorly conducting land. In these circumstances, electric currents tend to flow in the conducting sea, resulting in large vertical fields at the coast. This phenomenon has been called the coast effect (PARKINSON, 1964; PARKINSON and JONES, 1979). It should be noticed, however, that the effect depends on the polarization of inducing field. The coast effect should be seen most effectively for a polarization perpendicular to the coast line (the so-called E-polarization case), since in this case electric currents flow parallel to the coast.

One representative coast effect has been observed at the coast of California, U.S.A. (SCHMUCKER, 1964, 1970; GREENHOUSE, 1972). Figure 12.5 shows in-phase induction arrows for a period of 60 min at various sites in the California coast area (GREENHOUSE, 1972). The arrows in the Pacific Ocean were derived from measurements of three components of geomagnetic variation on the sea floor. The coast effect is clearly demonstrated by long arrows at coastal sites. Anomalies indicated by long arrows are also seen near the continental slope, reflecting another conductivity boundary due to a sharp transition in sea depth.

One can find, through a detailed study of the coast effect, that the details

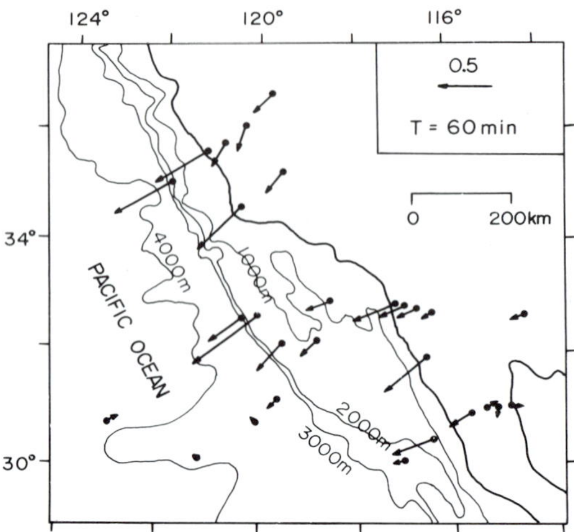

FIG. 12.5. Distribution of induction arrows for a period of 60 min at the southwestern coast of North America. (GREENHOUSE, 1972)

of the effect are somewhat different from one area to another. Since the conductivity structure in the crust and upper mantle affects the coast effect, the difference appears to be associated with regional differences in conductivity structure underneath coastal areas; for instance, a lateral change in underground conductivity structure, if it exists, gives rise to an anomalous Z field which must be superposed on the coast effect.

In an extreme case, the coast effect may be virtually canceled by an underground conductivity anomaly. This corresponds to the case in which a conducting layer lies closer to the earth's surface on the land side. The presence of such a structure could be the reason why no distinct coast effect has been observed in the Peruvian Andes (SCHMUCKER, 1969) or at the east coast of U.S.A. (EDWARDS and GREENHOUSE, 1975).

If a conducting layer lies at a shallower depth on the sea side, anomalous Z fields of deep origin have an effect similar to the coast effect. However, the effect of the deep conductivity structure is reflected in the decay pattern in the Z field amplitude inland (EVERETT and HYNDMAN, 1967; COCHRANE and HYNDMAN, 1970; HYNDMAN and COCHRANE, 1971; BENNETT and LILLEY, 1971; DRAGERT, 1973; WHITE and POLATAJKO, 1978).

Even if the underground conducting layer is laterally uniform in the coast area, its influence on the coast effect would be manifested as a depressed coast effect, especially at short periods, caused by reduction in the intensity of electric currents in the sea owing to electromagnetic coupling between the sea and the underground conducting layer.

In general, investigation of conductivity anomalies in the crust and upper mantle beneath coastal areas is no easy matter because of the coast effect there. This difficulty may be overcome by examining the frequency dependence of the distribution of anomalous Z fields, since differences in electromagnetic response between surface and deep conductivity anomalies would be reflected in differences in the frequency dependence of geomagnetic variation anomaly (HONKURA, 1974). In any case, highly conducting seas essentially obscure the electromagnetic response of the underground structure and hence high resolution cannot be expected.

12.2.2 Island effect

One of the most outstanding anomalies in short-period geomagnetic variations can be observed for a small island. It is not unusual for the polarities of the vertical component to be opposite at two sites on an island. This kind of anomaly is called the island effect. Figure 12.6 shows a typical example of the island effect obtained on Hachijo-jima Island, a small volcanic island south of Tokyo, Japan (HONKURA *et al.*, 1974). For a bay-type variation with a duration of 30 min or so, the Z component at the southern station (NA) is almost in phase with the H component, while the phase of Z at the northern station (EI) is nearly 180° different from that at NA.

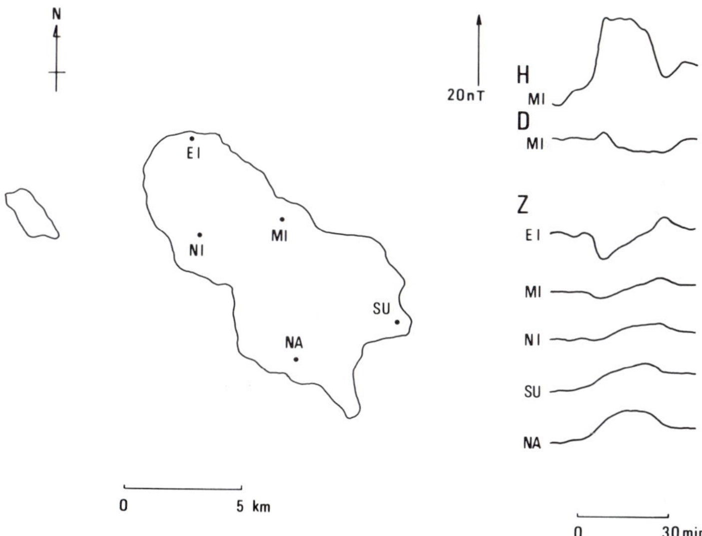

FIG. 12.6. A bay-like variation observed on Hachijo-jima Island. The horizontal components are nearly uniform and records at MI only are shown. (HONKURA *et al.*, 1974)

Since the discovery of the island effect on Oahu Island by MASON (1962), many examples have been reported (MASON, 1963, 1964; SASAI, 1967, 1968; HONKURA, 1971; HONKURA *et al.*, 1974, 1981; KLEIN, 1972, 1976; KLEIN and LARSEN, 1978). This effect has been understood as perturbation of induced electric currents flowing in the sea by a poorly conducting island. In fact, theoretical calculations based on a conducting thin-sheet model (see Section 11.4) could explain successfully the observed island effects on some islands (SASAI, 1968; HONKURA, 1971). Figure 11.5 shows the flow pattern of electric currents around Miyake-jima Island, as derived from calculations of perturbed electric currents, for a magnetic field varying in the north-south direction (HONKURA, 1971). In the figure, the distortion of the currents by the island is clearly demonstrated.

The island effect also provides information on the underground conductivity structure, since the intensity and phase of induced currents are affected by electromagnetic coupling between the sea and a conducting layer in the mantle. On the assumptions that the conductivity of the crust is much lower than that of the mantle conducting layer and that the conductivity structure is laterally uniform, the mantle conductivity structure could be inferred from the island effect observed on Miyake-jima (HONKURA, 1971), Hachijo-jima (HONKURA *et al.*, 1974), and Minami-daito Islands (HONKURA *et al.*, 1981), as will be described later. However, little information concerning the lateral heterogeneity in conductivity structure will be provided from studies of the island effect itself.

12.2.3 Peninsula effect

A vertical field of extremely large amplitude has often been observed at the tip of a peninsula. This phenomenon may be called the peninsula effect. The effect can be considered to be a combination of the coast and island effects. In a broad sense, the peninsula effect is a special case of the irregular coastline effect. This kind of anomaly is so striking that the amplitude of the Z component sometimes exceeds that of the H or D component with which the anomalous Z component is correlated.

Figure 12.7 shows an example of the peninsula effect obtained from observations in the Izu Peninsula area, Japan. As expected, the correlation between the Z and H components is very high at coastal stations. The amplitudes of Z at southern stations in the peninsula are extremely large and even comparable to the amplitude of H. They are larger by $20 \sim 30$ % than the amplitude of Z at a station, HAM, located at the coast west of the peninsula.

Quantitative interpretation of the peninsula effect is possible by combining the methods of numerical calculation for the coast and island effects.

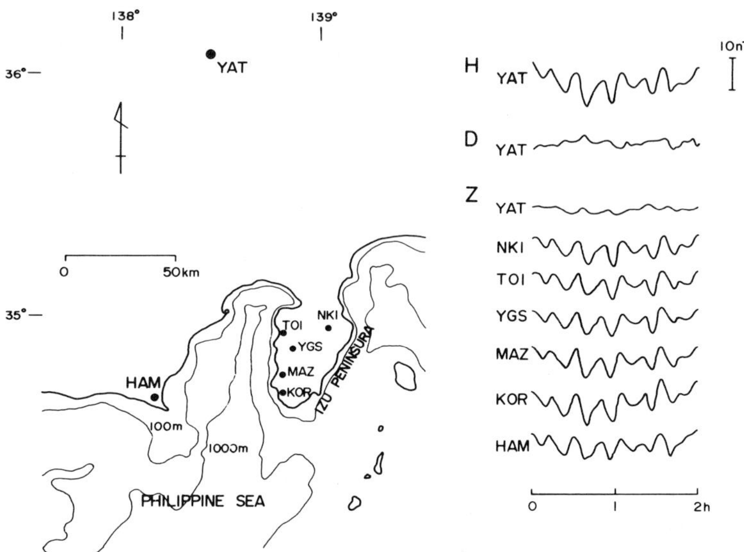

FIG. 12.7. Anomalous Z variations observed in the Izu Peninsula, Japan. The horizontal and vertical components at the Yatsugatake Observatory denoted by YAT are also shown.

Figure 12.8 shows the flow pattern of induced currents for a magnetic variation having a polarization perpendicular to the regional coastline in the vicinity of the Izu Peninsula region. Estimation of magnetic fields due to such distorted currents is in good agreement with the anomalous fields observed at sites in and around the Izu Peninsula. If this method is employed appropriately, observation of the peninsula effect may provide valuable information on the underground conductivity structure beneath coastal areas.

12.2.4 Channeling effect

The reverse of the island effect, that is, concentration of induced electric currents would occur in a narrow sea passage like a strait, which would result in anomalous Z fields of large amplitude. In this case, the polarity of Z at one side of a strait would be opposite to that at another. This phenomenon may be called the strait effect. In a broad sense, it is an example of the channeling effect.

One example of this sort is shown in Fig. 12.9 (H. YAMASHITA, personal communication, 1976). The induction arrow at a station TOI on the northern side of the Tsugaru Strait points to the south, while the arrow at OMA on the

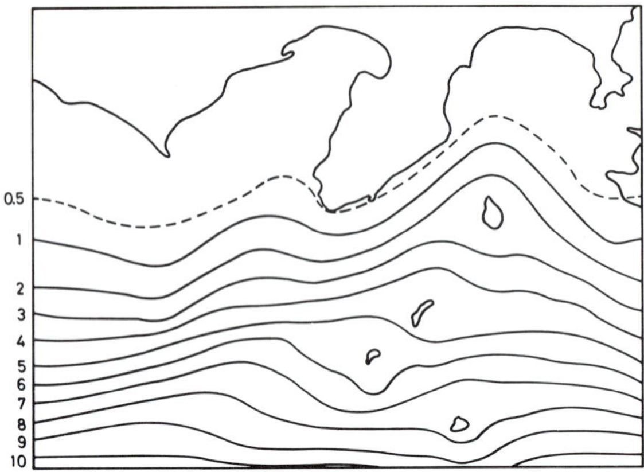

FIG. 12.8. Flow pattern of induced currents in the vicinity of the Izu Peninsula for a magnetic variation with a polarization perpendicular to the overall coastline.

southern side points to the north. Therefore, it is evident that concentration of electric currents indeed occurs for a magnetic field variation with a north-south polarization.

In Fig. 12.9, another anomaly seems to be implied by a small induction arrow at SKN located south of OMA. It is not obvious why the arrow is so small there, particularly in view of large arrows at sites further to the south. This anomaly can be interpreted in terms of another channeling effect; electric currents would flow between the Pacific Ocean and the Japan Sea through a sedimentary layer in the area around TGS and the bay south of SKN. Thus the two channeling effects cancel each other out at SKN.

As in the above case, highly conducting sediments often form a path for current channeling which connects one open sea with another. Another example of this kind has been obtained in western Hokkaido, Japan (NISHIDA, 1976). In this case, the channeling effect is best seen for the east-west polarization of magnetic variation. A notable geomagnetic variation anomaly found in the northern Pyrenees, southwestern France, can also be explained by the channeling effect (BABOUR *et al.*, 1976; VASSEUR *et al.*, 1977). In fact, VASSEUR and WEIDELT (1977) showed quantitatively that the anomaly can be interpreted quite well in terms of currents flowing in a sedimentary conduction path between the Atlantic Ocean and the Mediterranean Sea. The Alert anomaly in the Canadian Arctic may also be due to the channeling effect, as pointed out by DYCK and GARLAND (1969), NIBLETT and WHITHAM (1970), and NIBLETT *et al.* (1974).

The importance of conduction currents in geomagnetic variation anomalies has also been noticed with special reference to a thick sedimentary layer

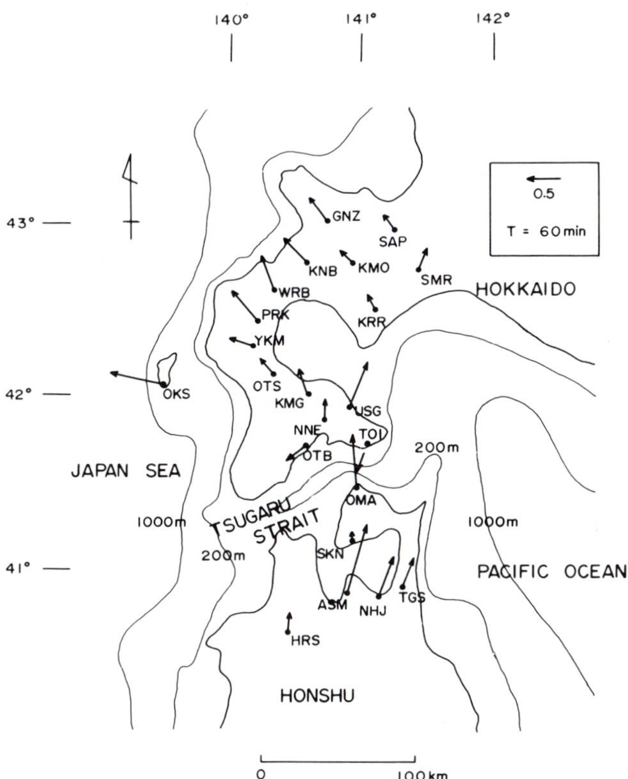

FIG. 12.9. Distribution of induction arrows for a period of 60 min in northern Japan. (H. Yamashita, personal communication, 1976)

covering a wide area. Typical examples of geomagnetic variation anomalies of this sort have been found in the North German basin (UNTIEDT, 1970; RITTER, 1975) and the southern Great Plains, U.S.A. (PORATH and DZIEWONSKI, 1971a, b). In these cases, however, electromagnetic induction is supposed to be operative in the sea with which the sedimentary layer is connected electrically.

12.3 Magnetotellurics

Induced electric fields observed on the earth's surface should contain some information on the underground conductivity structure. Hence observations of both electric and magnetic fields have been carried out to study a regional conductivity distribution in the crust and upper mantle (e.g. FOURNIER and ÁDÁM, 1971; HUTTON, 1976a; FOURNIER, 1980). This method is called *magnetotellurics* (often denoted by MT). In contrast to geomagnetic variation anomalies, usually characterized by anomalous vertical components, in magnetotellurics the horizontal components are used for analyses and the ratio of the electric to magnetic fields at a station yields estimates of the apparent resistivity, as will be shown later.

The magnetotelluric method provides a powerful tool, if it is properly employed in a region where a laterally homogeneous conductivity structure can be reasonably inferred; in such a case the conductivity structure can be determined, at least in principle, from electric and magnetic data at a single station. In practice, however, there is a severe restriction in MT; that is, the electric field is directly and greatly affected by perturbation due to local inhomogeneity in surface conductivity distribution. In this sense, MT should be employed carefully, particularly in a geologically complicated area.

Suppose that the conductivity is uniform in the earth and the inducing magnetic field is also spatially uniform, then (11.42) holds and consequently the conductivity σ can be determined from E_x/H_y or E_y/H_x at a certain period. In MT studies, (11.42) is usually written as

$$\rho_a = 0.2T|Z|^2 \tag{12.8}$$

where

$$Z = E_x/H_y \text{ or } E_y/H_x. \tag{12.9}$$

ρ_a is called the apparent resistivity (CAGNIARD, 1953). Provided that the period T is given in seconds, E in mV/km, and H in gammas (nanoteslas), the unit of ρ_a in (12.8) is ohm · m.

Plots of ρ_a values against T are conventionally used in MT data analysis, even though the above mentioned condition for MT is not satisfied. However, as long as the assumption of laterally uniform conductivity distribution is valid, (12.8) is subject to only slight modification and the essential point remains unchanged. In such a case, the apparent resistivity at a certain period can be considered as reflecting the average resistivity from the earth's surface to the skin depth corresponding to the period. Therefore, the dependence of the apparent resistivity on depth can be derived from data at various periods, and the conductivity distribution can be inferred as a function of depth.

In geologically complicated areas, the condition of lateral uniformity is no longer satisfied; that is, a two-dimensional or three-dimensional treatment is required. In general, the magnetotelluric relation between the observed electric and magnetic fields is expressed in terms of an impedance tensor:

$$\begin{pmatrix} E_x \\ E_y \end{pmatrix} = \begin{pmatrix} z_{11} & z_{12} \\ z_{21} & z_{22} \end{pmatrix} \begin{pmatrix} H \\ D \end{pmatrix} \tag{12.10}$$

where E_x and E_y denote the northward and eastward components, respectively, of the electric field with respect to the geomagnetic direction. In the one-dimensional case, $z_{11} = z_{22} = 0$ and $z_{12} = -z_{21}$, and (12.10) is reduced to (12.9).

In a strictly two-dimensional case, diagonal elements of the impedance tensor vanish if the measuring coordinates are rotated properly. It is then shown that one of the new axes is parallel to the strike of the two-dimensional structure and the other is perpendicular to it. Non-vanishing elements of the rotated tensor, z'_{12} and z'_{21} are used to estimate the apparent resistivity for the *H*- and *E*-polarization cases (e.g. HERMANCE, 1973). These two estimates of the apparent resistivity do not coincide with each other near the conductivity boundary.

Figure 12.10 provides a qualitative representation of the distribution of apparent resistivity estimates in the direction perpendicular to the strike for the *H*- and *E*-polarization cases. On the conductive side, the apparent resistivity gives a value higher than the real value in the *E*-polarization case. On the other hand, the *H*-polarization case results in a higher resistivity estimate on the resistive side. In order to determine a detailed distribution, impedance estimates are required for various periods and theoretical calculations must be made as described in Section 11.3.

In practice, however, it is not unusual that diagonal elements never vanish by rotating coordinates in any direction. In such a case, a parameter called skew may be helpful to judge whether a two-dimensional approximation is reasonable or not. The skew is defined by

$$\text{skew} = \frac{|z_{11}+z_{22}|}{|z_{12}-z_{21}|}. \tag{12.11}$$

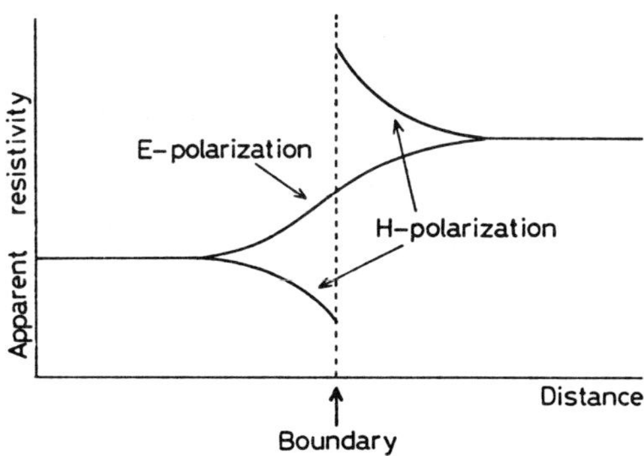

FIG. 12.10. Illustration of apparent resistivity distribution across a two-dimensional conductivity boundary for the *H*-and *E*-polarization cases.

In a strictly two-dimensional case, the following relations hold for any rotation (HERMANCE, 1973):

$$\begin{aligned} z_{11} &= -z_{22} \\ z_{12} - z_{21} &= \text{const} . \end{aligned} \tag{12.12}$$

Therefore, the skew must be small for the two-dimensional approximation to be valid

Although no strict criterion has been established, it is conventionally recommended that a two-dimensional approach should not be attempted if the skew is much larger than 0.2. In the case of the data shown in Fig. 12.1!, the skew is larger than 0.6 for the period range from 20 min to 240 min; hence in this case a two-dimensional treatment must be avoided. As for the three-dimensional case, no general method of analysis has been developed. One typical case, which would be encountered frequently, is a set of magnetotelluric fields strongly affected by a channeling effect.

In the above discussion, it has been assumed that a magnetic field variation is spatially uniform; in other words, the wavelength of the source field is considered to be infinite. However, it is not obvious whether or not this assumption is actually reasonable. In fact, WAIT (1954) and PRICE (1962) pointed out that the MT response depends on the wavelength of the source

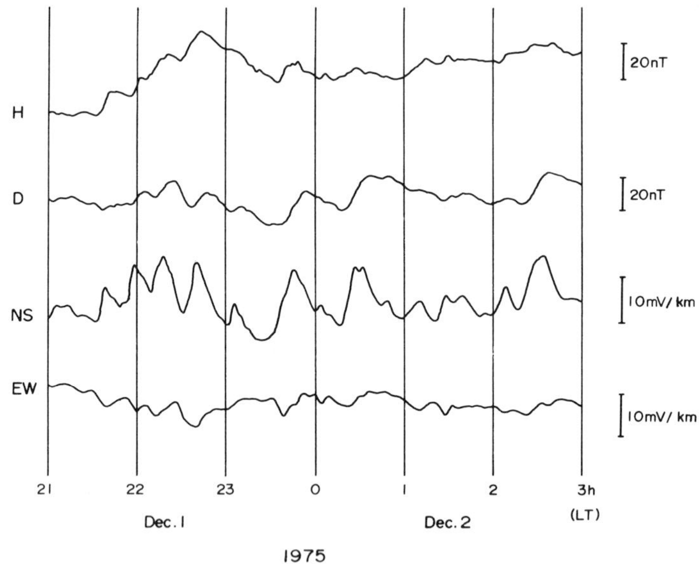

FIG. 12.11. An example of magnetotelluric fields observed at the Yatsugatake Observatory, central Japan. NS and EW denote the northward and eastward components of the telluric fields, respectively. (HONKURA and KOYAMA, 1979)

field. The effect of finite wavelength on apparent resistivity estimates may be quite large, particularly for a short wavelength.

In high-latitudes, substorm fields of large amplitude originate in the ionospheric electrojet (see Section 8.4) and contain components of short wavelength. However, HERMANCE and PELTIER (1970) pointed out that even for substorm fields in high-latitudes the source field effect is serious only near the electrojet region. In mid-latitudes, the wavelength of substorm fields is fairly long, several thousand kilometers (GOUGH and CAMFIELD, 1972) and the source field is unlikely to be significant.

As described in Section 11.2, the effect of finite wavelength appears through λ in the term $\lambda^2 + 4\pi\sigma\mu\omega i$. Therefore, the effect would be negligible if the magnitude of the second term is much larger than the first term. In mid-latitudes, $\lambda \approx 10^{-8}\ \mathrm{cm}^{-1}$ (GOUGH and CAMFIELD, 1972). If a value of 10^{-13} e.m.u. (0.01 S/m), which is not unreasonable as the conductivity of the crust and upper mantle, is assumed, the second term is comparable to the first term when $\omega \approx 10^{-4}\ \mathrm{sec}^{-1}$. It is concluded, therefore, that the source field effect is not serious in mid-latitudes as far as short-period events are used for analysis. However, if the MT method is employed for long-period events or in a highly resistive region, much attention must be paid to the source field effect.

12.4 Conductivity Anomalies in Continents

12.4.1 Anomalies at plate boundaries

(1) Spreading zones

Iceland is located just above the mid-Atlantic ridge and electromagnetic induction studies there would provide important information on the conductivity structure associated with the spreading process. In view of this, MT studies have been undertaken in the southwestern part (HERMANCE and GARLAND, 1968; HERMANCE and GRILLOT, 1970, 1974; HERMANCE, 1973) and also in the northeastern part of Iceland (BEBLO and BJÖRNSSON, 1978).

The estimated conductivity for southwestern Iceland ranges from 0.02 to 0.05 S/m at the base of the crust and 0.01 to 0.025 S/m down to 100 km or more in the mantle (HERMANCE and GRILLOT, 1974). Taking into account a high temperature-gradient there, the conducting layer at the base of the crust was interpreted as indicating a temperature of 1000°C or so at the crust-mantle boundary.

In northeastern Iceland, MT measurements were carried out at many sites arranged in an approximately east-west direction; this measuring line is perpendicular to the strike of the neovolcanic zone in northeastern Iceland (BEBLO and BJÖRNSSON, 1978). Figure 12.12 shows the conductivity structure across the neovolcanic zone. A fairly thin conducting layer is clearly recognized. The conductivity of this layer amounts to 0.1 S/m or so. There is a clear tendency for the depth to the conducting layer to increase with increasing distance from the spreading center corresponding to the neovolcanic zone. This conducting layer has been interpreted as indicating the existence of molten basalt produced as a result of partial melting of peridotite (see Section 12.6). It has also been speculated that because of this partial melting zone the lithosphere above it may easily move away from the spreading axis; such a movement is one of the fundamental requirements in the theory of plate tectonics.

The East African Rift Valley is considered to be a zone where the spreading process has just become activated, so that, as in Iceland, this region has drawn much attention. Thus intensive observations of geomagnetic and telluric variations have been carried out in the Gregory Rift Valley, Kenya, and also in the Afar Depression and the Main Rift, Ethiopia. An overall review of studies of geomagnetic variation anomalies and magnetotellurics in these regions was given by HUTTON (1976b).

In the Gregory Rift Valley, BANKS and OTTEY (1974) found an anomaly of geomagnetic variation as shown by the distribution of induction arrows at a period of 25 min in Fig. 12.13. The existence of a highly conducting layer is suggested by the reversal of induction arrows at sites across the valley. In fact, a two-dimensional model requires a conducting layer at a

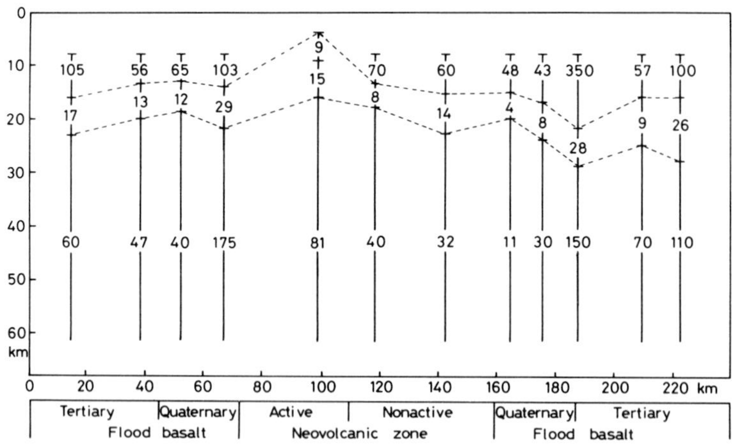

FIG. 12.12. Resistivity profile across the neovolcanic zone in northeastern Iceland. Numerals indicate resistivity values in ohm · m. (BEBLO and BJÖRNSSON, 1978)

depth of 20 km beneath the rift floor as shown in Fig. 12.13. Such a model was further confirmed by more extensive array observations in the same area (BEAMISH, 1977). This conducting layer beneath the valley does not seem to extend further southwards beyond a latitude of 1°S.

MT observations made in the same area also suggest that the conductivity is high below the valley (ROONEY and HUTTON, 1977). The estimated apparent resistivity ranges from 2 to 20 ohm · m, and in order to interpret the apparent resistivity curve as a function of period, a conducting layer is required at a depth less than 8 km, that is, within the crust. Long-period data suggest the existence of another conducting layer at a depth more than 30 km. ROONEY and HUTTON (1977) interpret their model of crustal conductivity anomaly in terms of conducting water and high temperature, while BANKS and OTTEY (1974) argue that their model of mantle conductivity anomaly reflects a partial melting of mantle material. Although different models have been derived from two different methods, it seems doubtless that a conducting layer having a conductivity of 0.1 S/m or so is located very close to the surface of the Gregory Rift Valley.

The result of two-dimensional array observations of geomagnetic variations is presented in a review paper by HUTTON (1976b), although it is rather

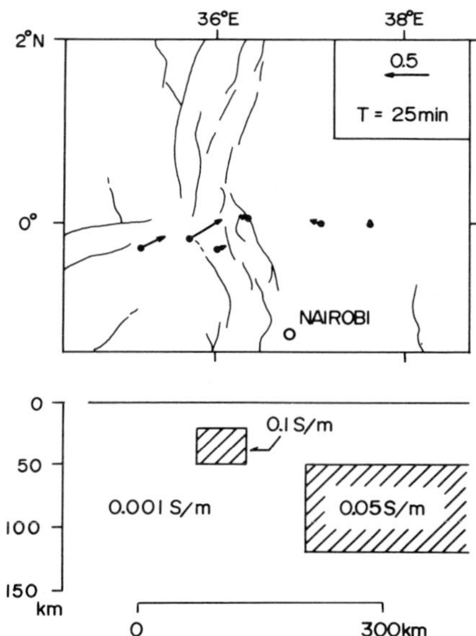

FIG. 12.13. Distribution of induction arrows for a period of 25 min in the Gregory Rift Valley, Kenya, and a two-dimensional model of underground conductivity structure. (BANKS and OTTEY, 1974)

a preliminary one. Nevertheless it is important to find a conducting layer having a conductivity higher than 0.1 S/m at a depth between 100 and 300 km beneath the Afar Depression. This layer seems to be inferred from large induction arrows for the period range from 10 min to 3 hours. However, a completely different result has been derived from MT measurements at 36 sites in the Afar Depression; a thin conducting layer, 0.02 ~ 0.1 S/m in conductivity and about 20 km in thickness, exists at a depth of 15 km and the conductivity below this layer remains lower than 0.01 S/m down to 300 km (HAAK, 1977).

Beneath the Gulf of California, there lies a spreading zone which is a northern extension of the East Pacific Rise. In an attempt to search for a conductivity structure associated with the spreading process, observations of geomagnetic variation were carried out at land stations on both sides of the gulf (WHITE, 1973a, b). Although there is a striking effect due to the sea water which primarily accounts for the induction arrows and obscures the resolution of underground conductivity structure, the existence of a conducting layer is suggested. The depth to and the conductivity of the layer are estimated as 20 ~ 30 km and 0.02 S/m, respectively, although these are by no means conclusive because of the poor resolution.

(2) Subduction zones

The Japanese Islands constitute a typical island arc behind a subduction zone in the western Pacific. It is expected, therefore, that CA studies in Japan would contribute to the understanding of an anomalous conductivity structure associated with the subduction process. In Japan, intensive observations of geomagnetic variation have been carried out since the early 1950's (e.g. RIKITAKE and YOKOYAMA, 1953; RIKITAKE, 1959, 1964a, 1966a); at that time a clear concept of plate tectonics had not yet been developed. The observations brought to light two major anomalies of short-period geomagnetic variation, the central and northeastern Japan anomalies, which were interpreted in terms of marked undulation of a mantle conducting layer beneath Japan, as shown in Fig. 12.14 (RIKITAKE, 1969).

One characteristic of the conductivity structure shown in Fig. 12.14 is a sharp contrast in the depth to the conducting layer between the subduction zone and the region behind the arc, as is clearly recognized in northeastern Japan. The Philippine Sea off southwestern Japan is considered to be a marginal sea behind the Izu-Bonin and Mariana trenches (see Fig. 12.23). Therefore, the existence of a conducting layer at shallow depth is in good agreement with our understanding of the tectonic environment in the Japanese region.

In Japan and its vicinity, so many measurements of heat flow have been made that the distribution of heat flow is well established. UYEDA and

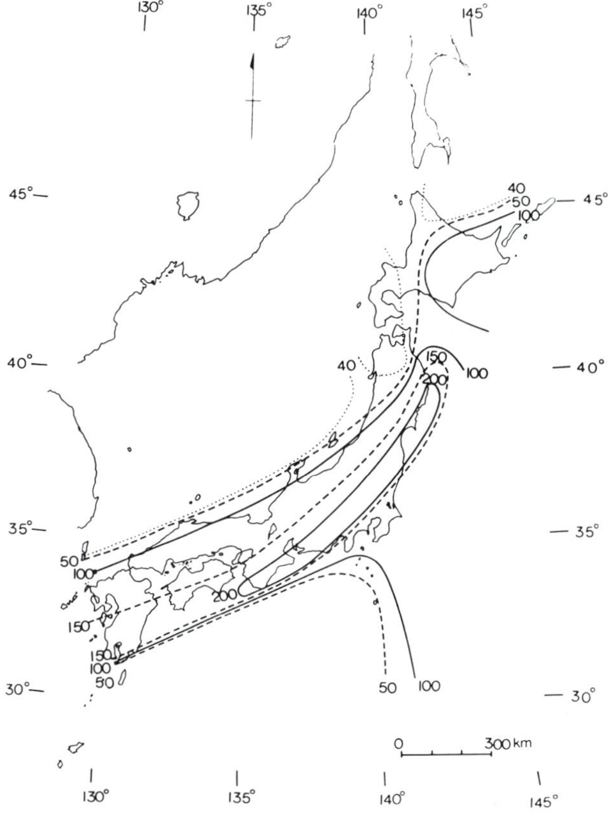

FIG. 12.14. A model of conductivity structure beneath Japan. Contours indicate the depth to the mantle conducting layer in kilometers. (RIKITAKE, 1969)

RIKITAKE (1970) pointed out that the inferred overall conductivity structure beneath the Japanese Islands is in good agreement with the heat flow distribution.

Since the early 1970's, more intensive observations of geomagnetic variation have been carried out, aiming at studying more detailed structures beneath various parts of Japan. In the Hokkaido region, the northern-most major island of Japan, NISHIDA (1976) examined the distribution of anomalous Z and H variations and pointed out that a conducting layer might exist at a depth of 40 km or so beneath the Japan Sea side.

Northeastern Japan, called the Tohoku District, is a typical trench - island arc - marginal sea system. Observations of geomagnetic variation were made at some land stations across the arc and induction arrows were derived as shown, for a period of 60 min, in Fig. 12.15 (HONKURA, 1974). According to two-dimensional model calculations, no conducting layer is required at shallow depth beneath the Pacific Ocean to account for the distribution of induction arrows. It was also suggested that the upper mantle beneath the Japan Sea might be highly conducting, although poor resolution makes such a conclusion less significant. In 1981 and 1982, more intensive observations were undertaken in the same area, including observations at some sites on the sea floor beneath the Pacific Ocean and the Japan Sea (T. YUKUTAKE, personal communication, 1982). Analyses of these data would provide more detailed information on the conductivity structure in this area.

In the central part of Japan, where a major anomaly called the Central Japan Anomaly has been found, induction arrows have been derived as shown for a period of 60 min in Fig. 12.16 (HONKURA, 1974). Arrows on islands were obtained by removing the island effect on each island. As mentioned in Subsection 12.2.1, the coast effect obscures the response of the underground

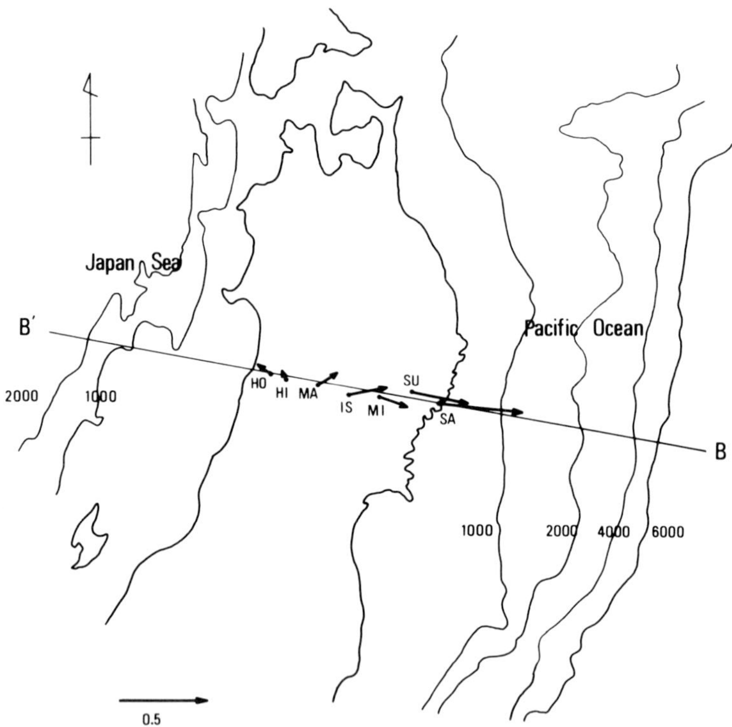

FIG. 12.15. Distribution of induction arrows for a period of 60 min in north-eastern Japan. (HONKURA, 1974)

conductivity structure. In order to overcome this difficulty, an attempt was made to account for the period dependence of the anomaly by making use of transfer functions for various periods. Then it was found that the two-dimensional model shown in Fig. 12.17 accounts reasonably well for the observed transfer functions at periods of 10, 30, 60 and 120 min, respectively, as demonstrated in Fig. 12.17. The model consists of highly conducting layers with a conductivity of 0.5 S/m, lying at 30 km depth beneath the Philippine and Japan Seas.

The above model is by no means unique. In fact, RIKITAKE (1975) showed that a depression model can also account for the observed transfer function.

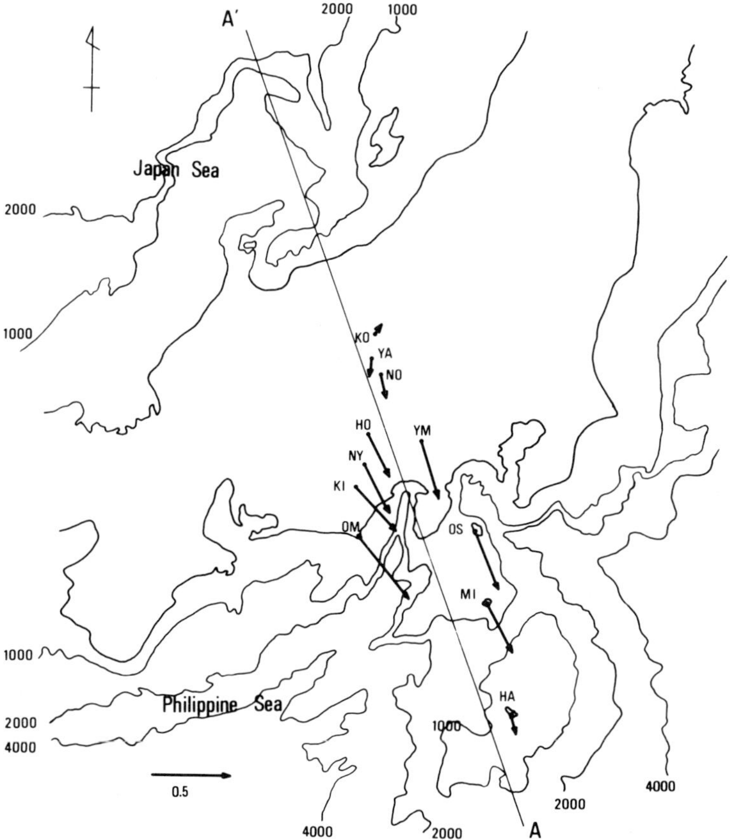

FIG. 12.16. Distribution of induction arrows for a period of 60 min in central Japan. Arrows at islands denoted by OS, MI, and HA are derived after the island effect is removed. (HONKURA, 1974)

(a)
0.6
0.4
0.2
-0.2
-0.4
-0.6

(b)
0.6
0.4
0.2
-0.2
-0.4
-0.6

(c)
0.6
0.4
0.2
-0.2
-0.4
-0.6

(d)
0.6
0.4
0.2
-0.2
-0.4
-0.6

A
A′
0
0.001 S/m
30
0.5 S/m
0.01 S/m
0.5 S/m
80
km
0
300 km

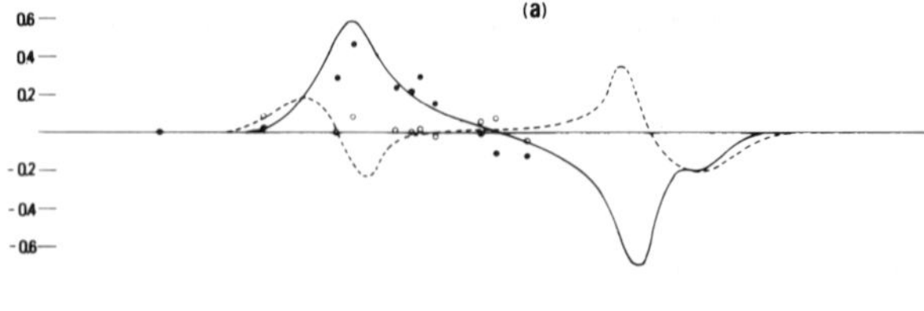

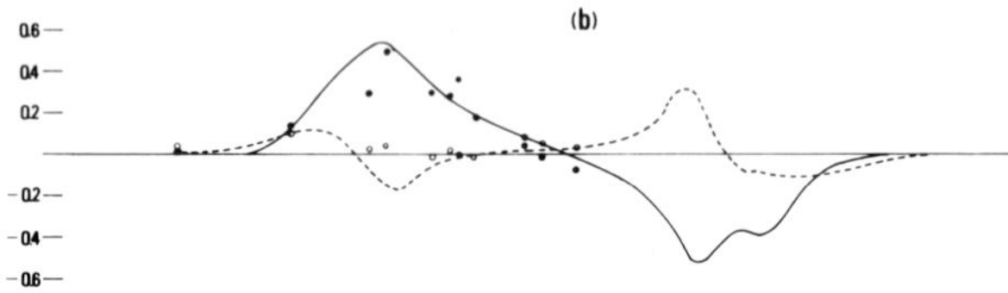

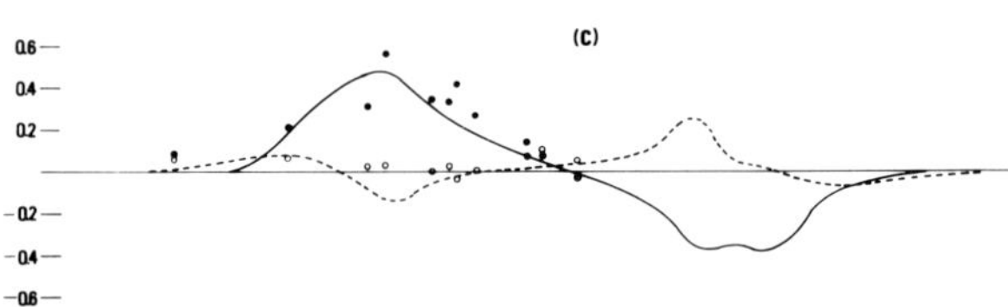

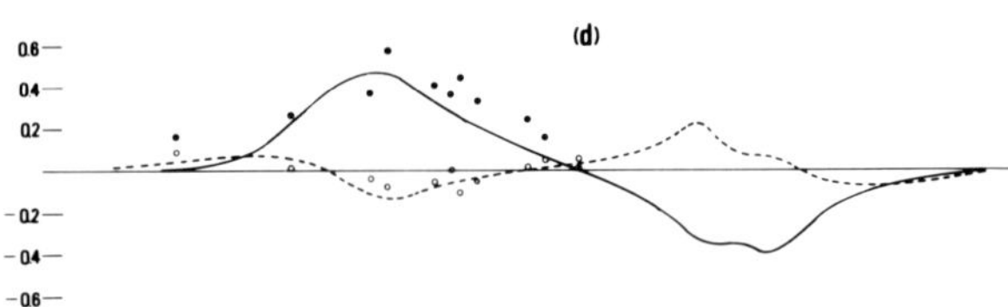

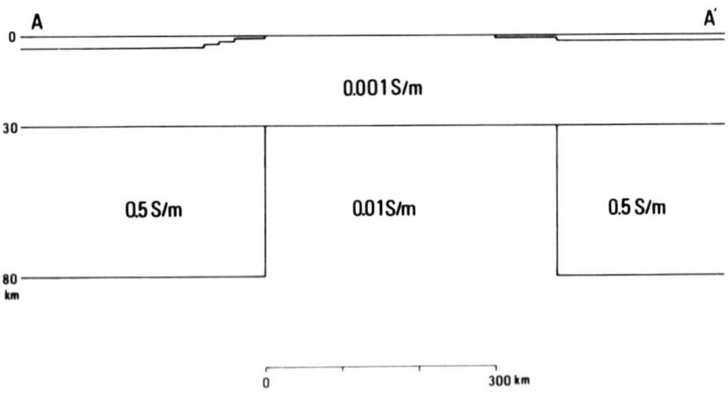

Fig. 12.17.

In the depression model, a mantle conducting layer having a conductivity of 0.1 S/m exists at a depth of 50 km or so beneath the Philippine and Japan Seas, whereas it is depressed below the central part of the Japanese mainland.

Studies of the peninsula effect in the central part of Japan support the depression model for the Philippine Sea region. As a conclusion, it can be claimed that highly conducting layers exist at a depth of 50 km or so beneath the Philippine and Japan Seas and their conductivity amounts to the order of 0.1 S/m.

In southwestern Japan, the main effort has been focused on investigation of the conductivity structure beneath the Japan Sea side (SUMITOMO, 1972; MIYAKOSHI, 1976, 1979). Figure 12.18 shows Parkinson vectors for periods of 15 sec, 6 min, and 60 min, respectively (MIYAKOSHI, 1979). The figure also shows the in-phase part of transfer function A, plotted against the distance from the coast. In an attempt to interpret the distribution of transfer functions, MIYAKOSHI (1976, 1979) presumed that a conducting layer existing at a depth of 30 km or so beneath the Japan Sea might extend further to the south, probably to a latitude of 35° in Fig. 12.18.

The detailed examination of local conductivity structure beneath various parts of Japan, as described above, generally confirms the overall model of conductivity structure beneath Japan as shown in Fig. 12.14, although some minor modifications may be required. However, no evidence has been obtained to support a lateral change in conductivity beneath the Izu-Bonin arc. To this end, it is essential to carry out seafloor observations. Since seafloor magnetometers are now available in Japan (J. SEGAWA, personal communication, 1981), the conductivity structure beneath the sea adjacent to Japan will soon be derived from seafloor observations. It is expected, therefore, that a more detailed structure, which may be typical for a subduction zone, will be resolved in the near future.

CA studies have also been undertaken at another subduction zone, the Pacific coast of South America. Observations of geomagnetic variation in Peru and Bolivia revealed a notable anomaly beneath the Andes (SCHMUCKER *et al.*, 1964, 1966; CASAVERDE *et al.*, 1969; ALDRICH *et al.*, 1972). As already described in Subsection 12.2.1, this anomaly cancels out the coast effect which should otherwise be observed at coastal stations.

The distributions of the H and Z components were examined along a profile in the N30°E direction, which is nearly perpendicular to the strike of

---

FIG. 12.17. A model of conductivity structure along a profile A-A' shown in Fig. 12.16. The conductivity of sea water is taken as 4 S/m. Closed and open circles denote the in-phase and out-of-phase parts of the observed transfer functions respectively, while solid and broken lines indicate the calculated ones for periods of (a) 10, (b) 30, (c) 60, and (d) 120 min, respectively. (HONKURA, 1974)

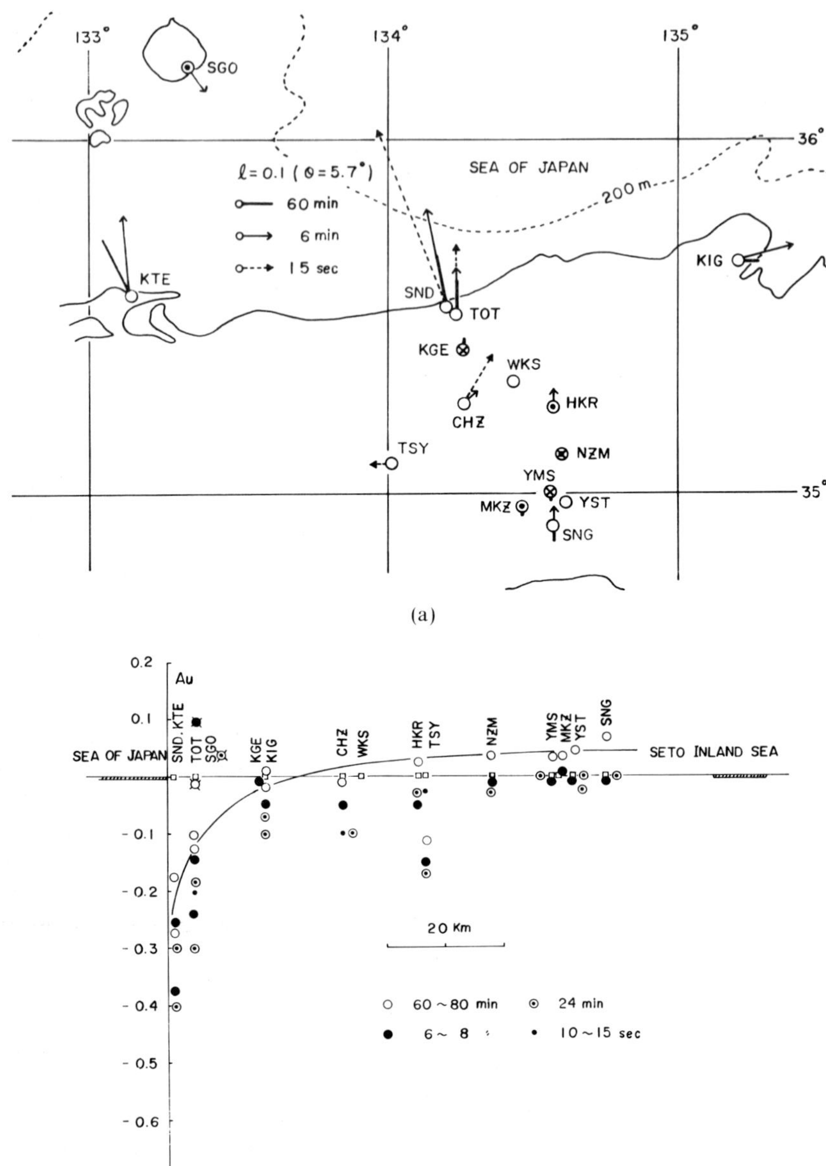

FIG. 12.18. (a)Parkinson vectors for periods of 15 sec, 6 min, and 60 min, respectively, in southwestern Japan. (b) Au-values for four period bands plotted against the distance from the Japan Sea coast. (MIYAKOSHI, 1979)

the Peruvian Andes. The result is as follows. The Z component vanishes just at the Andes, and its polarity is opposite on either side of the Andes, while the H component becomes the largest at locations where the Z component vanishes (SCHMUCKER, 1969). This Andean anomaly was interpreted in terms of a two-dimensional semi-cylindrical upheaval of a mantle conducting layer beneath the Andes. The radius of the semi-cylinder is estimated as 200 km and its top surface is located at a depth of 60 km.

An anomaly in the distribution of $Z - H$ transfer functions (corresponding to A) was also found in northern Chile and southern Bolivia (ALDRICH *et al.*, 1975a). On the other hand, in central Chile and Argentina the distribution of $Z - D$ transfer functions (corresponding to B) clearly shows a coast effect similar to the one observed at the California coast (ALDRICH *et al.*, 1975b); contours of $Z - D$ transfer functions which provide no clear evidence supporting the existence of a conductivity anomaly as notable as the one found beneath the Peruvian Andes. It should be noted that the coast effect is enhanced more at shorter periods. Such a period dependence could be an important clue to the elucidation of underground conductivity structure (HONKURA, 1974). However, more detailed analyses indicate a far more complex distribution of transfer functions (L. T. ALDRICH, personal communication, 1983).

(3) Collision zone

The microcontinent of India is supposed to be colliding at present with the Eurasian plate so that these two plates are greatly deformed at the boundary, as manifested by the creation of the Himalaya. In an attempt to discover any anomalous conductivity structure at this plate boundary, a two-dimensional array observation of geomagnetic variation was carried out in northwestern India (LILLEY *et al.*, 1981c). Analysis of an event having a nearly north-south polarization revealed a distinct anomaly in which even polarity reversal was found, in the Z component. This anomaly of geomagnetic variation is due to a seemingly narrow conducting zone in the crust. Contrary to that expected from the geometry of the plate boundary, however, the strike of the zone is not parallel but rather perpendicular to the boundary.

LILLEY *et al.* (1981c) further argue that electric currents flowing in the anomalous conducting zone and giving rise to a geomagnetic variation anomaly might be induced elsewhere, presumably in a conducting layer beneath the Tibetan plateau, although no direct evidence has been found to support the existence of such a coducting layer. The above conclusions were derived only for one particular polarization of magnetic variation and LILLEY *et al.* (1981c) suggest that a different kind of geomagnetic variation anomaly may be found, if analysis is made for different polarization. In any case, whether there is conductivity structure associated with plate collision remains, as yet, uncertain.

12.4.2 Anomalies in intraplate regions

(1) North America

SCHMUCKER (1964, 1970) initiated CA studies in the U.S.A. by carrying out observations of geomagnetic variation in the southwestern United States and found the Sierra and Rio Grande anomalies. At that time, these anomalies were interpreted as indicating local upheaval of a mantle conducting layer beneath the Sierra and Rio Grande areas. It was found that the conductivity structure model thus proposed is in good agreement with the distribution of heat flow (WARREN *et al.*, 1969). Consequently, the undulatory surface of the conducting layer was supposed to correspond to the 1200°C isotherm.

Meanwhile, Gough, Porath, and their co-workers started, for the first time, a series of two-dimensional array observations covering the western United States and also southwestern Canada (e.g. GOUGH, 1973, 1974; CAMFIELD and GOUGH, 1975; ALABI *et al.*, 1975). The first survey area was the central part of the western United States and anomalies in both vertical and horizontal components were found over the Wasatch Front and the southern Rockies (REITZEL *et al.*, 1970).

Separation of a substorm event recorded simultaneously at the array sites into external and internal origins confirmed that anomalous fields are really of internal origin (PORATH *et al.*, 1970). These anomalies could be reasonably accounted for by two completely different two-dimensional models as shown in Fig. 12.19(a) and (b). The model (a) indicates local upheaval of a mantle conducting layer due to high temperature beneath the anomalous zones, whereas the conducting zone in the model (b) coincides with a seismic low-velocity zone in the upper mantle (PORATH and GOUGH, 1971; PORATH, 1971).

The two-dimensional array observations were then extended further to the south. The survey area covers some of the sites where SCHMUCKER (1964) made measurements. The conductivity structure model proposed by SCHMUCKER was subject to only slight modification (PORATH and GOUGH, 1971); that is, a regional upheaval of a mantle conducting layer under the Basin and Range Province with a local conducting ridge beneath the Rio Grande Rift Belt. However, such a model is not unique and the observed anomaly of geomagnetic variation also seems to be accounted for by another model consisting of a low-velocity zone in the upper mantle (GOUGH, 1974).

In western Canada, observations of geomagnetic and telluric fields had been carried out before CAMFIELD *et al.* (1971) made a two-dimensional ar-

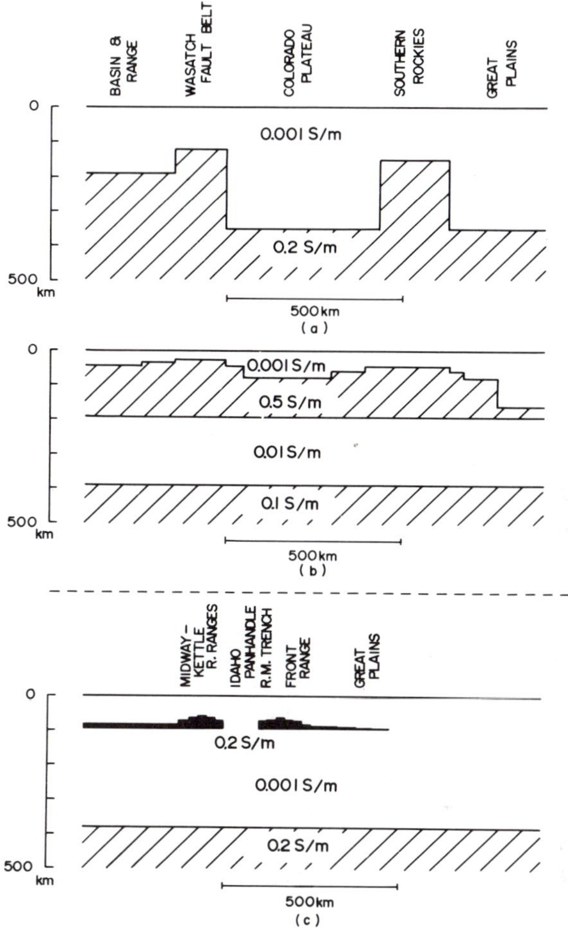

FIG. 12.19. (a), (b) Two models of conductivity structure beneath the central part of the western United States. (PORATH and GOUGH, 1971; PORATH, 1971) (c) a model beneath the northern part of the western United States. (CAMFIELD and GOUGH, 1975).

ray observation of geomagnetic variation (LAMBERT and CANER, 1965; CANER *et al.*, 1967, 1971; COCHRANE and HYNDMAN, 1970). The anomaly found in the Cordillera has been characterized by attenuation of the vertical component at periods shorter than two hours; that is, this anomaly is of high-cut ΔZ type. MT results in the anomalous region show that a conducting layer (0.2 S/m) exists at a depth of $10 \sim 15$ km and below it a less conducting layer extends; its conductivity is 0.02 S/m or so (CANER, 1970, 1971).

PORATH *et al.* (1971) attempted to interpret anomalies of geomagnetic variation as derived from two-dimensional array observations in the northwestern United States and southwestern Canada (CAMFIELD *et al.*, 1971). The proposed model consists of steps and ridges of a mantle conducting layer. However, the condition that ΔZ reduction is observed only for periods shorter than two hours turned out not to be satisfied in the above model. Hence a new model consisting of a thin conducting layer, 15 km in thickness and 0.2 S/m in conductivity, was proposed to remove this inconsistency (CAMFIELD and GOUGH, 1975). The new model is shown in Fig. 12.19(c). The depth to the conducting layer is not well resolved and its possible range is between 50 and 100 km.

DRAGERT and CLARKE (1977) reinvestigated the conductivity structure underneath the Cordillera and concluded that the conducting layer suggested by CANER (1971) and CAMFIELD and GOUGH (1975) lies at a depth of $40 \sim 50$ km. Although the depth to the conducting layer is different from one model to another, its thickness and conductivity are indeed consistent; about 20 km and 0.2 S/m. This conducting layer has been interpreted in terms of hydration and/or partial melting (CANER, 1970; DRAGERT and CLARKE, 1977).

Another very striking anomaly of local nature was found in the Black Hills area and has been called the North American Central Plains (NACP) anomaly (CAMFIELD *et al.*, 1971; GOUGH and CAMFIELD, 1972). In order to investigate further the NACP anomaly, ALABI *et al.* (1975) carried out two-dimensional array observations covering the area extending from the Black Hills to the border between the Great Plains and the Canadian Shield. This anomaly, of crustal origin, was found to extend along a narrow band further to the north, probably to the border of the Canadian Shield. It was noticed that this conducting band coincides closely with fold and fracture zones and also with metamorphic belts.

Figure 12.20 shows schematically the overall conductivity structure in the western United States and western Canada as derived from two-dimensional array observations of geomagnetic variation by Gough, Porath, and their co-workers (GOUGH, 1974; ALABI *et al.*, 1975). The most conducting layers lie beneath the southern and northern Rockies and also beneath the Wasatch Front. In the Basin and Range Province, also, there is a highly conducting layer in the upper mantle. The NACP anomaly is characterized by a narrow conducting zone in the crust. It extends for as much as 1000 km in a nearly north-south direction.

In addition to the above regional structure, local conductivity anomalies of crustal origin have been found beneath the Cascade Volcanic Belt (LAW *et al.*, 1980) and the Yellowstone region (LEARY and PHINNEY, 1974). These local

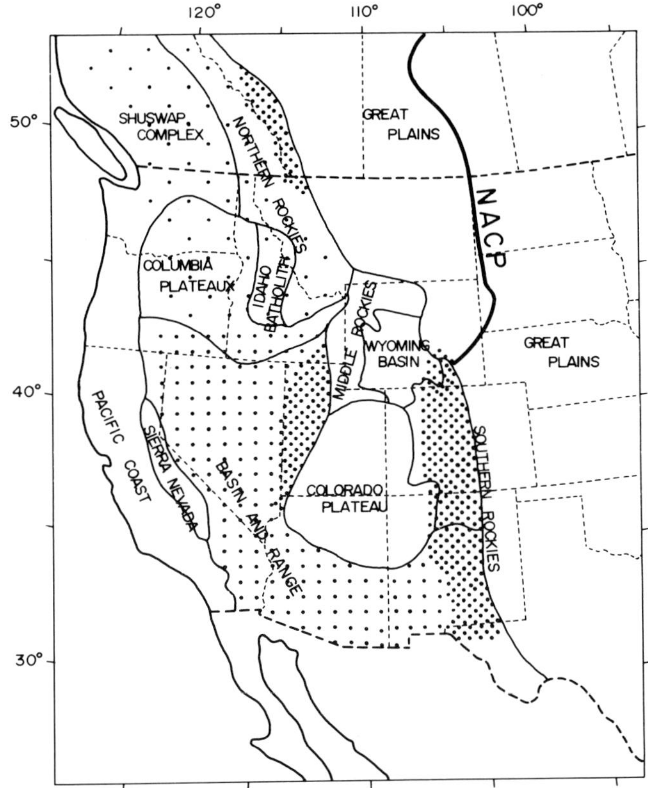

FIG. 12.20. Outline of conductivity anomalies in western North America. The density of dots indicates qualitative estimates of conductivity anomalies. (GOUGH, 1974) NACP denotes the North American Central Plains anomaly. (ALABI, *et al.*, 1976)

anomalies seem to be associated with local geothermal activity.

In the Atlantic coast region of North America, observations of geomagnetic and telluric fields have been undertaken and some crustal conductivity anomalies have been found (HYNDMAN and COCHRANE, 1971; COCHRANE and HYNDMAN, 1974; BAILEY *et al.*, 1974, 1978; EDWARDS and GREENHOUSE, 1975; KURTZ and GARLAND, 1976; HONKURA *et al.*, 1977; NEKUT *et al.*, 1977; COCHRANE and WRIGHT, 1977; CAMFIELD, 1981).

The crustal conducting layer extending along the Atlantic coast is called the Atlantic conductivity anomaly. Its presence on the continental side greatly reduces the coast effect there (HYNDMAN and COCHRANE, 1971; EDWARDS and GREENHOUSE, 1975). The anomalous layer lies in the lower crust and the estimated conductivity amounts to about 0.1 S/m. It has been pointed out that the layer might be related to the Appalachian system and the enhancement of conductivity would be due to hydration process associated with tectonic subsidence (COCHRANE and HYNDMAN, 1974). In some portions, the anomalous layer may be associated with a fracture zone or high temperature as inferred from its good correlation with an aeromagnetic anomaly or high heat flow (BAILEY *et al.*, 1978).

In the Canadian Arctic, it has been noticed that magnetograms at the Alert and Mould Bay Magnetic Observatories (see Table 1.1) are something unusual, probably showing anomalies of geomagnetic variation. Hence systematic studies of these anomalies have started and temporary observations have been carried out in and around the anomalous regions (e.g. WHITHAM, 1965; NIBLETT and WHITHAM, 1970).

The Alert anomaly, originally characterized by the NW-SE polarization of the horizontal magnetic field, was studied over Ellesmere Island and the path of electric currents causing the anomaly could be traced over a distance of 500 km or so (NIBLETT and WHITHAM, 1970). NIBLETT *et al.* (1974) carried out observations on the ice in the Lincoln Sea and found that the current path extends 100 km or more further to the northeast from Alert. The narrow conducting zone providing a conduction path for induced electric currents has been interpreted as hydration in the lower crust because no evidence showing high temperature in the anomalous area has so far been reported.

The Mould Bay anomaly, which is famous for its typical high-cut ΔZ type, was studied in detail by DELAURIER *et al.* (1974). The anomalous area extends over distances of 620 km and 290 km in the north-south and east-west directions, respectively. An unusually high value, 1 S/m or higher, is required for crustal conductivity to account for attenuation of short-period variation in the Z component. Heat flow is fairly low in this region, implying that temperature must be rather low in the crust. Consequently it is unlikely that such a high conductivity is of high-temperature origin. The cause of the Mould Bay anomaly has not yet been clearly understood.

The distribution of induction arrows, which is rather complicated because of the perturbation of induced currents flowing in the sea by irregular coastlines, shows no clear anomaly at the edge of the crustal conducting layer, whereas, in theory, a directional variation type anomaly should appear at the edge of a high-cut ΔZ type anomaly (UYEDA and RIKITAKE, 1970). In order to study those problems as yet unsolved, more observations have been carried out in regions surrounding the anomalous area (E. R. NIBLETT, personal communication, 1976).

(2) Eurasia

In Scotland, a geomagnetic variation anomaly characterized by the high gradient of ΔZ, including polarity reversal, was found as a result of an array observation (EDWARDS *et al.*, 1971; BAILEY and EDWARDS, 1976). This anomaly has been called the Eskdalemuir anomaly. It seems to be ascribed to a crustal conducting layer. Since the anomalous structure is correlated with the Iapetus suture, the conducting zone was interpreted as corresponding to an old oceanic crust (BAILEY and EDWARDS, 1976).

More intensive observations of geomagnetic and telluric variation were carried out in the Eskdalemuir anomaly region (JONES and HUTTON, 1979a, b; HUTTON and JONES, 1980). The distribution of anomalous Z variations turned out to be fairly complicated, although the Eskdalemuir anomaly was clearly recognized. On the other hand, MT data were mostly used to estimate the underground conductivity. It was then concluded that the Eskdalemuir anomaly is due to a conducting zone existing at a depth greater than 25 km; it is a mantle anomaly rather than the crustal anomaly proposed previously (HUTTON and JONES, 1980). In addition, crustal conductors seem to exist on both sides of the anomalous zone.

In north Germany, studies of geomagnetic variation anomaly were initiated by Bartels and his co-workers (e.g. BARTELS, 1954) and an anomaly characterized by polarity reversal in the Z component has been found; this is the north Germany anomaly (SCHMUCKER, 1959). It was then found that the anomaly extends further eastwards to Poland as shown by the distribution of Wiese vector in Fig. 12.21 (UNTIEDT , 1970). This anomaly is now well known as the North German - Polish anomaly and has been interpreted mainly in

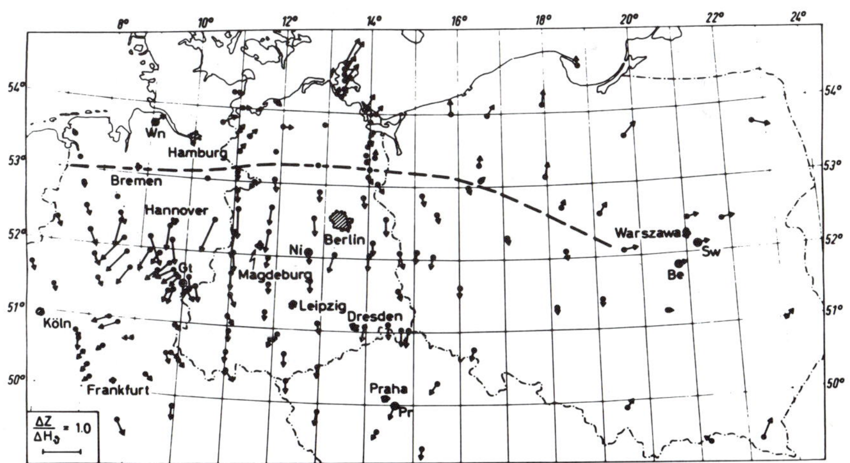

FIG. 12.21. Distribution of Wiese vectors showing the North German—Polish anomaly. Wiese vectors almost vanish along the dashed line. (UNTIEDT, 1970)

terms of conduction currents flowing in thick sediments. In fact, the anomalous region is well correlated with the sedimentary layer called the North German basin (RITTER, 1975).

Another local anomaly has been found along the Rhine graben. According to the results of analyses of geomagnetic variation anomaly and magnetotellurics, a highly conducting layer having a conductivity of 0.02 S/m lies at a depth range from 25 to 45 km and another layer of 0.04 S/m exists at 75 to 95 km (REITMAYR, 1975). The upper layer may correspond to a low-velocity zone in the lower crust and upper mantle as discussed by GARLAND (1975).

On the other hand, BABOUR and MOSNIER (1980) claim that the geomagnetic variation anomaly in the Rhine graben can be accounted for by conduction currents flowing in the sedimentary layer only and consequently no deep source is required. Their argument is based on simultaneous measurements of geomagnetic variation at a depth of 650 m and at the surface.

In central Europe, studies of geomagnetic variation anomaly and magnetotellurics have been extensively made and some anomalies have been brought to light (e.g. RITTER, 1975; ROKITYANSKY, 1980). One of the most striking anomalies would be the one found in the Carpathian region. A conducting layer seems to be required to account for the observed anomaly as represented by the distribution of induction arrows. The layer is supposed to lie in the lower crust and upper mantle, while its conductivity is estimated as 0.02 S/m or so (JANKOWSKI *et al.*, 1977).

CA studies in the U.S.S.R. are mostly based on MT, and the results in various parts of the U.S.S.R. are presented in many articles included in Chapter 5 of a monograph edited by ÁDÁM (1976). This chapter also includes some articles concerning CA studies in regions such as North Germany, Poland, Czechoslavakia, the Panovian basin, and the East Carpathians. Among many anomalies in the U.S.S.R., two notable ones will be described in the following.

In the Baikal Rift zone, eastern Siberia, a conducting layer was found at a depth range of 12 $\sim$ 25 km in the crust. Another conducting layer exists at a depth range of 60 $\sim$ 100 km. The deeper layer was interpreted as indicating high temperature, while the shallow one was supposed to be due to conducting water supplied by dehydration processes (POSPEEV and MIKHALEVSKY, 1976). On the other hand, BERDICHEVSKY *et al.* (1972) interpreted the crustal conducting layer in terms of molten granitic magma extracted during the process of metamorphism.

Highly conducting layers also exist at depth ranges of 40 $\sim$ 60 km and 100 $\sim$ 130 km, respectively, beneath the South-Caspian Depression

(AVAGINOV *et al.*, 1976). Although the resolution of the conductivity structure is rather poor, the conducting layers are supposed to be due to partial melting of mantle material.

In southern India, CA studies have been actively undertaken (SRIVASTAVA *et al.*, 1974; SRIVASTAVA and ABBAS, 1980). Conductivity anomalies are supposed to exist, although geomagnetic variations in this region are likely to be greatly affected by the sea effect.

CA studies have also been undertaken in China and conductivity anomalies are reported in the coastal area east of Beijing (CHEN, 1974) and in the Kansu Province (XU *et al.*, 1978).

(3) Africa

Two-dimensional array studies revealed notable anomalies of short-period geomagnetic variation in South Africa in addition to the coast effect there (GOUGH *et al.*, 1973; DE BEER and GOUGH, 1980). It seems most likely that an anomalous conductor is located in the crust or upper-most mantle, as inferred from the distribution of anomalous horizontal and vertical components. This anomalous conductor coincides well with the Cape Fold Belt along which a static magnetic anomaly is observed. The combination of conductivity and magnetic anomalies suggests the existence of an oceanic crustal layer which is usually highly magnetized (see Subsection 6.2.1). The layer is also likely to contain highly conducting sea-water in its pores, resulting in a high bulk conductivity of the layer.

The usual interpretation of an anomalous conductor in terms of high temperature, particularly combined with partial melting of mantle material, seems inappropriate in the above case, since heat flow is not high in the Cape Fold Belt (DE BEER and GOUGH, 1980). Even though heat flow is low, the source of the magnetic anomaly would be located in the crust; in the mantle, temperature is likely to exceed the Curie point and strong magnetization cannot be expected. It is concluded, therefore, that the conducting layer lies in the crust rather than in the mantle.

Another array observation of geomagnetic variation was undertaken in South-West Africa, Botswana, and northwestern Rhodesia (DE BEER *et al.*, 1976), resulting in the discovery of another conductivity anomaly. As shown in Fig. 12.22, the anomalous conductor, presumably located in the crust, elongates in the east-west direction in central South-West Africa. It then extends eastwards to Botswana, where location of the conductor is not well identified, presumably because it splits into branches. The conducting zone can then be traced northeastwards to Rhodesia and probably further to the Rift Valley of East Africa.

(4) Australia

Two-dimensional array studies were undertaken in the southeastern and

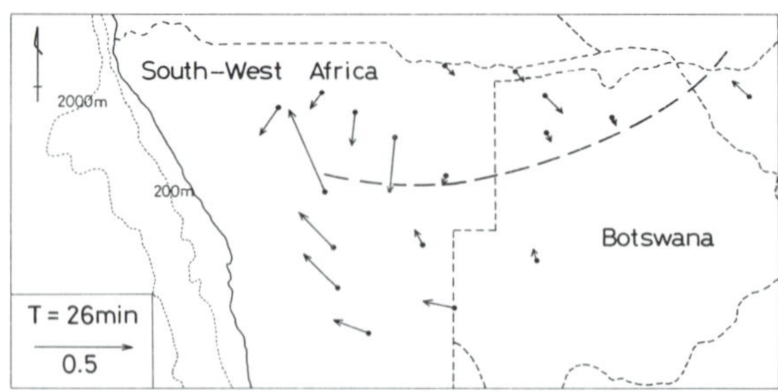

FIG. 12.22. Distribution of induction arrows for a period of 26 min in the south-western part of Africa. The dashed line indicates an approximate elongation of an anomalous conductor. (DE BEER *et al.*, 1976)

central parts of Australia as intensively as in the western part of North America (LILLEY and BENNETT, 1972; GOUGH *et al.*, 1974; LILLEY, 1976; WOODS and LILLEY, 1979, 1980). Contrary to the expectation that no striking conductivity anomaly would exist in a stable shield region such as the Australian Continent, the existence of some highly conducting layers has been confirmed from the observed geomagnetic variation and magnetotelluric data.

In southeastern Australia, conductivity anomalies were found in the Otway Ranges and the Adelaide Geosyncline (LILLEY and TAMMEMAGI, 1972; GOUGH *et al.*, 1974; LILLEY, 1976). The anomalous conductors seem to lie in the lower crust or upper mantle. In view of high seismicity in both regions, the conductors are supposed to be related to shear zones associated with tectonics in Australia or local weak zones, presumably containing molten magma.

In the central part of Australia a local geomagnetic variation anomaly was also found which could be interpreted as being due to a conducting sedimentary layer in the crust (WOODS and LILLEY, 1979). Apart from this anomaly, geomagnetic variation data could be used to estimate the deep conductivity structure beneath the Australian Continent. As a result it was found that a mantle conducting layer exists at a depth of 300 ~ 400 km and its conductivity amounts to 0.1 ~ 1.0 S/m. This result is in quite good agreement with the global conductivity structure as described in Chapter 10.

A more detailed analysis was performed for magnetic quiet daily variations recorded simultaneously at array sites in the central and southeastern parts of Australia (LILLEY *et al.*, 1981a, b). The method of analysis is as follows. A parameter c is defined as

$$c=\Delta Z/(\partial\Delta X/\partial x+\partial\Delta Y/\partial y) \tag{12.13}$$

where ΔZ, ΔX, and ΔY are the downward, northward, and eastward components of geomagnetic variation, respectively. For a laterally uniform structure, it is easily shown that the apparent resistivity ρ_a is derived from c as

$$\rho_a = 0.8\pi^2 |c|^2 / T. \tag{12.14}$$

The unit of ρ_a is given in ohm · m, if T and c are measured in seconds and kilometers, respectively.

The results of the above analysis show that, in central Australia, the conductivity rises sharply at 500 km depth. This would be considered as a normal structure beneath the shield area, since a similar structure has been found beneath the Baltic Shield and the East European Platform (e.g. VANYAN, 1981). On the other hand, a mantle conducting layer appears at 200 ~ 300 km depth in southeastern Australia. LILLEY *et al.* (1981b) interprets the existence of a conducting layer at rather shallow depth as implying a partial melting of 5% or so below the above depth. As will be seen in the next section, it is rather usual for a highly conducting layer to exist at a depth of 100 km or so beneath oceans. In most cases, such a conducting layer has been interpreted in terms of partial melting. In this sense, the conductivity structure beneath southeastern Australia may reflect a transition zone between the ocean and the shield.

12.5 Conductivity Anomalies Beneath Oceans

12.5.1 Ocean effects and the mantle conductivity structure

Ocean effects on geomagnetic variations, such as the island, coast, and peninsula effects, contain some information on the mantle conductivity structure, as described in Section 12.2. Therefore, detailed studies of such effects would be significant in investigating the conductivity structure beneath oceans, although not all the oceanic areas can be covered by such a method. In fact, information on the conductivity structure beneath the Philippine Sea, which is a representative marginal sea in th western Pacific, has been obtained from studies of the above effects, as will be described in the following.

The response function derived from the island effect observed on Miyake-jima and Hachijo-jima Islands made it possible to propose a model of the conductivity structure beneath the Izu-Bonin arc to which both the islands belong. The proposed model is characterized by a layer having a conductivity of 0.01 S/m or so and existing at the depth range of 30 to 100 km (HONKURA, 1972; HONKURA *et al.*, 1974). Similarly, a conductivity model for the West Philippine Basin was derived from the island effect observed on Minami-daito Island; its location is shown in Fig. 12.23. In this area, a highly conducting

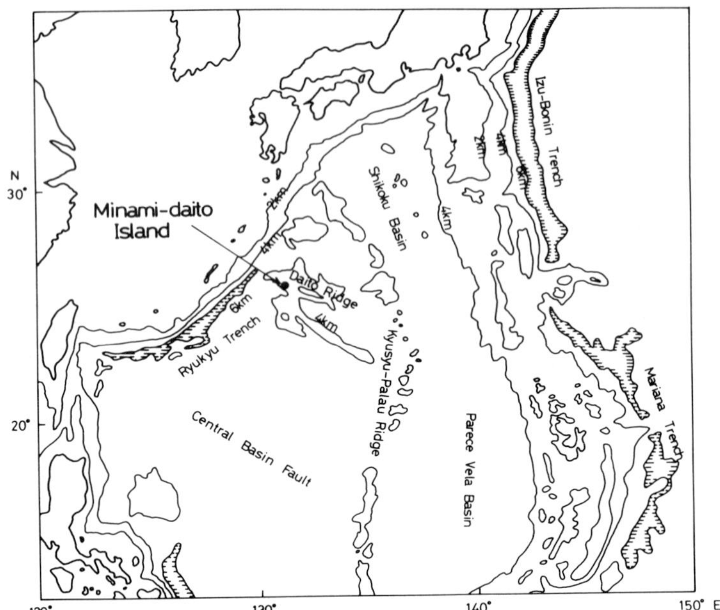

FIG. 12.23. Location of Minami-daito Island and some tectonic features in the Philippine Sea.

layer having a conductivity of 0.5 S/m or thereabouts seems to exist at a depth of 100 km or so (HONKURA *et al.*, 1981).

The geomagnetic variation anomaly found in central Japan has been interpreted as an indication of the existence of a conducting layer beneath the northern-most portion of the Shikoku Basin (RIKITAKE, 1969, 1975; HONKURA, 1974). A detailed study of the peninsula effect observed at the Kii Peninsula provided additional information on the conductivity structure. As a result, it was concluded that a conducting layer exists at a depth of 50 km and its conductivity amounts to 0.1 S/m or so.

The conductivity models for the areas studied in the Philippine Sea are summarized in Fig. 12.24. The age of seafloor is estimated as 20 m.y. and 80 m.y. for the areas studied in the Shikoku Basin and the West Philippine Basin, respectively. There seems to be a tendency for the mantle conducting layer to be located at shallower depth beneath younger seafloor. Such a relation will be confirmed later by more information derived from seafloor observations of geomagnetic and telluric variations in the Pacific Ocean.

Similar studies were attempted for the island effect observed on Oahu and Hawaii Islands. KLEIN (1972) pointed out that no conducting layer exists between the surface and 200 km depth beneath Oahu. More detailed analyses

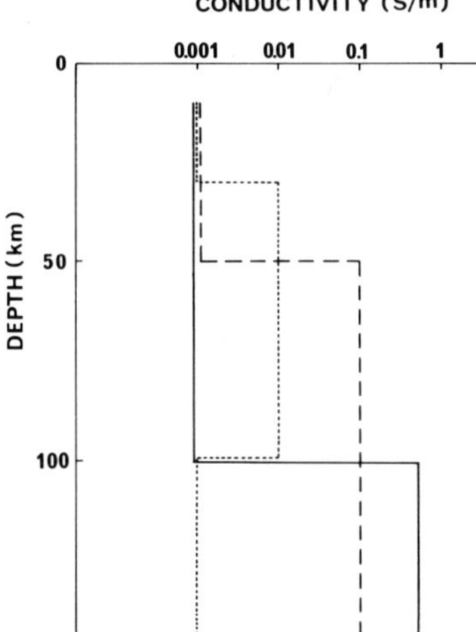

FIG. 12.24. Models of conductivity structure for the West Philippine Basin, the Shikoku Basin, and the Izu-Bonin Arc, as denoted by solid, dashed, and dotted lines, respectively.

made for the island effect observed on Hawaii disclosed that the conductivity amounts to the order of 0.01 S/m down to 200 km depth and then increases to 0.1 S/m or so (KLEIN, 1976; KLEIN and LARSEN, 1978). The above result would give a more reliable estimate for the conductivity at shallow depth than the model proposed by LARSEN (1975); in Larsen's study magnetotelluric data of long period, 4 hours to 10 days, are used for analysis, while the model of KLEIN (1976) is based on geomagnetic variations having a period range from 50 min to 12 hours.

12.5.2 Seafloor observation of geomagnetic and telluric variations

The most powerful means of investigating the mantle conductivity structure in ocean areas is certainly the observation of geomagnetic and telluric fields at the seafloor. Conditions for a magnetotelluric study are usually very good at the seafloor of a deep ocean, since difficulties in MT studies arising through surface effects, which often affect telluric fields on land to a considerable extent, can be avoided.

FILLOUX (1967) succeeded for the first time in making observations of geomagnetic and telluric fields at the seafloor off the California coast. Since then, considerable improvements have been made to devices for seafloor observation. Using several sets of seafloor instruments, FILLOUX (1977, 1980a, b, 1981) carried out extensive magnetotelluric experiments in the Northern

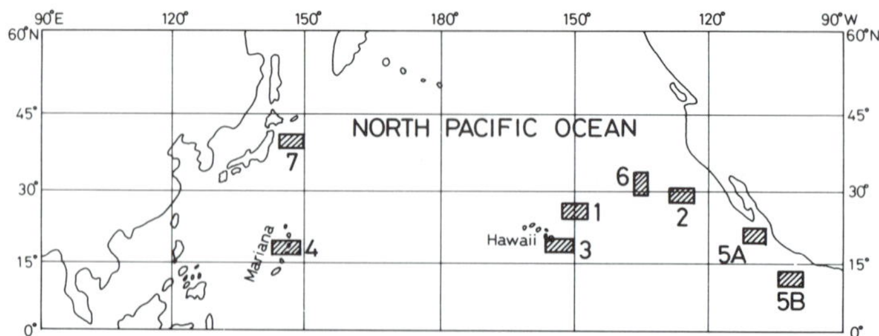

FIG. 12.25. Locations of magnetotelluric experiments in the North Pacific Ocean. (FILLOUX, 1981)

Pacific. The areas investigated are shown in Fig. 12.25. Among them, the experiment denoted by 5B was planned to investigate the conductivity structure near the East Pacific Rise, while experiments 4 and 7 were carried out, at least partly, in relation to the trench—island arc system.

The results of experiments 1, 2, and 5B provide us with significant information which may be closely related to the concepts of plate tectonics. In these areas, a conducting layer was found to exist in the upper mantle. Its conductivity does not vary much from one area to another, amounting to about 0.1 S/m or slightly less. However, it was noticed that the depth to the conducting layer varies from one area to another. Moreover, a clear pattern can be recognized in that the depth to the conducting layer tends to increase as the age of seafloor (plate age) increases, and in the case of experiment 4, no distinct conducting-layer has been found below the sea floor east of Mariana, where the plate is much older than in the other areas. The relation between the plate age and the depth to the conducting layer will be reviewed later.

In the area northeast of Hawaii, where experiment 3 was carried out (FILLOUX, 1977), several sets of instruments were deployed at the same time and another data set was used for analysis by CHAVE *et al.* (1981). Their conductivity model is slightly different from the model proposed by FILLOUX (1977), but in view of the rather poor resolution of the conductivity structure, the difference is unlikely to be significant and the models of FILLOUX (1977) and CHAVE *et al.* (1981) will be combined in later discussion about lithosphere thickening.

Seafloor observations were also carried out in the northwestern Atlantic ocean (BENNETT and FILLOUX, 1975; POEHLS and VON HERZEN, 1976; COX, 1980). Comparing geomagnetic variations observed at the seafloor with those at a nearby island, POEHLS and VON HERZEN (1976) examined attenuation in the amplitude of the horizontal component due to conducting sea-water. Since the attenuation depends on the underground conductivity structure through electromagnetic coupling between the sea and the underground conductor, analysis of amplitude attenuation may possibly provide information on the conductivity structure. In principle, non-attenuated geomagnetic variations must be observed at the sea surface. In practice, however, sea-surface observation is not possible at present and hence it was replaced by the observation on the island.

Such a treatment can be justified, in an approximate sense, only for low frequencies. Therefore, the reliability of a conductivity model would not be so good as that for magnetotellurics based on seafloor data. In any case, POEHLS and VON HERZEN (1976) put forward a model consisting of a conducting layer at a depth range 50–100 km. Its conductivity amounts to 0.1 S/m or so. The age of the seafloor in the area investigated is about 130 m.y. This result seems to be inconsistent with the relation between the plate age and the depth to the conducting layer for the North Pacific, although such a relation has not yet been generally established.

COX *et al.* (1980) reported more detailed accounts which include the results of magnetotellurics applied to seafloor data as well as the analysis of data obtained on islands. Although the proposed conductivity models are different from one method of analysis to another, a common feature can be found in all the models, as will be summarized in the following. A conducting layer about 30 km thick exists at a depth of 70 km. Below this layer lies another conducting layer at a depth of 250 km. The conductivity of these layers is slightly higher than 0.1 S/m. In later discussion about the relation between the plate age and the depth to the conducting layer, the upper conducting layer will be taken into account.

Seafloor observations of geomagnetic variation were also carried out by LAW and GREENHOUSE (1981) at the Juan de Fuca Ridge in the north-eastern Pacific. Their method of data analysis is similar to the one used by POEHLS and VON HERZEN (1976); that is, the horizontal variations at the seafloor were compared with those at the Victoria Observatory located on the west coast of Canada and amplitude attenuation due to conducting sea-water was examined. A conductivity structure model which accounts for the estimated attenuation shows a conducting layer located at a depth range from 55 to 75 km.

The seafloor is about one million years old and as young as the East Pacific Rise west of Mexico, where FILLOUX (1981) made seafloor observations. In the case of the East Pacific Rise, however, a conducting layer appeared at 20 km depth. It is not certain at present whether or not such discrepancy in the depth to the conducting layer is really significant.

12.5.3 Relationship between the plate age and the depth to the conducting layer

Information on the conductivity structure in ocean areas has accumulated as a result of seafloor observations in various regions. It now appears that the existence of a conducting layer having a conductivity of 0.1 S/m or so in the upper mantle is a common feature for major oceans. Moreover, a rather systematic variation in the depth to the conducting layer can be recognized in relation to the age of the seafloor.

Presently available data are summarized in Fig. 12.26, which shows the depth to the conducting layer plotted against the age of the seafloor for three regions; the Pacific Ocean, the Atlantic Ocean, and the Philippine Sea. The conducting layer has usually been interpreted as indicating a partial melting zone (see next section). From the viewpoint of plate tectonics, the partial melting zone lies below the lithosphere and constitutes the asthenosphere. Therefore, the conducting layer corresponds to the asthenosphere and consequently the depth to the conducting layer must be identical to the thickness of the lithosphere. In this sense, a curve representing the plate growth proposed by YOSHII (1975) is also shown in the figure.

Except in the case of the conducting layer found beneath the seafloor of 130 m.y. age in the Atlantic Ocean, the depth to the conducting layer tends to be greater than expected from the plate-growth curve. An attempt to explain such a tendency will be made in the next section. Alternatively, it is certainly possible to ascribe the discrepancy to poor resolution in determination of the depth to the conducting layer.

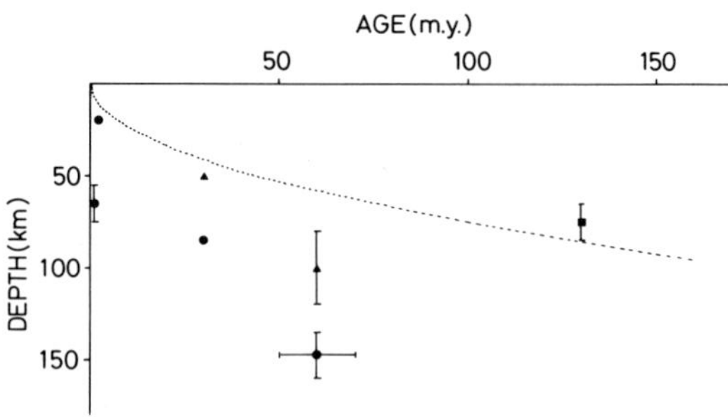

FIG. 12.26. The depth to the conducting layer plotted against the seafloor age. Circles, squares, and triangles indicate the data for the Pacific Ocean, the Atlantic Ocean, and the Philippine Sea, respectively. The dotted line shows a model of plate growth proposed by YOSHII (1975).

12.6 Geophysical Implication of Conductivity Anomaly

The electrical conductivity of the earth's constituents depends strongly on the temperature (see Section 10.9). Therefore, high temperature is naturally considered as a cause of high conductivity. Reliable experimental results for olivine show that the temperature must exceed at least 1200°C to yield a conductivity as high as 0.1 S/m. However, at such a high temperature mantle material is subject to partial melting (KUSHIRO *et al.*, 1968). Once partial melting occurs, basaltic magma comes into existence and the bulk conductivity is controlled mostly by molten basalt. It is concluded, therefore, that the conductivity value of the order of 0.1 S/m corresponds to that of material containing molten basaltic magma.

The conductivity of molten basalt is much higher, at least two orders of magnitude, than that of olivine at the same temperature (WATANABE, 1970; PRESNALL *et al.*, 1972; WAFF and WEILL, 1975; RAI and MANGHNANI, 1978). For instance, the conductivity value of 0.1 S/m is possible at 1500°C or higher for olivine (DUBA *et al.*, 1974), whereas the conductivity of molten basalt exceeds 0.1 S/m at 1200°C (RAI and MANGHNANI, 1978).

It should be pointed out here that inclusion of molten magma does not necessarily imply high conductivity. It is well known that the bulk conductivity depends strongly on the degree of interconnection of the liquid phase. Figure 12.27 shows two extreme cases; in (a) the liquid phase exists in isolated pockets, while in (b) it is interconnected to each other through narrow passages. The bulk conductivity is expressed for case (a) as

$$\sigma = \sigma_s\{\sigma_m + 2\sigma_s + 2f(\sigma_m - \sigma_s)\}/\{\sigma_m + 2\sigma_s - f(\sigma_m - \sigma_s)\} \quad (12.15)$$

and for case (b) as

$$\sigma = \sigma_m\{3\sigma_s - 2f(\sigma_s - \sigma_m)\}/\{3\sigma_m + f(\sigma_s - \sigma_m)\} \quad (12.16)$$

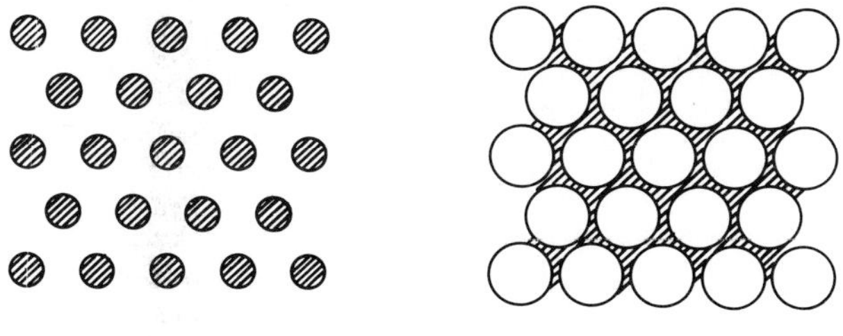

(a) (b)

FIG. 12.27. Two models of partial melting. Shaded zones indicate liquid phase. (HONKURA, 1975)

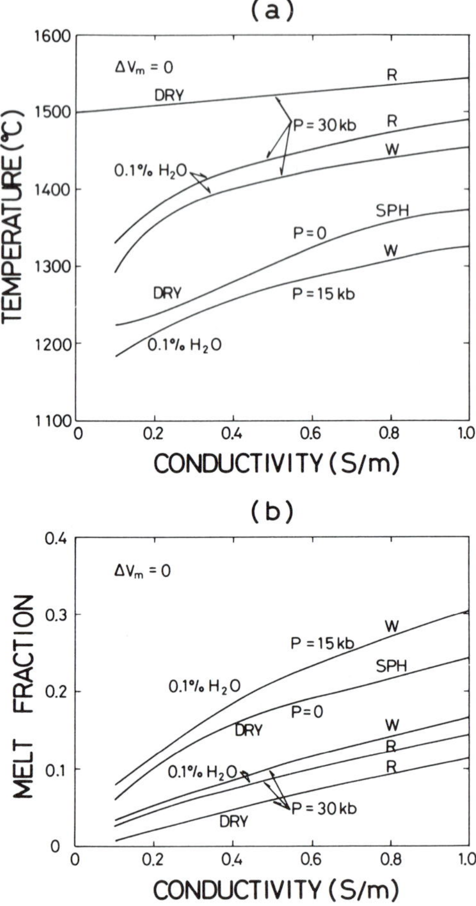

FIG. 12.28. The relationships (a) between temperature and conductivity and (b) between melt fraction and conductivity for various mantle models denoted by R, W, and SPH. ΔV_m and P denote the activation volume and pressure, respectively. (SHANKLAND and WAFF, 1977)

where σ_m and σ_s denote the conductivity of liquid and solid phases, respectively, and f is the melt fraction (WAFF, 1974; HONKURA, 1975). The above expressions correspond to the lower and upper bounds, respectively, imposed on the bulk conductivity (HASHIN and SHTRIKMAN, 1962).

SHANKLAND and WAFF (1977) extended this discussion and pointed out that the melt fraction and temperature can be determined independently if the case (b) in Fig. 12.27 is assumed to be applicable to partial melting zones in the mantle. Figure 12.28 shows the relationships between the temperature and conductivity and between the melt fraction and conductivity for some mantle models and pressures of 0, 15, and 30 kb, respectively. It must be remembered, however, that this argument holds only if the case (b) in Fig. 12.27 reflects the circumstances in the earth's mantle.

The above diagram of SHANKLAND and WAFF (1977) was used by OLDENBURG (1981) to interpret the conducting layers found beneath the Juan de Fuca Ridge, the 30 m.y. old seafloor off the California coast, and the 72 m.y. old seafloor northeast of Hawaii. Oldenburg claims that a partial melting zone, derived from CA studies, coincides closely with a zone of low S-wave velocity for the various regions. Low velocity zones are usually considered to correspond to the asthenosphere. Consequently, this supports the proposition that the conducting layer indicates the asthenosphere.

In the meantime, the possibility of interconnection of the liquid phase was investigated theoretically and also experimentally by WAFF and BULAU (1979) and BULAU *et al.* (1979). They examined theoretically a condition under which the state of interconnection is realized. Their experimental result clearly shows that the condition imposed theoretically is satisfied and the state of interconnection is indeed possible. These discussions support the applicability of the diagram of SHANKLAND and WAFF (1977) to the earth's mantle.

WAFF (1980a, b) further argues that the gravity effect has an important bearing on the state of interconnection. According to his argument, at a certain height from the base of the partial melting zone, the condition necessary for the state of interconnection will be broken because of an increase in the surface tension of the liquid due to buoyancy forces and passages will be closed, leaving the liquid phase pockets isolated from each other. Once the state of interconnection is broken, buoyancy forces are no longer transmitted upwards and the condition necessary for interconnection will be recovered. This series of processes will be realized along a vertical direction from the base of a partial melting zone to its top. In such a stratified structure, the electrical conductivity must be strongly anisotropic. Such a structure will also have an important bearing on magma transport problems, although they are beyond the scope of this book.

Now we return to the problem concerning the dependence of the lithospheric thickness on its age. As pointed out by SHANKLAND *et al.* (1981), only a few percent melt fraction is sufficient to significantly reduce seismic velocity, provided that voids containing liquid are of flattened shape, while several percent is required to increase the bulk conductivity to the order of 0.1 S/m at a temperature of 1200°C or so. A clue to interpreting the discrepancy between seismic and electromagnetic estimates of the thickness of the lithosphere is suggested in the above discussion. That is to say, the discrepancy may possibly be ascribed to the depth dependence of the degree of partial melting. At the onset of partial melting, the melt fraction would be small yet enough to reduce the seismic velocity, whereas the conductivity remains unchanged because of insufficient interconnection. At greater depth, the partial melting process will develop and enhance the melt fraction, resulting in effective interconnection of liquid phase.

As a final remark in this section, it should be pointed out that the order-disorder mechanism at a temperature slightly below the melting temperature may be an alternative cause of high conductivity, as experimentally demonstrated for an albite sample by PIWINSKII and DUBA (1974). It should also be noted that TOZER (1979) attributes high conductivity to conducting water associated with amphibole dehydration. If this is the case, the temperature at the conducting layer is near 600°C and much lower than the solidus of mantle material.

12.7 Time-Dependent Conductivity Anomaly

12.7.1 Experimental studies of pressure dependence of the conductivity

Crustal conductivity is controlled mainly by saline water contained in cracks and pores. Since these voids would be affected by pressure, the conductivity may change in association with changes in tectonic stress. In fact, BRACE *et al.* (1965) and BRACE and ORANGE (1968b) showed experimentally that the conductivity of rock samples depends on pressure through changes in water content in cracks and pores. The pattern of change is controlled by the degree of water saturation in cavities.

Moreover, many rocks exhibit a phenomenon called dilatancy at one-third or two-thirds of fracture stress (BRACE *et al.*, 1966). Such dilatancy is a result of creation of new cracks at high stress. At the stage of dilatancy, the conductivity is expected to increase, provided that water is available to saturate newly created cracks. The experiment of BRACE and ORANGE (1968a) clearly demonstrated changes in conductivity, by as much as one order of magnitude in some cases, in association with dilatancy.

The above micro-crack dilatancy is a characteristic of intact rocks and it occurs repeatedly after many cycles of stress to within 60 to 90% of the fracture stress (BRACE, 1977). Therefore, the crustal conductivity is expected to change before rock fracture resulting in the occurrence of an earthquake, as pointed out by SCHOLZ *et al.* (1973).

In view of repeated occurrences of earthquakes along an active fault, a mechanism of stick-slip seems to be more important than dilatancy. WANG *et al.* (1975) showed that the conductivity also changes in association with stick-slip occurring along the pre-existing fault, although the amount of change is much smaller than that associated with dilatancy of an intact rock. Thus it would not be unusual for the crustal conductivity to change in association with the accumulation of tectonic stress which usually results in occurrences of earthquakes.

12.7.2 Crustal conductivity changes associated with earthquake occurrences

As one method of earthquake prediction in the Garm area, U.S.S.R., artificial electric currents have been driven into the ground through a pair of electrodes about one kilometer apart and the electric potential due to the currents has been measured to determine the apparent resistivity of the crust. Such measurements, called the dipole-dipole method, have been carried out repeatedly and a time-dependence of crustal conductivity has been derived as shown in Fig. 12.29 (BARSUKOV, 1974). The observed changes in apparent resistivity estimates correlate well with occurrences of earthquakes of considerable magnitude. Moreover, these changes seem to precede earthquake occurrences.

Measurements of apparent resistivity similar to those in the Garm area have also been carried out in the San Andreas fault area, U.S.A. (MAZZELLA and MORRISON, 1974). On one occasion a change amounting to more than 10% at its maximum was observed and after this anomaly, which lasted for about 60 days, an earthquake of magnitude 3.9 occurred in the measurement area. This anomaly in resistivity was interpreted as indicating a crustal conductivity change preceding the earthquake occurrence.

In China, apparent resistivity measurements based on artificial current sources have been undertaken so actively that many examples of apparent resistivity changes associated with earthquakes have been reported (e.g. HONKURA, 1981); some of them seem to have contributed to practical

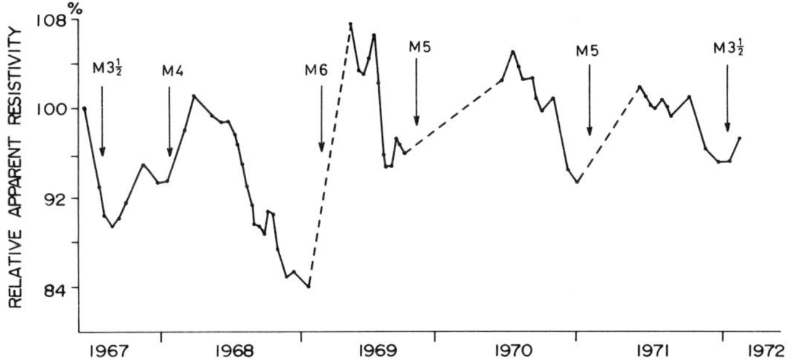

FIG. 12.29. Changes in apparent resistivity in percent observed at Garm, U.S.S.R. Arrows indicate the occurrences of earthquakes with their respective magnitudes. (BARSUKOV, 1974)

prediction of some earthquake occurrences. In the case of the destructive Tangshan earthquake of magnitude 7.8 that occurred in 1978, distinct changes appeared prior to its occurrence (ZHAO and QIAN, 1978).

All the reports mentioned above provide clear evidence that the crustal conductivity is a physical parameter which is subject to changes, particularly in relation to tectonic stress changes as manifested in earthquake occurrences. It is therefore expected that electromagnetic response of the crust will also be time-dependent, and studies of the time-dependent nature of electromagnetic induction parameters may turn out to be a useful means of earthquake prediction.

12.7.3 Time-dependence of electromagnetic induction parameters

(1) changes in transfer functions

YANAGIHARA (1972) reported that transfer functions characterizing an anomaly of short-period geomagnetic variation at the Kakioka Magnetic Observatory, Japan, have undergone a marked secular change since 1900. For instance, transfer function A decreased until around 1920, then increased to a maximum at about 1940, and has shown a decreasing trend thereafter. The difference between the maximum and minimum in A-value amounts to as much as 0.2. If such a large change in A-value is really of sub-surface origin, a large conductivity change must have occurred in the crust or upper mantle. Since there is no evidence to support an external origin for such a change in transfer function A, YANAGIHARA (1972) attributed the change in A to a conductivity change associated with the Kanto earthquake of magnitude 7.9 which occurred in 1923.

Yanagihara's report is based on a simple method of analysis for ssc's (sudden storm commencement), si's (sudden impulse), and similar variations of short period. A more detailed study based on the method of spectral analysis confirms in general the result of YANAGIHARA (1972); that is, the time-dependence of Au (in-phase part of transfer function A) agrees well with that of the A-value reported previously (SHIRAKI and YANAGIHARA, 1977).

In addition to the long-term change described above, short-term changes in transfer functions have also been observed at Kakioka. YANAGIHARA and NAGANO (1976) pointed out a good correlation between changes in transfer functions and occurrences of earthquakes of magnitude greater than 5 in the vicinity of Kakioka. In view of these interesting results, the time-dependence of transfer functions has been monitored in great detail at Kakioka, as shown in Fig. 12.30 (SHIRAKI, 1980).

Similar studies of the time-dependence of transfer functions were undertaken with special reference to the 1966 Tashkent earthquake of magnitude 5.5 (MIYAKOSHI, 1975) and the 1972 Sitka earthquake of magni-

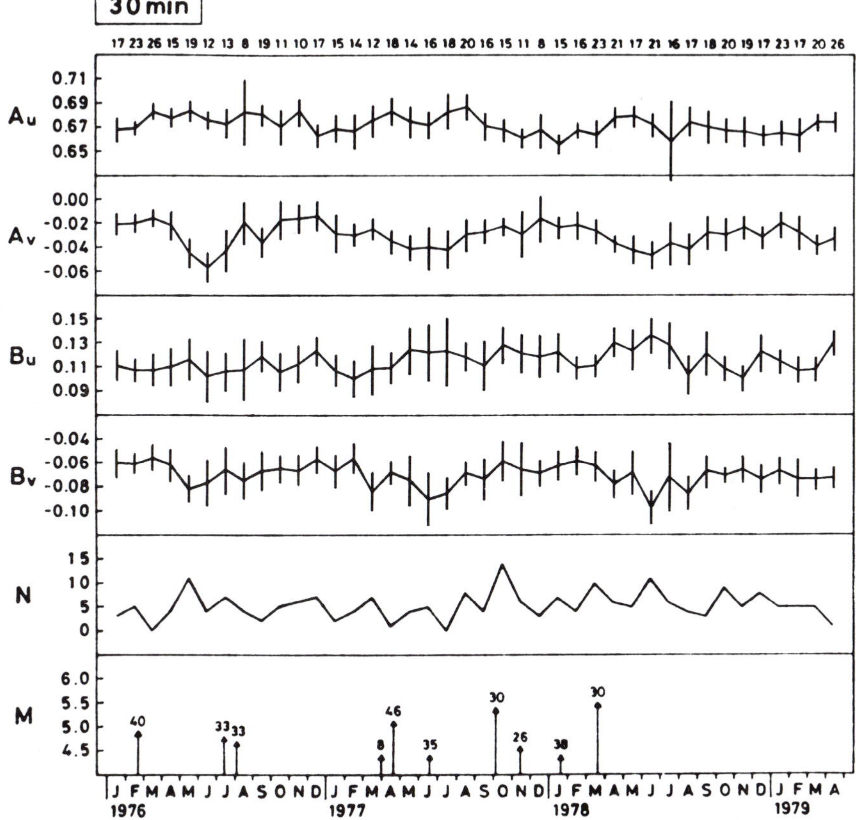

FIG. 12.30. Secular changes in transfer functions at the Kakioka Magnetic Observatory. *N* denotes the number of data sets used for monthly analysis. *M* indicates the magnitude of earthquakes shown by arrows. Numerals attached to the arrows denote the epicentral distances in kilometers for the respective earthquakes. (SHIRAKI, 1980)

tude 7.2 (RIKITAKE, 1979). Data used for analysis are geomagnetic variations of short period recorded at the Tashkent Observatory in Central Asia and at the Sitka Observatory, Alaska. In both cases, it was reported that precursory changes appeared about one year before the earthquake ocurrence.

There are some other reports on the time-dependence of transfer functions associated with seismic activity. Most of them can be found in review articles by NIBLETT and HONKURA (1980) and HONKURA (1981).

(2) Changes in the amplitude of horizontal variation

It is well known that the amplitude of the horizontal component of short-period geomagnetic variations is enhanced over a conducting body as a result of the concentration of induced electric currents. It is useful, therefore, to monitor the amplitude of the horizontal component as a means of detecting an underground conductivity change. An amplitude change can be monitored by calculating the ratio of the amplitude at a station of interest to that at another remote station, used as the reference station. If the separation between the two stations is much less than the scale length of the external source field, the amplitude of the external field can be regarded as the same at both stations, and changes in the amplitude ratio, if any, can be attributed to those of internal origin.

HONKURA and KOYAMA (1978) used this method to examine whether or not any changes in crustal conductivity can be detected in the Izu Peninsula, Japan, where anomalous crustal uplift and high microearthquake activity have been observed. As the remote reference station, the Yatsugatake Magnetic Observatory, which is about 140 km distant from the site in the Izu Peninsula, was chosen and the amplitude ratio has been determined every half a month for the *H* and *D* components, respectively, of short-period geomagnetic variations.

During the observation period, two major earthquakes occurred within 30 km of the station: *M* 7.0 and *M* 6.7 events in 1978 and 1980, respectively. In addition to these, many earthquakes of small magnitude have taken place. As shown in Fig. 12.31, some changes exceeding the error estimates have been

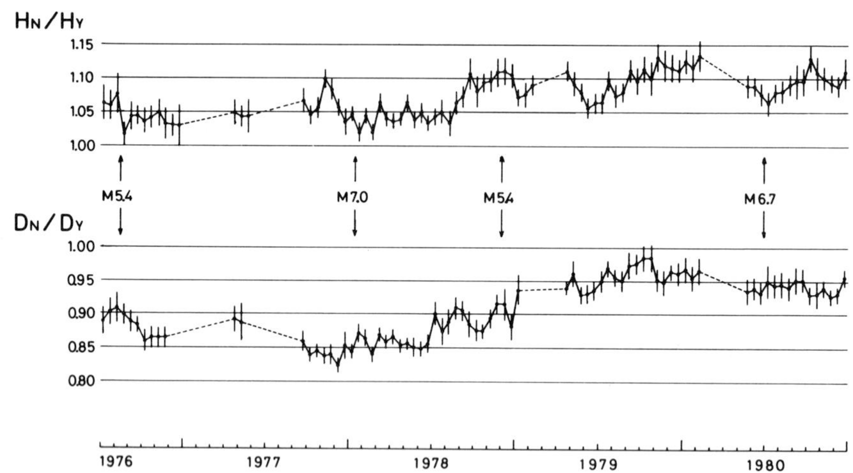

FIG. 12.31. Secular changes in the amplitude ratio for the *H* and *D* components, respectively, as observed at an observation site in the Izu Peninsula. Arrows indicate occurrences of earthquakes and their respective magnitudes.

observed, but it seems difficult to form any difinite conclusion concerning the relationship between changes in the amplitude ratio and seismic events.

Unusual crustal uplift has also persisted near the observation site. Since dilatancy is considered to be one of the likely causes of the crustal uplift, the time-dependence of the amplitude ratio was compared with time-dependent nature of crustal uplift. Figure 12.32 shows leveling data at bench marks BM 9341 and BM 003–012 near the observation site, as well as the difference in sea level between a tide station (Ito) located in the uplift area and a remote reference station (Aburatsubo). A remarkable correspondence can be found between the uplift, particularly as derived from the sea level data, and the amplitude ratio, particularly for the *D* component.

(3) Changes in magnetotelluric parameters

In a seismically active region northeast of Quebec City, Canada, magnetotelluric observations have been carried out in an attempt to detect a

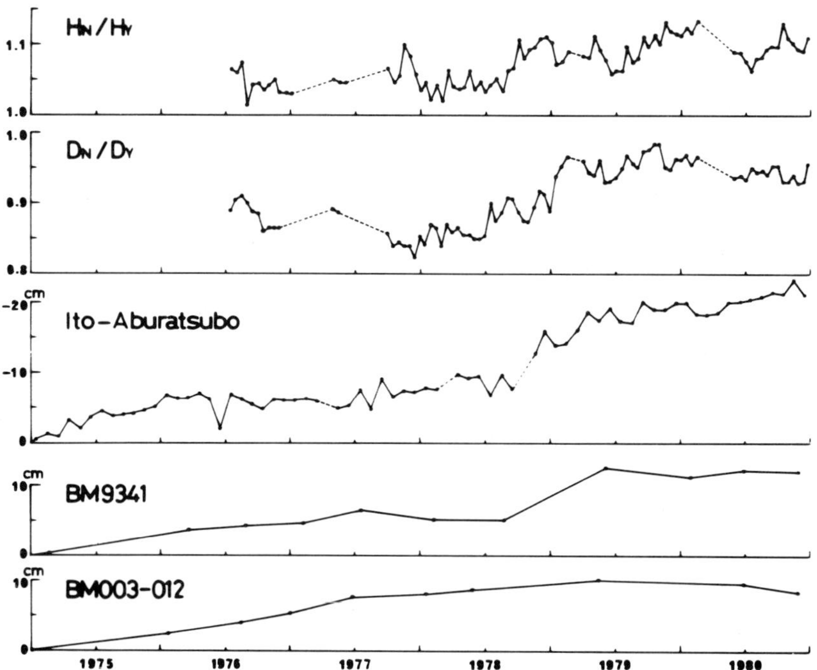

FIG. 12.32. Secular changes in the amplitude ratio for the *H* and *D* components, respectively, and time-dependent crustal uplift as represented by the difference in sea level between Ito and Aburatsubo and also by leveling data at bench marks BM9341 and BM003–012 in the Izu Peninsula.

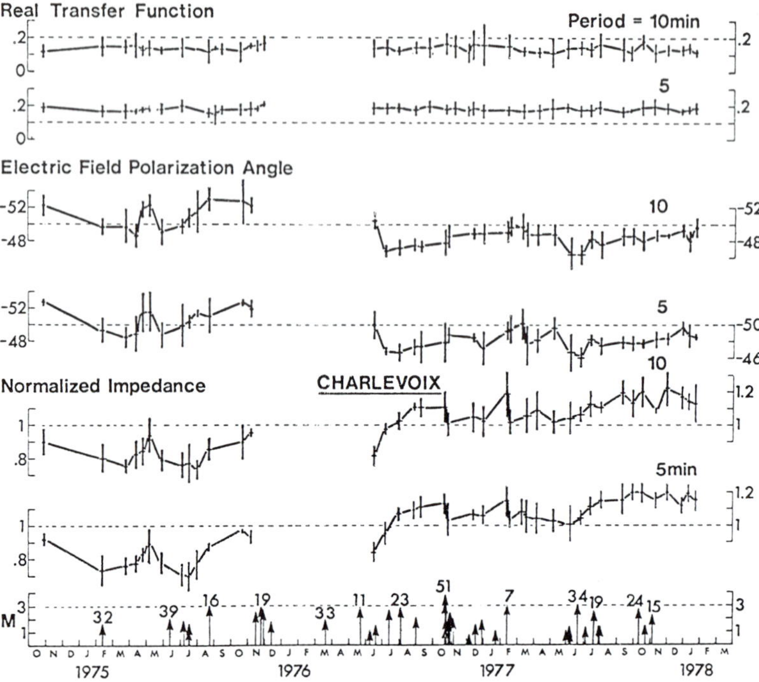

FIG. 12.33. Secular changes in the in-phase transfer function, the angle of electric field polarization, and the normalized impedance, respectively, as observed at Charlevoix. M denotes the magnitude of earthquakes shown by arrows. Numerals attached to the arrows denote the epicentral distances in kilometers for the respective earthquakes. (NIBLETT and HONKURA, 1980)

possible change in the impedance tensor reflecting a crustal conductivity change (HONKURA *et al.*, 1976; KURTZ and NIBLETT, 1978). The results obtained at one of the sites, Charlevoix, are shown in Fig. 12.33 for principal impedance and electric field polarization, and in-phase transfer function ($\sqrt{Au^2 + Bu^2}$ in this case) at periods of 5 and 10 min, respectively (NIBLETT and HONKURA, 1980). Changes exceeding error estimates could be detected in the normalized impedance and electric field polarization. However, no large earthquake has occurred near the Charlevoix station. KURTZ and NIBLETT (1978) suggest that the above changes may be an indication of a change in crustal conductivity in association with a possible accumulation of strain.

Similar magnetotelluric observations were carried out in southern

California (Reddy *et al.*, 1976; Phillips *et al.*, 1977). In this case, high frequency components, $0.03 \sim 0.04$ Hz and $7 \sim 8$ Hz bands, were used to examine the time-dependence of crustal conductivity. Some changes exceeding the uncertainty in the impedance estimates were reported, but their association with tectonic events is not clear.

REFERENCES

ACUÑA, M. H. and N. F. NESS, The main magnetic field of Jupiter, *J. Geophys. Res.*, **81**, 2917–2922, 1976.

ACUÑA, M. H. and N. F. NESS, The magnetic field of Saturn: Pioneer 11 observations, *Science*, **207**, 444–446, 1980.

ÁDÁM, A, *Geoelectric and Geothermal Studies*, 752 pp., Akadémiai Kiadó, Budapest, 1976.

ADAMS, J. C., The theory of terrestrial magnetism, giving the expressions of the magnetic forces on the earth's surface, the earth being regarded as a sphere, *Sci. Papers, Cambridge Univ. Press*, **2**, 401–640, 1900.

ADAMS, W. G., An account of the late Professor John Couch Adams' determination of the Gaussian magnetic constants, Brit. Assoc. Rept. Brestol Meeting, 1898.

AFANASIEVA, W. I., The spherical harmonic analysis of the earth's magnetic field for the epoch 1945, *Terr. Mag. Atmos. Electr.*, **51**, 19–30, 1946.

AKIMOTO, S., The system MgO–FeO–SiO_2 at high pressures and temperatures—phase equilibria and elastic properties, Tectonophysics, **13**, 161–187, 1972.

AKIMOTO, S. and H. FUJISAWA, Demonstration of the electric conductivity jump produced by the olivine-spinel transition, *J. Geophys. Res.*, **70**, 443–449, 1965.

ALABI, A. O., P. A. CAMFIELD, and D. I. GOUGH, The North American Central Plains conductivity anomaly, *Geophys. J. R. astr. Soc.*, **43**, 815–833, 1975.

ALDRICH, L. T., M. CASAVERDE, R. SALGUEIRO, J. BANNISTER, F. VOLPONI, S. DEL POZO, L. TAMAYO, L. BEACH, D. RUBIN, R. QUIROGA, and E. TRIEP, Electrical conductivity studies in the Andean Cordillera, *Carnegie Inst. Washington Yearbook*, 317, 1972.

ALDRICH, L. T., J. R. BANNISTER, S. DEL POZO, R. SALGUEIRO, and L. BEACH, Electrical conductivity studies in South America: Chile-Bolivia, *Carnegie Inst. Washington Yearbook*, 292–293, 1975a.

ALDRICH, L. T., M. CASAVERDE, J. R. BANNISTER, and L. BEACH, Electrical conductivity studies in South America: Argentina-Central Chile, *Carnegie Inst. Washington Yearbook*, 291–292, 1975b.

ALLAN, D. W., On the behaviour of systems of coupled dynamos, *Proc. Cambridge Phil. Soc.*, **58**, 671–693, 1962.

ANDERSSEN, R. S., On the inversion of global electromagnetic induction data, *Phys. Earth Planet. Inter.*, **10**, 292–298, 1975.

ANDERSSEN, R. S. and E. SENETA, New analysis for the geomagnetic Dst field of the magnetic storm on June 18–19, 1936, *J. Geophys. Res.*, **74**, 2768–2773, 1969.

ASHOUR, A. A., The induction of electric currents in a uniform circular disk, *Quart. J. Mech. Appl. Math.*, **3**, 119–128, 1950.

ASHOUR, A. A., Electromagnetic induction in finite thin sheets, *Quart. J. Mech. Appl. Math.*, **18**, 73–86, 1965.

ASHOUR, A. A., On a transformation of coordinates by inversion and its application to electromagnetic induction in a thin perfectly conducting hemispherical shell, *Proc. London Math. Soc.*, (3) **15**, 557–576, 1965a.

ASHOUR, A. A., The coast-line effect on rapid geomagnetic variations, *Geophys. J. R. astr. Soc.*, **10**, 147–161, 1965b.

ASHOUR, A. A. and S. CHAPMAN, The magnetic field of electric currents in an unbounded plane sheet, uniform except for a circular area of different uniform conductivity, *Geophys. J. R.* astr. *Soc.*, **10**, 31–44, 1965.

ASHOUR, A. A. and A. T. PRICE, The induction of electric currents in a non-uniform ionosphere, *Proc. Roy. Soc. Lond. A*, **195**, 198-224, 1948.

ATWATER, T., Implications of plate tectonics for the Cenozoic tectonic evolution of western North America, *Geol. Soc. Am. Bull.*, **81**, 3513-3536, 1970.

ATWATER, T. M. and J. D. MUDIE, Detailed near-bottom geophysical study of the Gorda Rise, *J. Geophys. Res.*, **78**, 8665-8686, 1973.

AVAGIMOV, A. A., T. ASHIROV, V. G. DUBROVSKY, and K. NEPESOV, Deep magnetotelluric surveys in Turkmenia and Azerbaijan, in *Geoelectric and Geothermal Studies*, edited by A. Ádám, pp. 652-665, Akadémiai Kiadó, Budapest, 1976.

BABOUR, K. and J. MOSNIER, Direct determination of the characteristics of the currents responsible for the geomagnetic anomaly of the Rhinegraben, *Geophys. J. R. astr. Soc.*, **60**, 327-331, 1980.

BABOUR, K., J. MOSNIER, M. DAGNIERES, G. VASSEUR, J. L. LE MOUEL, and J. C. ROSSIGNOL, A geomagnetic variation anomaly in the Northern Pyrenees, *Geophys. J. R. astr. Soc.*, **45**, 583-600, 1976.

BACKUS, G., The axisymmetric self-excited fluid dynamo, *Astrophys. J.*, **125**, 500-524, 1957.

BACKUS, G., A class of self-sustaining dissipative spherical dynamos, *Ann. Phys.*, **4**, 372-447, 1958.

BACKUS, G. and S. CHANDRASEKHAR, On Cowling's theorem on the impossibility of self-maintained axisymmetric homogeneous dynamos, *Proc. Natl. Acad. Sci. U. S.*, **42**, 105-109, 1956.

BAILEY, R. C., Inversion of the geomagnetic induction problem, *Proc. Roy. Soc. Lond. A*, **315**, 185-194, 1970.

BAILEY, R. C., Electromagnetic induction over the edge of a perfectly conducting ocean: the H-polarization case, *Geophys. J. R. astr. Soc.*, **48**, 385-392, 1977.

BAILEY, R. C. and R. N. EDWARDS, The effect of source field polarization on geomagnetic variation anomalies in the British Isles, *Geophys. J. R. astr. Soc.*, **45**, 97-104, 1976.

BAILEY, R. C., R. N. EDWARDS, G. D. GARLAND, R. KURTZ, and D. PITCHER, Electrical conductivity studies over a tectonically active area in eastern Canada, *J. Geomag. Geoelectr.*, **26**, 125-146, 1974.

BAILEY, R. C., R. N. EDWARDS, G. D. GARLAND, and J. P. GREENHOUSE, Geomagnetic sounding of eastern North America and the White Mountain heat flow anomaly, *Geophys. J. R. astr. Soc.*, **55**, 499-502, 1978.

BANKS, R. J., Geomagnetic variations and the electrical conductivity of the upper mantle, *Geophys. J. R. astr. Soc.*, **17**, 457-487, 1969.

BANKS, R. J., The overall conductivity distribution of the earth, *J. Geomag. Geoelectr.*, **24**, 337-351, 1972.

BANKS, R. J. and P. OTTEY, Geomagnetic deep sounding in and around Kenya Rift Valley, *Geophys. J. R. astr. Soc.*, **36**, 321-335, 1974.

BANNISTER, J. R. and D. I. GOUGH, Development of a polar magnetic substorm: a two-dimensional magnetometer array study, *Geophys. J. R. astr. Soc.*, **51**, 75-90, 1977.

BANNISTER, J. R. and D. I. GOUGH, A study of two polar magnetic substorms with a two-dimensional magnetometer array, *Geophys. J. R. astr. Soc.*, **53**, 1-26, 1978.

BARRACLOUGH, D. R., Spherical harmonic models of the geomagnetic field, *Geomagn. Bull. Inst. Geol. Sci.*, No. 8, 1-66, 1978.

BARSUKOV, O. M., Variations in the electrical resistivity of rocks and earthquakes, in *Earthquake Precursors*, edited by M. A. Sadovsky, I. L. Nersesov and L. A. Latynina, Acad, Sci. U.S.S.R. (English edition, U.S. Geological Survey, 1974).

BARTELS, J., The eccentric dipole approximating the earth's magnetic field, *Terr. Mag. Atmos. Electr.*, **41**, 225-250, 1936.

Bartels, J., Erdmagnetisch erschliessbare lokale Inhomogenitäten der elektrischen Leitfähigkeit im Untergrund, Nachr. Akad. Wiss. Göttingen, II, *Math. Phys. Kl.*, 2a, 95–100, 1954.

Beahn, T. J., Geomagnetic field gradient measurements and noise reduction techniques in Colorado, *J. Geophys. Res.*, **81**, 6276–6280, 1976.

Beamish, D., The mapping of induced currents around the Kenya Rift: a comparison of techniques, *Geophys. J. R. astr. Soc.*, **50**, 311-332, 1977.

Beblo, M. and A. Björnsson, Magnetotelluric investigation of the lower crust and upper mantle beneath Iceland, *J. Geophys.*, **45**, 1–16, 1978.

Behannon, K. W., Intrinsic magnetic properties of the lunar body, *J. Geophys. Res.*, **73**, 7257–7268, 1968.

Benkova, N. P., Spherical harmonic analysis of the Sq-variations, May-August 1933, *Terr. Mag. Atmos. Electr.*, **45**, 425–432, 1940.

Bennett, D. J. and J. H. Filloux, Magnetotelluric deep electrical sounding and resistivity, *Rev. Geophys. Space Phys.*, **13**, 197–203, 1975.

Bennett, D. J. and F. E. M. Lilley, The effect of southeast coast of Australia on transient magnetic variations, *Earth Planet. Sci. Lett.*, **12**, 392–398, 1971.

Berdichevskiy, M. N. and V. I. Dmitriev, Distortion of magnetic and electrical fields by near-surface lateral inhomogeneities, *Acta Geodaet., Geophys. Montanist. Acad. Sci. Hung.*, **11**, 447–483, 1976a.

Berdichevsky, M. N. and V. I. Dmitriev, Basic principles of interpretation of magnetotelluric sounding curves, in *Geoelectric and Geothermal Studies*, edited by A. Adám, pp. 165–221, Akádemiai Kiadó, Budapest, 1976b.

Berdichevsky, M. N., L. L. Vanyan, I. S. Fel'dman, and G. Porstendorfer, Conducting layers in the Earth's crust and upper mantle, *Gerlands Beitr. Geophysik*, **81**, 187–196, 1972.

Bhattacharyya, B. K., Reduction and treatment of magnetic anomalies of crustal origin in satellite data, *J. Geophys. Res.*, **82**, 3379–3390, 1977.

Bhattacharyya, B. K. and K. C. Chan, Reduction of magnetic and gravity data on an arbitrary surface acquired in a region of high topographic relief, *Geophysics*, **42**, 1411–1430, 1977.

Bhattacharyya, B. K. and L. Leu, Analysis of magnetic anomalies over Yellowstone National Park: mapping of Curie point isothermal surface for geothermal reconnaissance, *J. Geophys. Res.*, **80**, 4461–4465, 1975.

Bird, J., *Plate Tectonics*, Am. Geophys. Union, Washington, 986 pp., 1980.

Blakely, R. J., An age-dependent, two-layer model for marine magnetic anomalies, in *The Geophysics of the Pacific Ocean Basin and Its Margin*, edited by G. H. Sutton, M. H. Manghnani, R. Moberly and E. U. McAfee, Geophys. Monogr. Ser. 19, pp. 227–234, Am. Geophys. Union, Washington, 1976.

Blakely, R. J. and S. C. Cande, Marine magnetic anomalies, *Rev. Geophys. Space Phys.*, **17**, 204–214, 1979.

Blakely, R. J. and R. L. Christiansen, The magnetization of Mount Shasta and implications for virtual geomagnetic poles determined from seamounts, *J. Geophys. Res.*, **83**, 5971–5978, 1978.

Blakely, R. J., K. D. Klitgord, and J. D. Mudie, Analysis of marine magnetic data, *Rev. Geophys. Space Phys.*, **13**, 182–185, 1975.

Blank, J. L. and W. R. Sill, Response of the moon to the time-varying interplanetary magnetic field, *J. Geophys. Res.*, **74**, 736–743, 1969.

Bott, M. H. P., Solution of the linear inverse problem in magnetic interpretation with application to oceanic magnetic anomalies, *Geophys. J. R. astr. Soc.*, **13**, 313–323, 1967.

BRACE, W. F., Recent laboratory studies of earthquake mechanics and prediction, *J. Phys. Earth*, **25**, Suppl., S185–S202, 1977.

BRACE, W. F. and A. S. ORANGE, Electrical resistivity changes in saturated rocks during fracture and frictional sliding, *J. Geophys. Res.*, **73**, 1433–1445, 1968a.

BRACE, W. F. and A. S. ORANGE, Further studies of the effects of pressure on electrical resistivity of rocks, *J. Geophys. Res.*, **73**, 5407–5420, 1968b.

BRACE, W. F., A. S. ORANGE, and T. R. MADDEN, The effect of pressure on the electrical resistivity of water-saturated crystalline rocks, *J. Geophys. Res.*, **70**, 5669–5678, 1965.

BRACE, W. F., B. W. PAULDING, Jr., and C. SCHOLZ, Dilatancy in the fracture of crystalline rocks, *J. Geophys. Res.*, **71**, 3939–3953, 1966.

BRADLEY, R.S., A. K. JAMIL, and D. C. MUNRO, The electrical conductivity of olivine at high temperatures and pressures, *Geochim. Cosmochim. Acta*, **28**, 1669–1678, 1964.

BRAGINSKII, S. I., Self-excitation of a magnetic field during the motion of a highly conducting fluid, *Soviet Phys. JETP*, **20**, 726–735, 1965.

BREWITT-TAYLOR, C. R. and J. T. WEAVER, On the finite difference solution of two-dimensional induction problems, *Geophys. J. R. astr. Soc.*, **47**, 375–396, 1976.

BUCHA, V., Changes of the earth's magnetic moment and radiocarbon dating, *Nature*, **224**, 681–683, 1969.

BUCHA, V., R. E. TAYLOR, R. BERGER, and E. W. HAURY, Geomagnetic intensity: changes during the past 3000 years in the western hemisphere, *Science*, **168**, 111–114, 1970.

BULAU, J. R., H. S. WAFF, and J. A. TYBURCZY, Mechanical and thermodynamic constraints on fluid distribution in partial melts, *J. Geophys. Res.*, **84**, 6102–6108, 1979.

BULLARD, E. C., The secular change in the earth's magnetic field, *Monthly Notices Roy. Astron. Soc., Geophys. Suppl.*, **5**, 248–257, 1948.

BULLARD, E. C., The magnetic field within the earth, *Proc. Roy. Soc. Lond. A,* **197**, 433–453, 1949.

BULLARD, E. C., The stability of a homopolar dynamo, *Proc. Cambridge Phil. Soc.,* **51**, 744–760, 1955.

BULLARD, E. C. and H. GELLMAN, Homogeneous dynamos and terrestrial magnetism, Phil. Trans. Roy. Soc. Lond. A, **247**, 213–278, 1954.

BULLARD, E. C. and R. L. PARKER, Electromagnetic induction in the oceans, in *The Sea*, edited by A. E. Maxwell, Wiley-Interscience, New York, pp. 637–693, 1970.

BULLARD, E. C., C. FREEDMAN, H. GELLMAN, and J. NIXON, The westward drift of the earth's magnetic field, *Phil. Trans. Roy. Soc. Lond. A*, **243**, 67–92, 1950.

BUSSE, F. H., Thermal instabilities in rapidly rotating systems, *J. Fluid Mech.*, **44**, 441–460, 1970.

BUSSE, F. H., A model of the geodynamo, *Geophys. J. R. astr. Soc.*, **42**, 437–459, 1975.

BUSSE, F. H., Generation of planetary magnetism by convection, *Phys. Earth Planet. Inter.*, **12**, 350–358, 1976.

BUSSE, F. H., An example of nonlinear dynamo action, *J. Geophys.*, **43**, 441–452, 1977.

BUTLER, R. F. and A. V. COX, A mechanism for producing magnetic remanence in meteorites and lunar samples by cosmic-ray exposure, *Science*, **172**, 939–941, 1971.

BUTLER, R. F. and N. D. OPDYKE, Magnetic polarity stratigraphy, *Rev. Geophys. Space Phys.*, **17**, 235–244, 1979.

BYERLY, P. E. and R. H. STOLTZ, An attempt to define the Curie point isotherm in northern and central Arizona, *Geophysics*, **42**, 1394–1400, 1977.

CAGNIARD, I., Basic theory of the magneto-telluric method of geophysical prospecting, *Geophysics*, **18**, 605–635, 1953.

Cagniard, M. L., Sur la nature des ondes séismiques capables de traverser le Noyau terrestre, *Compt. Rend. Acad. Sci. Paris*, **234**, 1705–1706, 1952.

Cain, J. C., S. J. Hendricks, R. A. Langel, and W. V. Hudson, A proposed model for the International Geomagnetic Reference Field—1965, *J. Geomag. Geoelectr.*, **19**, 335–355, 1967.

Camfield, P. A., Magnetometer array study in a tectonically active region of Quebec, Canada, *Geophys. J. R. astr. Soc.*, **65**, 553–570, 1981.

Camfield, P. A. and D. I. Gough, Anomalies in daily variation magnetic fields and structure under north-western United States and south-western Canada, *Geophys. J. R. astr. Soc.*, **41**, 193–218, 1975.

Camfield, P. A., D. I. Gough, and H. Porath, Magnetometer array studies in the northwestern United States and southwestern Canada, *Geophys. J. R. astr. Soc.*, **22**, 201–221, 1971.

Caner, B., Electrical conductivity structure in western Canada and petrological interpretation, *J. Geomag. Geoelectr.*, **22**, 113–129, 1970.

Caner, B., Quantitative interpretation of geomagnetic depth-sounding data in western Canada, *J. Geophys. Res.*, **76**, 7202–7216, 1971.

Caner, B., W. H. Cannon, and C. E. Livingstone, Geomagnetic depth-sounding and upper mantle structure in the Cordillera region of western North America, *J. Geophys. Res.*, **72**, 6335–6351, 1967.

Caner, B., D. R. Auld, H. Dragert, and P. A. Camfield, Geomagnetic depth-sounding and crustal structure in western Canada, *J. Geophys. Res.*, **76**, 7181–7201, 1971.

Cap, F. F., Possible production mechanisms of lunar magnetic fields, *J. Geophys. Res.*, **77**, 3328-3333, 1972.

Casaverde, M., A. A. Giesecke, Jr., R. Salgueiro, S. Del Pozo, L. Tamayo, M. A. Tuve, and L. T. Aldrich, Studies of conductivity anomaly under the Andes, *Carnegie Inst. Washington Yearbook*, **65**, 369–372, 1969.

Chapman, S., The solar and lunar diurnal variation of the earth's magnetism, *Phil. Trans. Roy. Soc. Lond. A*, **218**, 1–118, 1919.

Chapman, S. and J. Bartels, *Geomagnetism*, Oxford Univ. Press, London, 1049 pp., 1940.

Chapman, S. and A. T. Price, The electric and magnetic state of the interior of the earth as inferred from terrestrial magnetic variations, *Phil. Trans. Roy. Soc. Lond. A*, **229**, 427–460, 1930.

Chave, A. D., R. P. Von Herzen, K. A. Poehls, and C. S. Cox, Electromagnetic induction fields in the deep ocean north-east of Hawaii: implications for mantle conductivity and source fields, *Geophys. J. R. astr. Soc.*, **66**, 379–406, 1981.

Chen Po-Fang, Conductivity anomaly in west coast of Po Hai, *Acta Geophysica Sinica*, **17**, 169–172, 1974 (in Chinese with English abstract).

Cisowski, C. S., M. Fuller, M. F. Rose, and P. Wasilewski, Magnetic effects of explosive shocking of lunar soil, *Geochim. Cosmochim. Acta*, **37**, Suppl. 4, 3003–3017, 1973.

Cochrane, N. A. and R. D. Hyndman, A new analysis of geomagnetic depth-sounding data from western Canada, *Can. J. Earth Sci.*, **7**, 1208–1218, 1970.

Cochrane, N. A. and R. D. Hyndman, Magnetotelluric and magnetovariational studies in Atlantic Canada, *Geophys. J. R. astr. Soc.*, **39**, 385–406, 1974.

Cochrane, N. A. and J. A. Wright, Geomagnetic sounding near the northern termination of the Appalachian System, *Can. J. Earth Sci.*, **14**, 2858–2864, 1977.

Colburn, D. S., R. G. Currie, J. D. Mihalov, and C. P. Sonett, Diamagnetic solar wind cavity discovered behind moon, *Science*, **158**, 1040–1042, 1967.

Coleman, Jr., P. J. and C. T. Russell, The remanent magnetic field of the moon, *Phil. Trans. Roy. Soc. Lond. A*, **285**, 489–506, 1977.

COLEMAN, Jr., P. J., B. R. LICHTENSTEIN, C. T. RUSSELL, G. SCHUBERT, and L. R. SHARP, The particles and fields subsatellite magnetometer experiment, Apollo 16 Preliminary Science Report, NASA SP-315, 1972.

COLES, R. L. and G. V. HAINES, Long-wavelength magnetic anomalies over Canada, using polynomial and upward continuation techniques, *J. Geomag. Geoelectr.*, **31**, 545–566, 1979.

COLES, R. L., G. V. HAINES, and W. HANNAFORD, Large scale magnetic anomalies over western Canada and the Arctic: a discussion, *Can. J. Earth Sci.*, **13**, 790–802, 1976.

COLLINSON, D. W., S. K. RUNCORN, and A. STEPHENSON, Magnetic properties of Apollo 16 rocks, in *Lunar Science*, vol. 4, edited by J. Chamberlain and C. Watkins, Lunar Science Institute, Houston, pp. 155–157, 1973.

COOK, A. E. and P. H. ROBERTS, The Rikitake two-disk dynamo system, *Proc. Cambridge Phil. Soc.*, **68**, 547–569, 1970.

COWLING, T. G., The magnetic field of sunspots, *Monthly Notices Roy. Astron. Soc.*, **94**, 39–48, 1934.

COWLING, T. G., The dynamo maintenance of steady magnetic fields, *Quart. J. Mech. Appl. Math.*, **10**, 129–136, 1957.

COX, A., Geomagnetic reversals, *Science*, **163**, 237–245, 1969.

COX, A., Plate Tectonics and Geomagnetic Reversals, 702 pp., W. H. Freeman, San Francisco, 1973.

COX, A., R. R. DOELL, and G. B. DALRYMPLE, Reversals of the earth's magnetic field, *Science*, **144**, 1537–1543, 1964.

COX, C., Electromagnetic induction in the oceans and inferences on the constitution of the earth, *Geophys. Surveys*, **4**, 137–156, 1980.

COX, C. S., J. H. FILLOUX, D. I. GOUGH, J. C. LARSEN, K. A. POEHLS, R. P. VON HERZEN, and R. WINTER, Atlantic lithosphere sounding, *J. Geomag. Geoelectr.*, **32**, SI13–SI32, 1980.

CURRIE, R. G., The geomagnetic spectrum—40 days to 5.5 years, *J. Geophys. Res.*, **71**, 4579–4598, 1966.

CURRIE, R. G., Geomagnetic line spectra—2 to 70 years, *Astrophys. Space Sci.*, **21**, 425–438, 1973.

DAVIS, P. M., The computed piezomagnetic anomaly field for Kilauea volcano, Hawaii, *J. Geomag. Geoelectr.*, **28**, 113–122, 1976.

DAVIS, P. M., D. B. JACKSON, J. FIELD, and F. D. STACEY, Kilauea volcano, Hawaii: a search for the volcanomagnetic effect, *Science*, **180**, 73–74, 1973.

DAVIS, P. M., D. D. JACKSON, and M. J. S. JOHNSTON, Further evidence of localized geomagnetic field changes before the 1974 Thanksgiving Day earthquake, Hollister, California, *Geophys. Res. Lett.*, **7**, 513–516, 1980.

DAVIS, P. M., D. D. JACKSON, C. A. SEARLS, and R. L. MCPHERRON, Detection of tectonomagnetic events using multichannel predictive filtering, *J. Geophys. Res.*, **86**, 1731–1737, 1981.

DAWSON, T. W. and J. T. WEAVER, *H*-polarization induction in two thin half-sheets, *Geophys. J. R. astr. Soc.*, **56**, 419–438, 1979.

DE BEER, J. H. and D. I. GOUGH, Conductive structures in southernmost Africa: a magnetometer array study, *Geophys. J. R. astr. Soc.*, **63**, 479–495, 1980.

DE BEER, J. H., J. S. V. VAN ZIJL, R. M. J. HUYSSEN, P. L. V. HUGO, S. J. JOUBERT, and R. MEYER, A magnetometer array study in South-West Africa, Botswana and Rhodesia, *Geophys. J. R. astr. Soc.*, **45**, 1–17, 1976.

DELAURIER, J. M., L. K. LAW, E. R. NIBLETT, and F. C. PLET, Geomagnetic variation anomalies in the Canadian Arctic II. Mould Bay anomaly, *J. Geomag. Geoelectr.*, **26**, 223–245, 1974.

DODSON, R., J. R. DUNN, M. FULLER, I. WILLIAMS, H. ITO, V. A. SCHMIDT, and Y. WU, Paleomagnetic record of a late Tertiary field reversal, *Geophys. J. R. astr. Soc.*, **53**, 373–412, 1978.

DOELL, R. R., C. S. GROMMÉ, A. N. THORPE, and F. E. SENFTLE, Magnetic studies of Apollo 11 lunar samples, *Geochim. Cosmochim. Acta*, **34**, Suppl. 1, 2097–2120, 1970.

DOLGINOV, S. S., E. G. EROSHENKO, L. N. ZHUZGOV, and N. V. PUSHKOV, Investigation of the magnetic field of the moon, *Geomagn. Aeron.*, **1**, 18, 1961.

DOLGINOV, Sh. Sh., YE. G. YEROSHENKO, and L. N. ZHUZGOV, The magnetic field of Mars according to the data from the Mars 3 and Mars 5, *J. Geophys. Res.*, **81**, 3353–3362, 1976.

DOSS, S. S. and A. A. ASHOUR, Some results on the magnetic field of electric currents induced in a thin hemispherical shell of finite conductivity, with geomagnetic application, *Geophys. J. R. astr. Soc.*, **22**, 385–400, 1971.

DRAGERT, H., A transfer function analysis of a geomagnetic depth sounding-profile across central British Columbia, *Can. J. Earth Sci.*, **10**, 1089–1098, 1973.

DRAGERT, H. and G. K. C. CLARKE, A detailed investigation of the Canadian Cordillera geomagnetic transition anomaly, *J. Geophys.*, **42**, 373–390, 1977.

DUBA, A., Electrical conductivity of olivine, *J. Geophys. Res.*, **77**, 2483–2495, 1972.

DUBA, A., Are laboratory electrical conductivity data relevant to the earth?, *Acta Geodaet., Geophys. Montanist. Acad. Sci. Hung.*, **11**, 485–495, 1976.

DUBA, A. and I. A. NICHOLLS, The influence of oxidation state on the electrical conductivity of olivine, *Earth Planet. Sci. Lett.*, **18**, 59–64, 1973.

DUBA, A., J. ITO, and J. C. JAMIESON, The effect of ferric iron on the electrical conductivity of olivine, *Earth Planet. Sci. Lett.*, **18**, 279–284, 1973.

DUBA, A., H. C. HEARD, and R. N. SCHOCK, Electrical conductivity of olivine at high pressure and under controlled oxygen fugacity, *J. Geophys. Res.*, **79**, 1667–1673, 1974.

DUNLOP, D. J. and G. F. WEST, An experimental evaluation of single domain theories, *Rev. Geophys. Space Phys.*, **7**, 709–757, 1969.

DYAL, P. and C. W. PARKIN, The magnetism of the moon, *Sci. Am.*, **225**, 62–73, 1971.

DYAL, P. and C. W. PARKIN, Global electromagnetic induction in the moon and planets, *Phys. Earth Planet. Inter.*, **7**, 251–265, 1973.

DYAL, P., C. W. PARKIN, and C. P. SONETT, Apollo 12 magnetometer: measurement of a steady magnetic field on the surface of the moon, *Science*, **169**, 762–764, 1970a.

DYAL, P., C. W. PARKIN, and C. P. SONETT, Lunar surface magnetometer, *IEEE Trans. Geosci. Electr.*, GE-8, No. 4, 203–215, 1970b.

DYAL, P., C. W. PARKIN, C. P. SONETT, and D. S. COLBURN, Electrical conductivity and temperature of the lunar interior from magnetic transient response measurements, NASA Technical Memorandum X-62, 012, 42 pp., 1970c.

DYAL, P., C. W. PARKIN, C. P. SONETT, R. L. DUBOIS, and G. SIMMONS, Lunar portable magnetometer experiment, Apollo 14 Preliminary Science Report, NASA Spec. Publ., 272, 227, 1971.

DYAL, P., C. W. PARKIN, and W. D. DAILY, Magnetism and the interior of the moon, *Rev. Geophys. Space Phys.*, **12**, 568–591, 1974.

DYAL, P., C. W. PARKIN, and W. D. DAILY, Structure of the lunar interior from magnetic field measurements, *Proc. 7th Lunar Sci. Conf.*, 3077–3095, 1976.

DYCK, A. V. and G. D. GARLAND, A conductivity model for certain features of the Alert anomaly in geomagnetic variation, *Can. J. Earth Sci.*, **6**, 513–516, 1969.

DYSON, F. W. and H. FURNER, The earth's magnetic potential, *Monthly Notices Roy. Astron. Soc., Geophys.* Suppl., **1**, 76–88, 1923.

ECKHARDT, D., Geomagnetic inductions in a concentrically stratified earth, *J. Geophys. Res.*,

68, 6273-6278, 1963.

ECKHARDT, D., K. LARNER, and T. MADDEN, Long period magnetic fluctuations and mantle electrical conductivity estimates, *J. Geophys. Res.*, **68**, 6279-6286, 1963.

EDWARDS, R. N. and J. P. GREENHOUSE, Geomagnetic variations in the eastern United States: evidence for a highly conducting lower crust?, *Science*, **188**, 726-728, 1975.

EDWARDS, R. N., L. K. LAW, and A. WHITE, Geomagnetic variations in the British Isles and their relation to electrical currents in the ocean and shallow seas, *Phil. Trans. Roy. Soc. Lond. A*, **270**, 289-323, 1971.

ELSASSER, W. M., Induction effects in terrestrial magnetism 1: Theory, Phys. Rev., **69**, 106-116, 1946a.

ELSASSER, W. M., Induction effects in terrestrial magnetism 2: The secular variation, *Phys. Rev.*, **70**, 202-212, 1946b.

ELSASSER, W. M., The earth's interior and geomagnetism, *Rev. Mod. Phys.*, **22**, 1-35, 1950.

ELSASSER, W. M., Hydromagnetism 1: A review, *Am. J. Phys.*, **23**, 590-609, 1955.

EMILIA, D. A. and G. BODVARSSON, Numerical methods in the direct interpretation of marine magnetic anomalies, *Earth Planet. Sci. Lett.*, **7**, 194-200, 1969.

EMILIA, D. A. and R. L. MASSEY, Magnetization estimation for nonuniformly magnetized seamounts, *Geophysics*, **39**, 223-231, 1974.

ENGLAND, A. W., G. SIMMONS, and D. STRANGWAY, Electrical conductivity of the moon, *J. Geophys. Res.*, **73**, 3219-3226, 1968.

ERMAN, A. and H. PETERSEN, *Die Grundlagen der Gaussischen Theorie und die Erscheinungen des Erdmagnetismus im Jahre 1829*, Berlin, 1874.

EVERETT, J. E. and R. D. HYNDMAN, Geomagnetic variations and electrical conductivity structure in south-western Australia, *Phys. Earth Planet. Inter.*, **1**, 24-34, 1967.

FAINBERG, E. B., Electromagnetic induction in the world ocean, *Geophys. Surveys*, **4**, 157-171, 1980.

FILLOUX, J. H., Oceanic electric currents, geomagnetic variations and the deep electrical conductivity structure of ocean-continent transition of central California, Ph.D. Thesis, Univ. California, San Diego, La Jolla, 166 pp., 1967.

FILLOUX, J. H., Ocean-floor magnetotelluric sounding over North Central Pacific, *Nature*, **269**, 297-301, 1977.

FILLOUX, J. H., North Pacific magnetotelluric experiments, *J. Geomag. Geoelectr.*, **32**, SI33-SI43, 1980a.

FILLOUX, J. H., Magnetotelluric soundings over the northeast Pacific may reveal spatial dependence of depth and conductance of the asthenoshere, *Earth Planet. Sci. Lett.*, **46**, 244-252, 1980b.

FILLOUX, J. H., Magnetotelluric exploration of the North Pacific: progress report and preliminary soundings near a spreading ridge, *Phys. Earth Planet. Inter.*, **25**, 187-195, 1981.

FINCH, H. F. and B. R. LEATON, The earth's main magnetic field, epoch 1955.0, *Monthly Notices Roy. Astron. Soc., Geophys. Suppl.*, **7**, 314-317, 1957.

FISCHER, G., P. -A. SCHNEGG, and K. D. USADEL, Electromagnetic response of an ocean-coast model to *E*-polarization induction, *Geophys. J. R. astr. Soc.*, **53**, 599-616, 1978.

FITTERMAN, D. V., Electrokinetic and magnetic anomalies associated with dilatant regions in a layered earth, *J. Geophys. Res.*, **83**, 5923-5928, 1978.

FITTERMAN, D. V., Theory of electrokinetic-magnetic anomalies in a faulted half-space, *J. Geophys. Res.*, **84**, 6031-6040, 1979.

FOURNIER, H., Recent work in magnetotelluric soundings of the lower crust and uppermost mantle, *Geophys. Surveys*, **4**, 89-96, 1980.

FOURNIER, H. G. and A. ÁDÁM, Proposal for a first upper mantle magnetotelluric E-W profile across Europe, *Acta Geodaet., Geophys. Montanist. Acad. Hung.*, **6**, 459–477, 1971.

FRITSCHE, H., *Über die Bestimmung der koeffizienten der Gaussischen Allgemeinen Theorie des Erdmagnetismus für das Jahr 1885, etc.*, St. Petersburg, 1897.

FUJITA, N., Secular change of the geomagnetic total force in Japan for 1970.0, *J. Geomag. Geoelectr.*, **25**, 181–194, 1973.

FUKUSHIMA, N. and Y. KAMIDE, Partial ring current models for worldwide geomagnetic disturbances, *Rev. Geophys. Space Phys.*, **11**, 795–853, 1973.

FULLER, M., The remanent magnetization of lunar soils, *Science*, **178**, 154–156, 1972.

FULLER, M., Lunar magnetism, *Rev. Geophys. Space Phys.*, **12**, 23–70, 1974.

FULLER, M., I. WILLIAMS, and K. A. HOFFMAN, Paleomagnetic records of geomagnetic field reversals and the morphology of the transitional fields, *Rev. Geophys. Space Phys.*, **17**, 179–203, 1979.

GARLAND, G. D., Correlation between electrical conductivity and other geophysical parameters, *Phys. Earth Planet. Inter.*, **10**, 220–230, 1975.

GAUSS, C. F., Allgemeine Theorie des Erdmagnetismus, in *Resultate aus den Beobachtungen des Magnetischen Vereins im Jahre 1838*, edited by C. F. Gauss and W. Weber, pp. 1–57, Weidmann, Leipzig, 1839.

GIBSON, R. D., The Herzenberg dynamo I, *Quart. J. Mech. Appl. Math.*, **21**, 243–255, 1968a.

GIBSON, R. D., The Herzenberg dynamo II, *Quart. J. Mech. Appl. Math.*, **21**, 257–267, 1968b.

GIBSON, R. D., The Herzenberg dynamo, in *The Application of Modern Physics to the Earth and Planetary Interiors*, edited by S. K. Runcorn, pp. 571–576, Wiley-Interscience, London, 1969.

GIBSON, R. D. and P. H. ROBERTS, The Bullard-Gellman dynamo, in *The Application of Modern Physics to the Earth and Planetary Interiors*, edited by S. K. Runcorn, pp. 577–602, Wiley-Interscience, London, 1969.

GILBERT, W., *De Magnete, Magneticisque Corporibus, et de magno Magnete Tellure Physiologia nova, plurimis & argumentis, & experimentis demonstrata*, London, 1600: English translation, 368 pp. Wiley, New York, 1893; 368 pp., Dover, New York, 1958.

GOLDSTEIN, B. E., R. J. PHILLIPS, and C. T. RUSSELL, Magnetic permeability measurements and a lunar core, *Geophys. Res. Lett.*, **3**, 289–292, 1976.

GOUGH, D. I., The geophysical significance of geomagnetic variation anomalies, *Phys. Earth Planet. Inter.*, **7**, 379–388, 1973.

GOUGH, D. I., Electrical conductivity under western North America in relation to hear flow, seismology, and structure, *J. Geomag. Geoelectr.*, **26**, 105–123, 1974.

GOUGH, D. I. and J. R. BANNISTER, A polar magnetic substorm observed in the evening sector with a two-dimensional magnetometer array, *Geophys. J. R. astr. Soc.*, **55**, 435–450, 1978.

GOUGH, D. I. and P. A. CAMFIELD, Convergent geophysical evidence of a metamorphic belt through the Black Hills of South Dakota, *J. Geophys. Res.*, **77**, 3168–3170, 1972.

GOUGH, D. I., J. H. DE BEER, and J. S. V. VAN ZIJL, A magnetometer array study in southern Africa, *Geophys. J. R. astr. Soc.*, **34**, 421–433, 1973.

GOUGH, D. I., M. W. MCELHINNY, and F. E. M. LILLEY, A magnetometer array study in southern Australia, *Geophys. J. R. astr. Soc.*, **36**, 345–362, 1974.

GREEN, V. R. and J. T. WEAVER, Two-dimensional induction in a thin sheet of variable integrated conductivity at the surface of a uniformly conducting earth, *Geophys. J. R. astr. Soc.*, **55**, 721–736, 1978.

GREENHOUSE, J. P., Geomagnetic time variations on the sea floor off northern California,

Ph.D. Thesis, Univ. California, San Diego, 248 pp., 1972.

Gubbins, D., Numerical solutions of the kinematic dynamo problem, *Phil. Trans. Roy. Soc. Lond. A*, **274**, 493–521, 1973a.

Gubbins, D., On large magnetic Reynolds number dynamos, *Geophys. J. R. astr. Soc.*, **33**, 57–64, 1973b.

Gubbins, D., Numerical solutions of the hydromagnetic dynamo problem, *Geophys.* J. R. astr. Soc., **42**, 295–305, 1975.

Gubbins, D., Observational constraints on the generation process of the earth's magnetic field, *Geophys. J. R. astr. Soc.*, **47**, 19–39, 1976.

Gubbins, D., Energetics of the earth's core, *J. Geophys.*, **43**, 453–464, 1977.

Haak, V., The electrical resistivity of the upper 300 km of the Afar-depression in Ethiopia derived from magnetotelluric measurements, *Acta Geodaet., Geophys. Montanist. Acad. Sci. Hung.*, **12**, 7–10, 1977.

Haggerty, S. E., Mineralogical constraints on Curie isotherms in deep crustal magnetic anomalies, *Geophys. Res. Lett.*, **5**, 105–108, 1978.

Hagiwara, Y., Analysis of the results of the aeromagnetic surveys over valcanoes in Japan, *Bull. Earthq. Res. Inst., Univ. Tokyo*, **43**, 529–547, 1965.

Hagiwara, Y., Magnetic lines of force deformed by motion of a conducting fluid (I), *J. Geomag. Geoelectr.*, **19**, 229–238, 1967.

Hagiwara, Y., Pseudomagnetic anomaly derived from gravity observations in central Japan, *Bull. Earthq. Res. Inst., Univ. Tokyo*, **55**, 27–41, 1980.

Haigh, G., The process of magnetization by chemical change, *Phil. Mag.*, **3**, 267–286, 1958.

Hall, D. H., Long-wavelength aeromagnetic anomalies and deep crustal magnetization in Manitoba and northwestern Ontario, *J. Geophys.*, **40**, 403–430, 1974.

Hamano, Y., An experiment on the post-depositional remanent magnetization in artificial and natural sediments, *Earth Planet. Sci. Lett.*, **51**, 221–232, 1980.

Hamilton, R. M., Temperature variation at constant pressures of the electrical conductivity of periclase and olivine, *J. Geophys. Res.*, **70**, 5679–5692, 1965.

Harrison, C. G. A., Magnetization of the oceanic crust, *Geophys. J. R. astr. Soc.*, **47**, 257–283, 1976.

Hasegawa, M., Representation of the field of diurnal variations of terrestrial magnetism on quiet days by the method of graphical integration, *Proc. Imp. Acad. Tokyo*, **12**, 225–228, 1936.

Hasegawa, M. and M. Ota, On the magnetic field of *Sq* in the middle and lower latitudes during the Second Polar Year, *Trans. Oslo Meeting Intern. Assoc. Terr. Mag. Elec., Bull.*, **13**, 426–430, 1950.

Hashin, Z. and S. Shtrikman, A variational approach to the theory of the effective magnetic permeability of multiphase materials, *J. Appl. Phys.*, **33**, 3125–3133, 1962.

Heirtzler, J. R., X. le Pichon, and J. G. Baron, Magnetic anomalies over the Reykjanes Ridge, *Deep Sea Res.*, **13**, 427–443, 1966.

Heirtzler, J. R., G. O. Dickson, E. M. Herron, W. C. Pitman III, and X. le Pichon, Marine magnetic anomalies, geomagnetic field reversals, and motions of the ocean floor and continents, *J. Geophys. Res.*, **73**, 2119–2136, 1968.

Helsley, C. E., Magnetic properties of lunar 10022, 10069, 10084, and 10085 samples, *Geochim. Cosmochim. Acta*, **34**, Suppl. 1, 2213–2219, 1970.

Henderson, R. G., A comprehensive system of automatic computation in magnetic and gravity interpretation, *Geophysics*, **25**, 569–585, 1960.

Hermance, J. F., An electrical model for the sub-Icelandic crust, *Geophysics*, **38**, 3–13, 1973.

Hermance, J. F., Processing of magnetotelluric data, *Phys. Earth Planet. Inter.*, **7**, 349–364,

1973.

HERMANCE, J. F. and G. D. GARLAND, Magnetotelluric deep-sounding experiments in Iceland, *Earth Planet. Sci. Lett.*, **4**, 469–474, 1968.

HERMANCE, J. F. and L. R. GRILLOT, Correlation of magnetotelluric, seismic, and temperature data from southwest Iceland, *J. Geophys. Res.*, **75**, 6582–6591, 1970.

HERMANCE, J. F. and L. R. GRILLOT, Constraints on temperatures beneath Iceland from magnetotelluric data, *Phys. Earth Planet. Inter.*, **8**, 1–12, 1974.

HERMANCE, J. F. and W. R. PELTIER, Magnetotelluric fields of a line current, *J. Geophys. Res.*, **75**, 3351–3354, 1970.

HERZENBERG, A., Geomagnetic dynamos, *Phil. Trans. Roy. Soc. Lond. A*, **250**, 543–585, 1958.

HEWSON-BROWNE, R. C., V. C. L. HUTSON, P. C. KENDALL, and S. R. C. MALIN, New iterative methods of solution of oceanic induction problems, *Phys. Earth Planet. Inter.*, **7**, 431–436, 1973.

HIDE, R., Free hydromagnetic oscillations of the earth's core and the theory of geomagnetic secular variation, *Phil. Trans. Roy. Soc. Lond. A*, **259**, 615–650, 1966.

HIDE, R., On the earth's core-mantle interface, *Quart. J. Roy. Met. Soc.*, **96**, 579–590, 1970.

HIDE, R. and S. R. C. MALIN, Novel correlations between global features of the earth's gravitational and magnetic fields, *Nature*, **225**, 605–609, 1970.

HIGGINS, G. H. and G. C. KENNEDY, The adiabatic gradient and the melting point gradient of the core of the earth, *J. Geophys. Res.*, **76**, 1870–1878, 1971.

HILDE, T. W. C., N. ISEZAKI, and J. M. WAGEMAN, Mesozoic sea-floor spreading in the North Pacific, in *The Geophysics of the Pacific Ocean Basin and Its Margin*, edited by G. H. Sutton, M. H. Manghnani, R. Moberly and E. U. McAfee, Geophys. Mongr. Ser. 19, pp. 205–226, Am. Geophys. Union, Washington, 1976.

HILLHOUSE, J. and A. COX, Brunches-Matuyama polarity transition, *Earth Planet. Sci. Lett.*, **29**, 51–64, 1976.

HINZE, W. J., Continental magnetic anomalies, *Rev. Geophys. Space Phys.*, **17**, 257–273, 1979.

HIROOKA, K., Archaeomagnetic study for the past 2000 years in southwest Japan, *Mem. Fac. Sci. Kyoto Univ., Ser. Geol. Mineral.*, **38**, 167–207, 1971.

HOBBS. B. A., The calculation of geophysical induction effects using surface integrals, *Geophys. J. R. astr. Soc.*, **25**, 481–509, 1971.

HOBBS, B. A., Electromagnetic induction in the oceans, *Geophys. J. R. astr. Soc.*, **42**, 307–313, 1975a.

HOBBS, B. A., Analytical solutions to global and local problems of electromagnetic induction in the earth, *Phys. Earth Planet. Inter.*, **10**, 250–261, 1975b.

HOBBS, B. A. and A. T. PRICE, Surface integral formulae for geomagnetic studies, *Geophys. J. R. astr. Soc.*, **20**, 49–63, 1970.

HOBBS, B. A. and A. M. M. BRIGNALL, A method for solving general problems of electromagnetic induction in the oceans, *Geophys. J. R. astr. Soc.*, **42**, 307–313, 1976.

HOFFMAN, K. A., Polarity transition records and the geomagnetic dynamo, *Science*, **196**, 1329–1332, 1977.

HOFFMAN, K. A. and M. FULLER, Transitional field configurations and geomagnetic reversal, *Nature*, **273**, 715–718, 1978.

HOLMES, P., 'Strange' phenomena in dynamical systems and their physical implications, *Appl. Math. Modelling*, **1**, 362–366, 1977.

HONKURA, Y., Geomagnetic variation anomaly on Miyake-jima Island, *J. Geomag. Geoelectr.*, **23**, 307–333, 1971.

HONKURA, Y., Island effect and electrical conductivity structure beneath Miyake-jima Island, *J. Geomag. Geoelectr.*, **25**, 167–179, 1973.

HONKURA, Y., Electrical conductivity anomalies beneath the Japan arc, *J. Geomag. Geoelectr.*, **26**, 147–171, 1974.

HONKURA, Y., Partial melting and electrical conductivity anomalies beneath the Japan and Philippine Seas, *Phys. Earth Planet. Inter.*, **10**, 128–134, 1975.

HONKURA, Y., Electrical conductivity anomalies in the earth, *Geophys. Surveys*, **3**, 225–253, 1978a.

HONKURA, Y., On a relation between anomalies in the geomagnetic and telluric fields observed at Nakaizu and the Izu-Oshima-Kinkai earthquake of 1978, *Bull. Earthq. Res. Inst., Univ. Tokyo*, **53**, 931–937, 1978b (in Japanese with English abstract).

HONKURA, Y., Electric and magnetic approach to earthquake prediction, in *Current Research in Earthquake Prediction I*, edited by T. Rikitake, pp. 301–383, Center Acad. Publ. Japan, Tokyo/D. Reidel Publ. Co., Dordrecht, 1981.

HONKURA, Y. and S. KOYAMA, Observations of short-period geomagnetic variations at Nakaizu (1), *Bull. Earthq. Res. Inst., Univ. Tokyo*, **53**, 925–930, 1978 (in Japanese with English abstract).

HONKURA, Y. and S. KOYAMA, Electrical conductivity structure beneath the central part of Japan as inferred from magnetotelluric fields at the Yatsugatake Magnetic Observatory, *Bull. Earthq. Res. Inst., Univ. Tokyo*, **54**, 491–501, 1979.

HONKURA, Y. and T. RIKITAKE, Core motion as inferred from drifting and standing non-dipole fields, *J. Geomag. Geoelectr.*, **24**, 223–230, 1972.

HONKURA, Y., S. OSHIMA, and T. KONDO, Geomagnetic variation anomaly on Hachijo-jima Island, *J. Geomag. Geoelectr.*, **26**, 23–37, 1974.

HONKURA, Y., E. R. NIBLETT, and R. D. KURTZ, Changes in magnetic and telluric fields in a seismically active region of eastern Canada: preliminary results of earthquake prediction studies, *Tectonophysics*, **34**, 219–230, 1976.

HONKURA, Y., R. D. KURTZ, and E. R. NIBLETT, Geomagnetic depth sounding and magnetotelluric results from a seismically active region northeast of Quebec City, *Can. J. Earth Sci.*, **14**, 256–267, 1977.

HONKURA, Y., N. ISEZAKI, and K. YASKAWA, Electrical conductivity structure beneath the northeastern Philippine Sea as inferred from the island effect on Minami-daito Island, *J. Geomag. Geoelectr.*, **33**, 365–377, 1981.

HUTSON, V.C.L., P.C. KENDALL and S. R. C. MALIN, Computation of the solution of geomagnetic induction problems: a general method, with applications, *Geophys. J. R. astr. Soc.*, **28**, 489–498, 1972.

HUTSON, V. C. L., P. C. KENDALL, and S. R. C. MALIN, The modelling of oceans by spherical caps, *Geophys. J. R. astr. Soc.*, **33**, 377–387, 1973.

HUTTON, V. R. S., The electrical conductivity of the earth and planets, *Rep. Prog. Phys.*, **39**, 487–572, 1976a.

HUTTON, R., Induction studies in rifts and other active regions, *Acta Geodaet., Geophys. Montanist. Acad. Sci. Hung.*, **11**, 347–376, 1976b.

HUTTON, V. R. S. and A. G. JONES, Magnetovariational and magnetotelluric investigations in S. Scotland, *J. Geomag. Geoelectr.*, **32**, SI141–SI150, 1980.

HYNDMAN, R. D. and N. A. COCHRANE, Electrical conductivity structure by geomanetic induction at the continental margin of Atlantic Canada, *Geophys. J. R. astr. Soc.*, **25**, 425–446, 1971.

IAGA COMMISSION II, WORKING GROUP 4, International Geomagnetic Reference Field 1965. 0, *J. Geomag. Geoelectr.*, **21**, 569–571, 1969.

IAGA, DIVISION I, STUDY GROUP, Geomagnetic Reference Fields, International Geomagnetic Reference Field 1975, *J. Geomag. Geoelectr.*, **27**, 437–439, 1975.

IGRF EVALUATION GROUP OF JAPAN, Evaluation of IGRF models proposed for 1975.0, *J. Geomag. Geoelectr.*, **27**, 441–451, 1975.

IMAMITI, S., Secular variation of the magnetic declination in Japan, *Mem. Kakioka Mag. Obs.*, **7**, 49–55, 1956.

INGLIS, D. R., Dynamo theory of the earth's varying magnetic field, *Rev. Modern Phys.*, **53**, 481–496, 1981.

INTERNATIONAL GEOPHYSICAL DATA LIBRARY, SCIENCE COUNCIL OF JAPAN, Catalogue of Data in the International Geophysical Data Library, No. 19, 180–186, 1980.

IRVING, E., *Paleomagnetism and Its Application to Geological and Geophysical Problems*, 399 pp., J. Wiley, New York, 1964.

IRVING, E. and G. PULLAIAH, Reversals of the geomagnetic field, magnetostratigraphy, and relative magnitude of paleosecular variation in the Phanerozoic, *Earth Sci. Rev.*, **12**, 35–64, 1976.

ISEZAKI, N., Geomagnetic anomalies and tectonics around the Japanese Islands, *Oceanogr. Mag.*, **24**, 107–158, 1973.

ISHIDO, T. and H. MIZUTANI, Experimental and theoretical basis of electrokinetic phenomena in rock-water systems and its applications to geophysics, *J. Geophys. Res.*, **86**, 1763–1775, 1981.

ISHIKAWA, Y., Geomagnetic anomalies in central Japan based on results of aeromagnetic surveys, M.Sc. Thesis, Univ. Tokyo, 45 pp., 1979 (in Japanese).

ISHIKAWA, Y. and Y. SHONO, Reverse thermo-remanent magnetism in the $FeTiO_3$–Fe_2O_3 system, *J. Phys. Soc. Japan*, **17**, Suppl. B-I, 714–718, 1963.

ITO, K., Chaos in the Rikitake two-disc dynamo system, *Earth Planet. Sci. Lett.*, **51**, 451–456, 1980.

JACOBS, J. A., *Geomagnetic Micropulsations*, 179 pp., Springer-Verlag, Berlin, Heidelberg, 1970.

JANKOWSKI, J., A. SZYMANSKI, K. PĚČ, J. PĚČOVA, V. PETR, and O. PRAUS, Electromagnetic studies of the Carpathian conductivity anomaly, *Acta Geodaet., Geophys. Montanist. Acad. Sci. Hung.*, **12**, 99–109, 1977.

JAPANESE NATIONAL COMMITTEE FOR UPPER MANTLE PROJECT, The Crust and Upper Mantle of the Japanese Area, Part I. Geophysics, 119 pp. 1972.

JOHNSON, B. D., C. MCA. POWELL, and J. J. VEEVERS, Spreading history of the eastern Indian Ocean and greater India's northward flight from Antarctica and Australia, *Geol. Soc. Am. Bull.*, **87**, 1560–1566, 1976.

JOHNSON, F. S. and J. E. MIDGLEY, Notes on the lunar magnetosphere, *J. Geophys. Res.*, **73**, 1523–1532, 1968.

JOHNSON, H. P., Magnetization of the oceanic crust, *Rev. Geophys. Space Phys.*, **17**, 215–226, 1979.

JOHNSON, H. P. and R. T. MERRILL, A direct test of the Vine-Matthews hypothesis, *Earth Planet. Sci. Lett.*, **40**, 263–269, 1978.

JOHNSTON, M. J. S. and F. D. STACEY, Volcano-magnetic effect observed on Mt. Ruapehu, New Zealand, *J. Geophys. Res.*, **74**, 6541–6544, 1969.

JONES, A. G. and R. HUTTON, A multi-station magnetotelluric study in southern Scotland—I. Fieldwork, data analysis and results, *Geophys. J. R. astr. Soc.*, **56**, 329–349, 1979a.

JONES, A. G. and R. HUTTON, A multi-station magnetotelluric study in southern Scotland—II. Monte-Carlo inversion of the data and its geophysical and tectonic implications, *Geophys. J. R. astr. Soc.*, **56**, 351–368, 1979b.

JONES, F. W. and J. E. LOKKEN, Irregular coastline and channeling effects in three-dimensional geomagnetic perturbation models, *Phys. Earth Planet. Inter.* **10**, 140–150, 1975.

JONES, F. W. and L. J. PASCOE, A general computer program to determine the perturbation of alternating electric currents in a two-dimensional model of a region of uniform conductivity with an embedded inhomogeneity, *Geophys. J. R. astr. Soc.*, **24**, 3–30, 1971.

JONES, F. W. and L. J. PASCOE, The perturbation of alternating geomagnetic fields by three-dimensional conductivity inhomogeneities, *Geophys. J. R. astr. Soc.*, **27**, 479–485, 1972.

JONES, F. W. and A. T. PRICE, The perturbation of alternating geomagnetic fields by conductivity anomalies, *Geophys. J. R. astr. Soc.*, **20**, 317–334, 1970.

JONES, H. S. and P. J. MELOTTE, The harmonic analysis of the earth's main magnetic field, for epoch 1942, *Monthly Notices Roy. Astron. Soc., Geophys. Suppl.*, **6**, 409–430, 1953.

KAHLE, A. B., E. H. VESTINE, and R. H. BALL, Estimated surface motions of the earth's core, *J. Geophys. Res.*, **72**, 1095–1108, 1967.

KAHLE, A. B., R. H. BALL, and J. C. CAIN, Prediction of geomagnetic secular change confirmed, *Nature*, **223**, 165, 1969.

KAIKKONEN, P., A finite element program package for electromagnetic modeling, *J. Geophys. Res.*, **43**, 179–192, 1977.

KAKUTA, C., The magnetic torque on the impulsive change of the rotation of the earth, *Publ. Astron. Soc. Japan*, **13**, 361–368, 1961.

KAKUTA, C., Note on the relaxation time of magnetic coupling in the earth's rotation, *Publ. Astron. Soc. Japan*, **17**, 337–338, 1965.

KAMIDE, Y. and N. FUKUSHIMA, Positive geomagnetic bays in evening high-latitudes and their possible connection with partial ring current, *Rep. Ionosph. Space Res., Japan*, **26**, 79–101, 1972.

KAMIDE, Y., F. YASUHARA, and S.-I. AKASOFU, A model current system for the magnetospheric substorm, *Planet. Space Sci.*, **24**, 215–222, 1976.

KARIG, D. E., Origin and development of marginal basins in the western Pacific, *J. Geophys. Res.*, **76**, 2542–2559, 1971.

KEAN, W. F., R. DAY, M. FULLER, and V. A. SCHMIDT, The effect of uniaxial compression on the initial susceptibility of rocks as a function of grain size and composition of their constituent titanomagnetites, *J. Geophys. Res.*, **81**, 861–872, 1976.

KENDALL, P. C., Oceanic induction and shifting the spectrum, *Geophys. J. R. astr. Soc.*, **52**, 201–204, 1978.

KENNEDY, G. C. and G. H. HIGGINS, The core paradox, *J. Geophys. Res.*, **78**, 900–904, 1973.

KING, E. R. and ZIETZ, The New York—Alabama Lineament; geophysical evidence for a major crustal break in the basement beneath the Appalachian Basin, *Geology*, **6**, 312–318, 1978.

KIRSCHVINK, J. L., The Precambrian-Cambrian boundary problem: paleomagnetic directions from the Amadeus Basin, Central Australia, *Earth Planet. Sci. Lett.*, **40**, 91–100, 1978.

KISABETH, J. L., Substorm fields in and near the auroral zone, *Phys. Earth Planet. Inter.*, **10**, 214–249, 1975.

KITAZAWA, K., Intensity of the geomagnetic field in Japan for the past 10,000 years, *J. Geophys. Res.*, **75**, 7403–7411, 1970.

KITAZAWA, K. and K. KOBAYASHI, Intensity variation of the geomagnetic field during the past 4000 years in South America, *J. Geomag. Geoelectr.*, **20**, 7–19, 1968.

KLEIN, D. P., Geomagnetic time-variations, the island effect, and electromagnetic depth sounding on oceanic islands: results from analysis of data obtained in the frequency range of 0.5 to 10 cycles per hour on Oahu, Hawaii, M.Sc. Thesis, Univ. Hawaii, 96 pp., 1972.

KLEIN, D. P., Magnetic variations (2–30 cpd) on Hawaii Island and mantle electrical conductivity, Ph.D. Thesis, Univ. Hawaii, 82 pp., 1976.

KLEIN, D. P. and J. C. LARSEN, Magnetic induction fields (2–30 cpd) on Hawaii Island and their implications regarding electrical conductivity in the oceanic mantle, *Geophys. J. R. astr. Soc.*, **53**, 61–77, 1978.

KNOPOFF, L., The interaction between elastic wave motions and magnetic field in electrical conductors, *J. Geophys. Res.*, **60**, 441–456, 1955.

KNOPOFF, L. and G. J. F. MACDONALD, The magnetic field and the central core of the earth, *Geophys. J. R. astr. Soc.*, **1**, 216–223, 1958.

KOBAYASHI, K., Chemical remanent magnetization of ferromagnetic minerals and its application to rock magnetism, *J. Geomag. Geoelectr.*, **10**, 99–117, 1959.

KOBAYASHI, K. and M. NAKADA, Magnetic anomalies and tectonic evolution of the Shikoku Inter-arc Basin, *J. Phys. Earth*, **26**, Suppl., S391-S402, 1978.

KOBAYASHI, Y. and H. MARUYAMA, Electrical conductivity of olivine single crystals at high temperature, *Earth Planet. Sci. Lett.*, **11**, 415–419, 1971.

KODAMA, K. and S. UYEDA, Magnetization of Izu Islands with special reference to Oshima volcano, *J. Volcanol. Geotherm. Res.*, **6**, 353–373, 1979.

KONTIS, A. L., Aeromagnetic field test of total intensity upward continuation, *Geophysics*, **36**, 418–425, 1971.

KRAUSE, F. and K. -H. RÄDLER, *Mean-Field Megnetohydrodynamics and Dynamo Theory*, 271 pp., Pergamon Press, Oxford, New York, 1980.

KRAUT, E. A., Free radial modes in a compressible conducting fluid sphere containing a uniform internal magnetic field, *J. Geophys. Res.*, **70**, 3927–3933, 1965.

KRUTIKHOVSKAYA, Z. A. and I. K. PASHKEVICH, Magnetic model for the earth's crust under the Ukrainian Shield, *Can. J. Earth Sci.*, **14**, 2718–2728, 1977.

KURTZ, R. D. and G. D. GARLAND, Magnetotelluric measurements in eastern Canada, *Geophys. J. R. astr. Soc.*, **45**, 321–347, 1976.

KURTZ, R. D. and E. R. NIBLETT, Time dependence of magnetotelluric fields in a tectonically active region in eastern Canada, *J. Geomag. Geoelectr.*, **30**, 561–577, 1978.

KUSHIRO, I., Y. SHONO, and S. AKIMOTO, Melting of a peridotite nodule at high pressures and high water pressures, *J. Geophys. Res.*, **73**, 6023–6029, 1968.

LAHIRI, B. N. and A. T. PRICE, Electromagnetic induction in non-uniform conductors, and the determination of the conductivity of the earth from terrestrial magnetic variations, *Phil. Trans. Roy. Soc. Lond. A*, **237**, 509–540, 1939.

LAMBERT, A. and B. CANER, Geomagnetic depth-sounding and the coast effect in western Canada, *Can. J. Earth Sci.*, **2**, 485–509, 1965.

LANGEL, R. A., Near-earth satellite magnetic field measurements; a prelude to Magsat, *EOS Trans. A. G. U.*, **60**, 667–668, 1979.

LANGEL, R. A., R. L. COLES, and M. A. MAYHEW, Comparison of magnetic anomalies of lithospheric origin measured by satellite and airborne magnetometers over Western Canada, *Can. J. Earth Sci.*, **17**, 876–887, 1980a.

LANGEL, R. A., R. H. ESTES, G. D. MEAD, E. B. FABIANO, and E. R. LANCASTER, Initial geomagnetic field model from Magsat vector data, *Geophys. Res. Lett.*, **7**, 793–796, 1980b.

LARSEN, J. C., Low frequency (0.1–6.0 cpd) electromagnetic study of deep mantle electrical conductivity beneath the Hawaiian Islands, *Geophys. J. R. astr. Soc.*, **43**, 17–46, 1975.

LARSON, E. E., D. E. WATSON, and W. JENNINGS, Regional comparison of a Miocene geomagnetic transition in Oregon and Nevada, *Earth Planet. Sci. Lett.*, **11**, 391–400, 1971.

LARSON, R. L. and T. W. C. HILDE, A revised time scale of magnetic reversals for the early Cretaceous and late Jurassic, *J. Geophys. Res.*, **80**, 2586–2594, 1975.

LARSON, R. L. and W. C. PITMAN III, World-wide correlation of Mesozoic magnetic anomalies, and its implications, *Geol. Soc. Am. Bull.*, **83**, 3645-3662, 1972.

LAW, L. K. and J. P. GREENHOUSE, Geomagnetic variation sounding of the asthenosphere beneath the Juan de Fuca Ridge, *J. Geophys. Res.*, **86**, 967-978, 1981.

LAW, L. K., D. R. AULD, and J. R. BOOKER, A geomagnetic variation anomaly coincident with the Cascade volcanic belt, *J. Geophys. Res.*, **85**, 5297-5302, 1980.

LAWVER, L. A. and J. W. HAWKINS, Diffuse magnetic anomalies in marginal basins: their possible tectonic and petrologic significance, *Tectonophysics*, **45**, 323–339, 1978.

LEARY, P. and R. A. PHINNEY, A magnetotelluric traverse across the Yellowstone region, *Geophys. Res. Lett.*, **1**, 265–268, 1974.

LEATON, B. R., S. R. C. MALIN, and M. J. EVANS, An analytical representation of the estimated geomagnetic field and its secular change for the epoch 1965.0, *J. Geomag. Geoelectr.*, **17**, 187-194, 1965.

LE PICHON, X., Sea-floor spreading and continental drift, *J. Geophys. Res.*, **73**, 3661-3697, 1968.

LE PICHON, X., J. FRANCHETEAU, and J. BONNIN, *Plate Tectonics*, 300 pp., Elsevier, Amsterdam, 1973.

LEVY, E. H., Magnetic dynamo in the moon: a comparison with the earth, *Science*, **178**, 52-53, 1972.

LIDIAK, E. G., Buried Precambrian rocks of South Dakota, *Bull. Geol. Soc. Am.*, **82**, 1411-1420, 1971.

LIENERT, B. R., J. H. WHITCOMB, R. J. PHILLIPS, I. K. REDDY, and R. A. TAYLOR, Long term variations in magnetotelluric apparent resistivities observed near the San Andreas Fault in Southern California, *J. Geomag. Geoelectr.*, **32**, 757-775, 1980.

LILLEY, F. E. M., On kinematic dynamos, *Proc. Roy. Soc. Lond. A*, **316**, 153-167, 1970.

LILLEY, F. E. M., A magnetometer array study across southern Victoria and the Bass Strait area, Australia, *Geophys. J. R. astr. Soc.*, **46**, 165-184, 1976.

LILLEY, F. E. M. and D. J. BENNETT, An array experiment with magnetic variometers near the coasts of south-east Australia, *Geophys. J. R. astr. Soc.*, **29**, 49-64, 1972.

LILLEY, F. E. M. and D. J. BENNETT, Linear relationships in geomagnetic variation studies, *Phys. Earth Planet. Inter.*, **7**, 9–14, 1973.

LILLEY, F. E. M. and C. M. CARMICHAEL, Electromagnetic damping of elastic waves: experimental results, *Can. J. Earth Sci.*, **5**, 825-829, 1968.

LILLEY, F. E. M. and C. M. CARMICHAEL, Electromagnetic damping of elastic waves: a simple theory, *Can. J. Earth Sci.*, **7**, 1304-1307, 1970.

LILLEY, F. E. M. and H. Y. TAMMEMAGI, Magnetotelluric and geomagnetic depth sounding methods compared, *Nature Phys. Sci.*, **240**, 184-187, 1972.

LILLEY, F. E. M., D. V. WOODS, and M. N. SLOANE, Electrical conductivity from Australian magnetometer arrays using spatial gradient data, *Phys. Earth Planet. Inter.*, **25**, 202–209, 1981a.

LILLEY, F. E. M., D. V. WOODS, and M. N. SLOANE, Electrical conductivity profiles and implications for the absence or presence of partial melting beneath central and southeast Australia, *Phys. Earth Planet. Inter.*, **25**, 419–428, 1981b.

LILLEY, F. E. M., B. P. SINGH, B. R. ARORA, B. J. SRIVASTAVA, S. N. PRASAD, and M. N. SLOANE, A magnetometer array study in northwest India, Phys. Earth Planet. Inter., **25**, 232–240, 1981c.

LINES, L. R. and F. W. JONES, The perturbation of alternating geomagnetic fields by three-dimensional island structures, *Geophys. J. R. astr. Soc.*, **32**, 133-154, 1973.

LOPER, D. E., The gravitationally powered dynamo, *Geophys. J. R. astr. Soc.*, **54**, 389-404,

1978.

Lourenco, J. S. and H. F. Morrison, Vector magnetic anomalies derived from measurements of a single component of the field, *Geophysics*, **38**, 359–368, 1973.

Lowes, F. J. and I. Wilkinson, Geomagnetic dynamo: a laboratory model, *Nature*, **198**, 1158–1160, 1963.

Lowes, F. J. and I. Wilkinson, Geomagnetic dynamo: an improved laboratory model, *Nature*, 219, 717–718, 1968.

Lund, S. P. and S. K. Banerjee, Paleosecular variations from lake sediments, *Rev. Geophys. Space Phys.*, **17**, 244–249, 1979.

Maeda, R., T. Rikitake, and T. Nagata, Sudden commencements of geomagnetic storms and their local irregularities, *J. Geomag. Geoelectr.*, **17**, 69–93, 1965.

Malkus, W. V. R., Precessional torques as the cause of geomagnetism, *J. Geophys. Res.*, **68**, 2871–2886, 1963.

Malkus, W. V. R., Precession of the earth as the cause of geomagnetism, *Science*, **160**, 259–264, 1968.

Martin III, R. J., Is piezomagnetism influenced by microcracks during cyclic loading?, *J. Geomag. Geoelectr.*, **32**, 741–755, 1980.

Martin III, R. J. and M. Wyss, Magnetism of rocks and volumetric strain in uniaxial failure tests, *Pure Appl. Geophys.*, **113**, 51–61, 1975.

Martin III, R. J., R. E. Habermann, and M. Wyss, The effect of stress cycling and inelastic volumetric strain on remanent magnetization, *J. Geophys. Res.*, **83**, 3485–3496, 1978.

Maruyama, T., Statical elastic dislocations in an infinite and semi-infinite medium, *Bull. Earthq. Res. Inst., Univ. Tokyo*, **42**, 289–368, 1964.

Mason, R. G., Spatial dependence of time variations of the geomagnetic field on Oahu, Hawaii, *Trans. Am. Geophys. Union*, **44**, 40, 1962.

Mason, R. G., Spatial dependence of time variations of the geomagnetic field in the range 24 hr to 3 min on Christmas Island, *Geophys. Dept. Imp. Coll. Sci. Technol., Lond., Publ.*, **63–3**, 1–20, 1963.

Mason, R. G., Magnetic effects at Canton Island of the 1962 high altitude nuclear tests at Johnston Island, *Geophys. Dept. Imp. Coll. Sci. Technol., Lond., Publ.*, **64–1**, 1–12, 1964.

Matsushita, S., Solar quiet and lunar daily variation fields, in *Physics of Geomagnetic Phenomena* 1, edited by S. Matsushita and W. H. Campbell, pp. 301–424, Academic Press, New York, 1967.

Matsushita, S., Morphology of slowly-varying geomagnetic external fields—a review, *Phys. Earth Planet. Inter.*, **10**, 299–312, 1975.

Matsushita, S. and H. Maeda, On the geomagnetic solar quiet daily variation field during the IGY, *J. Geophys. Res.*, **70**, 2535–2558, 1965a.

Matsushita, S. and H. Maeda, On the geomagnetic lunar daily variation field, *J. Geophys. Res.*, **70**, 2559–2578, 1965b.

Matsushita, S., J. D. Tarpley, and W. H. Campbell, IMF sector structure effects on the quiet geomagnetic field, *Radio Sci.*, **8**, 963–972, 1973.

Matuyama, M., On the direction of magnetization of basalts in Japan, Tyosen and Manchuria, *Proc. Japan Acad.*, **5**, 203–205, 1929.

Mauersberger, P., Mathematische Beschreibung und Statistische Untersuchung des Hauptfeldes und der Säkularvariation, in *Geomagnetismus und Aeronomie*, edited by G. Fanselau, 3, pp. 95–213, V. E. B. Deutscher Verlag der Wiss., Berlin, 1959.

Mayhew, M. A., Inversion of satellite magnetic anomaly data, *J. Geophys.*, **45**, 119–128, 1979.

MAZZELLA, A. and H. F. MORRISON, Electrical resistivity variations associated with earthquakes on the San Andreas fault, *Science*, **185**, 855–857, 1974.

McDONALD, K. L., Penetration of the geomagnetic secular field through a mantle with variable conductivity, *J. Geophys. Res.*, **62**, 117–141, 1957.

McELHINNY, M. W., *Paleomagnetism and Plate Tectonics*, 358 pp., Cambridge Univ. Press, London, 1973.

McKENZIE, D. P. and W. J. MORGAN, Evolution of triple junctions, *Nature*, **224**, 125–133, 1969.

McKENZIE, D. P. and R. L. PARKER, The north Pacific—an example of tectonics on a sphere, *Nature*, **216**, 1276–1280, 1967.

MINDLIN, R. D. and D. H. CHENG, Nuclei of strain in the semi-infinite solid, *J. Appl. Phys.*, **21**, 926–930, 1950.

MIYAKOSHI, J., Secular variation of Parkinson vectors in a seismically active region of Middle Asia, *J. Fac. Gen. Educ., Tottori Univ.*, **8**, 209–218, 1975.

MIYAKOSHI, J., Electrical conductivity structure beneath the Japan Island Arc by geomagnetic induction study, *Tech. Rep., Geol. Survey, Japan*, 31–40, 1976 (in Japanese).

MIYAKOSHI, J., Electrical conductivity structure beneath the Japan Island Arc by geomagnetic induction study, *J. Phys. Earth*, **27**, S153–S161, 1979.

MIZUTANI, H. and T. ISHIDO, A new interpretation of magnetic field variation associated with the Matsushiro earthquakes, *J. Geomag. Geoelectr.*, **28**, 179–188, 1976.

MIZUTANI, H. and H. KANAMORI, Electrical conductivity of rock forming minerals at high temperature, *J. Phys. Earth*, **15**, 25–31, 1967.

MIZUTANI, H., T. ISHIDO, Y. YOKOKURA, and S. OHNISHI, Electrokinetic phenomena associated with earthquakes, *Geophys. Res. Lett.*, **3**, 365–368, 1976.

MOFFATT, H. K., *Magnetic Field Generation in Electrically Conducting Fluids*, 343 pp., Cambrige Univ. Press, London, 1978.

MOGI, K., Relations between eruptions of various volcanoes and the deformations of the ground surfaces around them, *Bull. Earthq. Res. Inst., Univ. Tokyo*, **36**, 99–134, 1958.

MORGAN, W. J., Rises, trenches, great faults, and crustal blocks, *J. Geophys. Res.*, **73**, 1959–1982, 1968.

MORI, T. and T. YOSHINO, Local difference in variations of the geomagnetic total intensity in Japan, *Bull. Earthq. Res. Inst., Univ. Tokyo*, **48**, 893–922, 1970.

MURTHY, V. R. and S. K. BANERJEE, Lunar evolution: how well do we know it now?, *Moon*, **7**, 149, 1973.

MUTH, L. A. and E. R. BENTON, On the frozen flux velocity field at the surface of earth's core necessary to account for the poloidal main magnetic field and its secular variation, *Phys. Earth Planet. Inter.*, **24**, 245–252, 1981.

NAGATA, T., *Rock Magnetism*, 350 pp., Maruzen, Tokyo, 1961.

NAGATA, T., Two main aspects of geomagnetic secular variation—westward drift and non-drifting components, in *Proc. Benedum Earth Magnetism Symp.*, edited by T. Nagata, pp. 39–55, Univ. Pittsburgh Press, Pittsburgh, 1963.

NAGATA, T., Tectonomagnetism, *Int. Assoc. Geomag. Aeron., Bull.*, **27**, 12–43, 1969.

NAGATA, T., Effects of a uniaxial compression on remanent magnetizations of igneous rocks, *Pure Appl. Geophys.*, **78**, 100–109, 1970a.

NAGATA, T., Basic magnetic properties of rocks under the effects of mechanical stresses, *Tectonophysics*, **9**, 167–195, 1970b.

NAGATA, T., Notes on the amplification of magnetic field variations on the lunar surface, *J. Geomag. Geoelectr.*, **23**, 101–106, 1971.

NAGATA, T. and K. KOBAYASHI, Experimental studies on the generation of remanent magnetization of ferromagnetic minerals by chemical reaction, *Proc. Japan Acad.*, **34**, 269–273, 1958.

NAGATA, T. and M. OZIMA, Paleomagnetism, in *Physics of Geomagnetic Phenomena 1*, edited by S. Matsushita and W. H. Campbell, pp. 103–180, Academic Press, New York, 1967.

NAGATA, T. and T. RIKITAKE, Geomagnetic secular variation and poloidal magnetic fields produced by convectional motions in the earth's core, *J. Geomag. Geoelectr.*, **13**, 42–53, 1961.

NAGATA, T. and T. RIKITAKE, The northward shifting of the geomagnetic dipole and stability of the axial quadrupole of the earth, *J. Geomag. Geoelectr.*, **15**, 213–220, 1963.

NAGATA, T., K. AKASHI, and T. RIKITAKE, The natural remanent magnetism of sedimentary rocks (preliminary note), *Bull. Earthq. Res. Inst., Univ. Tokyo*, **21**, 276–297, 1943.

NAGATA, T., S. AKIMOTO, and S. UYEDA, Reverse thermoremanent magnetism, *Proc. Japan Acad.*, **27**, 643–645, 1951.

NAGATA, T., S. UYEDA, and S. AKIMOTO, Self-reversal of thermo-remanent magnetism of igneous rocks, *J. Geomag. Geoelectr.*, **4**, 22–38, 1952.

NAGATA, T., T. OGUTI, and H. MAEKAWA, Model experiments of electromagnetic induction within the earth, *Bull. Earthq. Res. Inst., Univ. Tokyo*, **33**, 561–569, 1955.

NAGATA, T., Y. ISHIKAWA, H. KINOSHITA, M. KONO, Y. SYONO, and R. M. FISHER, Magnetic properties and natural remanent magnetization of lunar materials, *Geochim. Cosmochim. Acta*, **34**, Suppl. 1, 2325–2340, 1970.

NAGATA, T., T. RIKITAKE, and M. KONO, Electrical conductivity and the age of the moon, *Space Res.*, **11**, 85–88, 1971.

NAGATA, T., R. M. FISHER, and F. C. SCHWERER, Lunar rock magnetism, *Moon*, **4**, 160–186, 1972a.

NAGATA, T., R. M. FISHER, F. C. SCHWERER, M. D. FULLER, and J. R. DUNN, Rock magnetism of Appolo 14 and 15 materials, *Geochim. Cosmochim. Acta*, Suppl. **3**, 2423–2447, 1972b.

NAGATA, T., R. M. FISHER, and F. C. SCHWERER, Some characteristic magnetic properties of lunar materials, *Moon*, **9**, 63–77, 1974.

NÉEL, L., L'inversion de l'aimantation permanente des rockes, *Ann. Géophys.*, **7**, 90–102, 1951.

NEKUT, A., J. E. P. CONNERNEY, and A. K. KUCKES, Deep crustal electrical conductivity; evidence for water in the lower crust, *Geophys. Res. Lett.*, **4**, 239–242, 1977.

NESS, N. F., The electrical conductivity and internal temperature of the moon, NASA-GSFC Rept. X-616-69-191, 1969.

NESS, N. F., K. W. BEHANNON, C. S. SCEARCE, and S. C. CANTARANS, Early results from the magnetic field experiment on Lunar Explorer 35, *J. Geophys. Res.*, **72**, 5769–5778, 1967.

NESS, N. F., K. W. BEHANNON, and R. P. LEPPING, The magnetic field of Mercury, 1, *J. Geophys. Res.*, **80**, 2708–2716, 1975.

NESS, N. F., M. H. ACUÑA, R. P. LEPPING, J. E. P. CONNERNEY, K. W. BEHANNON, L. F. BURLAGA, and F. M. NEUBAUER, Magnetic field studies by Voyager 1: preliminary results at Saturn, *Science*, **212**, 211–217, 1981.

NEUMAYER, G. and H. PETERSEN, Über das gegenwärtig vorliegende Material für die Erd-und Weltmagnetische Forschung, Verhandl. 8 Deut. Geographentages Berlin, Berlin, 1889.

NIBLETT, E. R. and Y. HONKURA, Time-dependence of electromagnetic transfer functions and their association with tectonic activity, *Geophys. Surveys*, **4**, 97–114, 1980.

NIBLETT, E. R. and K. WHITHAM, Multi-disciplinary studies of geomagnetic variation anomalies in the Canadian Arctic, *J. Geomag. Geoelectr.*, **22**, 99–111, 1970.

NIBLETT, E. R., J. M. DELAURIER, L. K. LAW, and F. C. PLET, Geomagnetic variation anomalies in the Canadian Arctic I. Ellesmere Island and Lincoln Sea, *J. Geomag. Geoelectr.*, **26**, 203–221, 1974.

NICOLL, M. A. and J. T. WEAVER, H-polarization induction over an ocean edge coupled to the mantle by a conducting crust, *Geophys. J. R. astr. Soc.*, **49**, 427–442, 1977.

NISHIDA, Y., Conductivity anomalies in the southern half of Hokkaido, Japan, *J. Geomag. Geoelectr.*, **28**, 375–394, 1976.

NOMURA, M., Marine geomagnetic anomalies with intermediate wavelengths in the western Pacific region, *Bull. Ocean Res. Inst., Univ. Tokyo*, No. 11, 1–42, 1979.

NORITOMI, K., The electrical conductivity of rock and the determination of the electrical conductivity of the earth's interior, *J. Mining Coll., Univ. Akita*, **1**, 27–59, 1961.

NOVOSELOVA, M. R., Magnetic anomalies of the Baikal rift zone and adjacent areas, *Tectonophysics*, **45**, 95–100, 1978.

OHSHIMAN, N., Local geomagnetic changes associated with fault activity, M.Sc. Thesis, Tokyo Inst. Tech., 178 pp., 1980.

OLDENBURG, D. W., One-dimensional inversion of natural source magnetotelluric observations, *Geophysics*, **44**, 1218–1244, 1979.

OLDENBURG, D. W., Conductivity structure of oceanic upper mantle beneath the Pacific plate, *Geophys. J. R. astr. Soc.*, **65**, 359–394, 1981.

OPDYKE, N. D., Paleomagnetism of deep-sea cores, *Rev. Geophys. Space Phys.*, **10**, 213–249, 1972.

OTOFUJI, Y. and S. SASAJIMA, A magnetization process of sediments: laboratory experiments on post-depositional remanent magnetization, *Geophys. J. R. astr. Soc.*, **66**, 241–259, 1981.

PARKER, E. N., Hydromagnetic dynamo models, *Astrophys. J.*, **122**, 293–314, 1955.

PARKER, R. L., The inverse problem of electrical conductivity in the mantle, *Geophys. J. R. astr. Soc.*, **22**, 121–138, 1971.

PARKIN, C. W., W. D. DAILY, and P. DYAL, Iron abundance and magnetic permeability of the moon, Proc. 5th Lunar Conf., *Geochim. Cosmochim. Acta.*, **3**, Suppl. 5, 2761–2778, 1974.

PARKINSON, W. D., Directions of rapid geomagnetic fluctuations, *Geophys. J. R. astr. Soc.*, **2**, 1–14, 1959.

PARKINSON, W. D., The influence of continents and oceans on geomagnetic variations, *Geophys. J. R. astr. Soc.*, **4**, 441–449, 1962.

PARKINSON, W. D., Conductivity anomalies in Australia and the ocean-effect, *J. Geomag. Geoelectr.*, **15**, 222–226, 1964.

PARKINSON, W. D. and B. A. HOBBS, Conditions for the geomagnetic induction relationship, *Phys. Earth Planet. Inter.*, **19**, P5–P6, 1979.

PARKINSON, W. D. and F. W. JONES, The geomagnetic coast effect, *Rev. Geophys. Space Phys.*, **17**, 1999–2015, 1979.

PASCOE, L. J. and W. JONES, Boundary conditions and calculation of surface values for the general two-dimensional electromagnetic induction problem, *Geophys. J. R. astr. Soc.*, **27**, 179–193, 1972.

PAYNE, M. A., SI and Gaussian CGS units, conversions and equations for use in geomagnetism, *Phys. Earth Planet. Inter.*, **26**, P10–P16, 1981.

PEDDIE, N. W. and E. B. FABIANO, A model of the geomagnetic field for 1975, *J. Geophys. Res.*, **81**, 2539–2542, 1976.

PHILLIPS, R. J., I. K. REDDY, and J. H. WHITCOMB, Investigation into the relationship between resistivity changes and tectonic activity in Southern California, *Trans. Am. Geophys. Union*, **58**, 731, 1977.

PITMAN III, W. C. and D. HAYES, Sea-floor spreading in the Gulf of Alaska, *J. Geophys. Res.*,

73, 6571–6580, 1968.

PITMAN III, W. C. and M. TALWANI, Sea-floor spreading in the North Atlantic, *Geol. Soc. Am. Bull.*, **83**, 619–646, 1972.

PIWINSKII, A. J. and A. DUBA, High temperature electrical conductivity of albite, *Geophys. Res. Lett.*, **1**, 209–211, 1974.

POEHLS, K. A. and D. D. JACKSON, Tectonomagnetic event detection using empirical transfer functions, *J. Geophys. Res.*, **83**, 4933–4940, 1978.

POEHLS, K. A. and R. P. VON HERZEN, Electrical resistivity structure beneath the North-west Atlantic Ocean, *Geophys. J. R. astr. Soc.*, **47**, 331–346, 1976.

PORATH, H., Magnetic variation anomalies and seismic low-velocity zone in the western United States, *J. Geophys. Res.*, **76**, 2643–2648, 1971.

PORATH, H. and A. DZIEWONSKI, Crustal electrical conductivity anomalies in the Great Plains Province of the United States, *Geophysics,* **36**, 382–395, 1971a.

PORATH, H. and A. DZIEWONSKI, Crustal resistivity anomalies from geomagnetic deep sounding studies, *Am. Geophys. Union Monog.*, **9**, 891–915, 1971b.

PORATH, H. and D. I. GOUGH, Mantle conductive structures in the western United States from magnetometer array studies, *Geophys. J. R. astr. Soc.*, **22**, 261–275, 1971.

PORATH, H., D. M. OLDENBURG, and D. I. GOUGH, Separation of magnetic variation fields and conductive structures in the western United States, *Geophys. J. R. astr. Soc.*, **19**, 237–260, 1970.

PORATH, H., D. I. GOUGH, and P. A. CAMFIELD, Conductive structures in the north-western United States and south-west Canada, *Geophys. J. R. astr. Soc.*, **23**, 387–398, 1971.

POSPEEV, V. I. and V. I. MIKHALEVSKY, Deep magnetotelluric surveys of the south of the Siberian Platform and in the Baikal Rift Zone, in *Geoelectric and Geothermal Studies*, edited by A. Ádám, pp. 673–681, Akadémiai Kiadó, Budapest, 1976.

PRESNALL, D. C., C. L. SIMMONS, and H. PORATH, Changes in electrical conductivity of a synthetic basalt during melting, *J. Geophys. Res.*, **77**, 5665–5672, 1972.

PRICE, A. T., The induction of electric currents in non-uniform thin sheets and shells, *Quart. J. Mech. Appl. Math.*, **2**, 283–310, 1949.

PRICE, A. T., Electromagnetic induction in a semi-infinite conductor with a plane boundary, *Quart. J. Mech. Appl. Math.*, **3**, 385–410, 1950.

PRICE, A. T., The theory of magnetotelluic methods when the source field is considered, *J. Geophys. Res.*, **67**, 1907–1918, 1962.

PRICE, A. T., Electromagnetic induction within the earth, in *Physics of Geomagnetic Phenomena* 1, edited by S. Matsushita and W. H. Campbell, pp. 235–298, Academic Press, New York, 1967.

PRICE, A. T. and G. A. WILKINS, New methods for the analysis of geomagnetic fields and their application to the Sq field of 1932–1933, *Phil. Trans. Roy. Soc. Lond. A,* **256**, 31–98, 1963.

RÄDLER, K.-H., On the electrodynamics of conducting fluids in turbulent motion I: The principles of mean field electrodynamics, *Z. Naturforsch.*, **23a,** 1841–1851, 1968 (English translation, Roberts and Stix, 1971).

RAI, C. S. and M. H. MANGHNANI, Electrical conductivity of ultramafic rocks to 1820 Kelvin, *Phys. Earth Planet. Inter.*, **17**, 6–13, 1978.

RAICHE, A. P., An integral equation approach to three-dimensional modelling, *Geophys. J. R. astr. Soc.*, **36**, 363–376, 1974.

RAMASWAMY, V., F. W. JONES, and H. W. DOSSO, A numerical study of the topographic effect on electromagnetic fields in a three-dimensional conductivity model, *Pure Appl. Geophys.*, **114**, 653–662, 1976.

REDDY, I. K. and D. RANKIN, Magnetotelluric response of laterally inhomogeneous and anisotropic media, *Geophysics*, **40**, 1035–1045, 1975.

REDDY, I. K., R. J. PHILLIPS, J. H. WHITCOMB, D. M. COLE, and R. A. TAYLOR, Monitoring of time-dependent resistivity by magnetotellurics, *J. Geomag. Geoelectr.*, **28**, 165–178, 1976.

REDDY, I. K., D. RANKIN, and R. J. PHILLIPS, Three-dimensional modelling in magnetotelluric and magnetic variational sounding, *Geophys. J. R. astr. Soc.*, **51**, 313–325, 1977.

REGAN, R. D., The reduction and analysis of satellite magnetometer data, *Geophys. Surveys*, **3**, 331–349, 1979.

REGAN, R. D., J. C. CAIN, and W. M. DAVIS, A global magnetic anomaly map, *J. Geophys. Res.*, **80**, 794–802, 1975.

REITMAYR, G., An anomaly of the upper mantle below the Rhine Graben, studied by the inductive response of natural electromagnetic fields, *J. Geophys.*, **41**, 651–658, 1975.

REITZEL, J. S., D. I. GOUGH, H. PORATH, and C. W. ANDERSON III, Geomagnetic deep sounding and upper mantle structure in the western United States, *Geophys. J. R. astr. Soc.*, **19**, 213–235, 1970.

RICHARDS, M. L., V. VACQUIER, and G. D. VAN VOORHIS, Calculation of the magnetization of uplifts from combining topographic and magnetic surveys, *Geophysics*, **32**, 678–707, 1967.

RIKITAKE, T., Electromagnetic induction within the earth and its relation to the electrical state of the earth's interior. 1(1), *Bull. Earthq. Res. Inst., Univ. Tokyo*, **28**, 45–100, 1950a.

RIKITAKE, T., Electromagnetic induction within the earth and its relation to the electrical state of the earth's interior. 1(2), *Bull. Earthq. Res. Inst., Univ. Tokyo*, **28**, 219–262, 1950b.

RIKITAKE, T., Electromagnetic induction within the earth and its relation to the electrical state of the earth's interior. 2, *Bull. Earthq. Res. Inst., Univ. Tokyo*, **28**, 263–283, 1950c.

RIKITAKE, T., Electromagnetic induction within the earth and its relation to the electrical state of the earth's interior. 3, *Bull. Earthq. Res. Inst., Univ. Tokyo*, **29**, 61–69, 1951a.

RIKITAKE, T., The distribution of magnetic dip in Ooshima (Oo-sima) Island and its change that accompanied the eruption of Volcano Mihara, 1950, *Bull. Earthq. Res. Inst., Univ. Tokyo*, **29**, 161–181, 1951b.

RIKITAKE, T., On the electrical conductivity in the earth's core, *Bull. Earthq. Res. Inst., Univ. Tokyo*, **30**, 191–205, 1952.

RIKITAKE, T., Oscillations of a system of disk dynamos, *Proc. Cambridge Phil. Soc.*, **54**, 89–105, 1958.

RIKITAKE, T., Anomaly of geomagnetic variations in Japan, *Geophys. J. R. astr. Soc.*, **2**, 276–287, 1959.

RIKITAKE, T., Electromagnetic induction in a hemispherical ocean by *Sq*, *J. Geomag. Geoelectr.*, **11**, 65–79, 1960.

RIKITAKE, T., The effect of the ocean on rapid geomagnetic changes, *Geophys. J. R. astr. Soc.*, **5**, 1–15, 1961a.

RIKITAKE, T., *Sq* and ocean, *J. Geophys. Res.*, **66**, 3245–3254, 1961b.

RIKITAKE, T., Supplement to paper "*Sq* and ocean", *J. Geophys. Res.*, **67**, 2588–2591, 1962a.

RIKITAKE, T., Poleward increase of the secular drift of geomagnetic field, Geomagnetica (publication for commemoration of the 50-year anniversary of S. Miguel Magnetic Observatory), Servico Meterologico Nacional, Lisbon, pp. 5–13, 1962b.

RIKITAKE, T., Outline of the anomaly of geomagnetic variations in Japan, *J. Geomag. Geoelectr.*, **15**, 181–184, 1964a.

RIKITAKE, T., Electromagnetic induction in a nearly spherical conductor, *Bull. Earthq. Res. Inst., Univ. Tokyo*, **42**, 621-629, 1964b.

RIKITAKE, T., Electromagnetic induction in a semi-infinite conductor having an undulatory surface, *Bull. Earthq. Res. Inst., Univ. Tokyo*, **43**, 161–166, 11965.

RIKITAKE, T., *Electromagnetism and the Earth's Interior*, 308 pp., Elsevier, Amsterdam, 1966a.

RIKITAKE, T., Elimination of non-local changes from total intensity values of the geomagnetic field, *Bull. Earthq. Res. Inst., Univ. Tokyo*, **44**, 1041–1070, 1966b.

RIKITAKE, T., Non-dipole field and fluid motion in the earth's core, *J. Geomag. Geoelectr.*, **19**, 129-142, 1967a.

RIKITAKE, T., Electromagnetic induction within non-uniform plane and spherical sheets, *Bull. Earthq. Res. Inst., Univ. Tokyo*, **45**, 1229-1294, 1967b.

RIKITAKE, T., Geomagnetism and earthquake prediction, *Tectonophysics*, **6**, 59-68, 1968a.

RIKITAKE, T., Theoretical magnetograms for s.s.c. when the ocean effect is considered, *Geophys. J. R. astr. Soc.*, **15**, 79-90, 1968b.

RIKITAKE, T., Resonant behavior of electric currents electromagnetically induced within a non-uniform sheet of conductor, *J. Geophys. Res.*, **73**, 7019–7029, 1968c.

RIKITAKE, T., The undulation of an electrically conductive layer beneath the islands of Japan, *Tectonophysics*, **7**, 257-264, 1969.

RIKITAKE, T., A theory of frequency-dependent effect of an island on geomagnetic variations, *Bull. Earthq. Res. Inst., Univ. Tokyo*, **48**, 189-203, 1970.

RIKITAKE, T., Electric conductivity anomaly in the earth's crust and mantle, *Earth-Sci. Rev.*, **7**, 35-65, 1971.

RIKITAKE, T., The mechanism of geomagnetic field reversal, *Phys. Earth Planet. Inter.*, **6**, 340-345, 1972.

RIKITAKE, T., Non-steady geomagnetic dynamo models, *Geophys. J. R. astr. Soc.*, **35**, 277-284, 1973a.

RIKITAKE, T., Global electrical conductivity of the earth, *Phys. Earth Planet. Inter.*, **7**, 245-250, 1973b.

RIKITAKE, T., A model of the geoelectric structure beneath Japan, *J. Geomag. Geoelectr.*, **27**, 233-244, 1975.

RIKITAKE, T., *Earthquake Prediction*, 357 pp., Elsevier, Amsterdam, 1976.

RIKITAKE, T., Changes in the direction of magnetic vector of short-period geomagnetic variations before the 1972 Sitka, Alaska, earthquake, *J. Geomag. Geoelectr.*, **31**, 441-448, 1979.

RIKITAKE, T. and Y. HAGIWARA, Magnetic anomaly over a magnetized circular core, *Bull. Earthq. Res. Inst., Univ. Tokyo*, **43**, 509-527, 1965.

RIKITAKE, T. and Y. HAGIWARA, Non-steady Bullard-Gellman dynamo model (1), *J. Geomag. Geoelectr.*, **20**, 57-65, 1968a.

RIKITAKE, T. and Y. HAGIWARA, Non-steady Bullard-Gellman dynamo model (2), *J. Geomag. Geoelectr.*, **20**, 415-422, 1968b.

RIKITAKE, T. and Y. HONKURA, Recent Japanese studies on conductivity anomalies, *Phys. Earth Planet. Inter.*, **7**, 203-212, 1973.

RIKITAKE, T. and S. SATO, The geomagnetic Dst field of the magnetic storm on June 18-19, 1936, *Bull. Earthq. Res. Inst., Univ. Tokyo*, **35**, 7-21, 1957.

RIKITAKE, T. and I. YOKOYAMA, Anomalous relations between H and Z components of transient geomagnetic variations, *J. Geomag. Geoelectr.*, **5**, 59-65, 1953.

RIKITAKE, T. and I. YOKOYAMA, Volcanic activity and changes in geomagnetism, *J. Geophys. Res.*, **60**, 165–172, 1955a.

RIKITAKE, T. and I. YOKOYAMA, The anomalous behaviour of geomagnetic variations of short period in Japan and its relation to the subterranean structure. The 6th report. (The results of further observations and some considerations concerning the influences of the sea on geomagnetic variations), *Bull. Earthq. Res. Inst., Univ. Tokyo*, **33**, 297-331, 1955b.

RIKITAKE, T., I. YOKOYAMA, A. OKADA, and Y. HISHIYAMA, Geomagnetic studies of Volcano Mihara. The 3rd paper. (Magnetic survey and continuous observation of changes in geomagnetic declination), *Bull. Earthq. Res. Inst., Univ. Tokyo*, **29**, 583-604, 1951.

RIKITAKE, T., Y. HONKURA, H. TANAKA, N. OHSHIMAN, Y. SASAI, Y. ISHIKAWA, S. KOYAMA, M. KAWAMURA, and K. OHCHI, Changes in the geomagnetic field associated with earthquakes in the Izu Peninsula, Japan, *J. Geomag. Geoelectr.*, **32**, 721-739, 1980.

RITTER, E., Results of geoelectromagnetic deep soundings in Europe, *Gerlands Beitr. Geophysik*, **84**, 261-273, 1975.

ROBERTS, P. H., *An Introduction to Magnetohydrodynamics*, 264 pp., Longmans, Green, London, 1967.

ROBERTS, P. H., Dynamo theory, in *Mathematical Problems in the Geophysical Sciences*, edited by W. H. Reid, pp. 129-206, Am. Math. Soc., Providence, Rhode Island, 1971.

ROBERTS, P. H., Electromagnetic core-mantle coupling, *J. Geomag. Geoelectr.*, **24**, 231-259, 1972.

ROBERTS, P. H. and M. STIX, *The Turbulent Dynamo*: a translation of a series of papers by F. Krause, K. -H. Rädler and M. Steenbeck, Tech. Note IA-60, National Center for Atmospheric Research, Boulder, 318 pp., 1971.

ROCHESTER, M. G., Geomagnetic westward drift and irregularities in the earth's rotation, *Phil. Trans. Roy. Soc. Lond. A*, **252**, 531-555, 1960.

ROCHESTER, M. G., Geomagnetic core-mantle coupling, *J. Geomag. Geoelectr.*, **67**, 4833-4836, 1962.

ROCHESTER, M. G., Perturbation in the earth's rotation and geomagnetic core-mantle coupling, *J. Geomag. Geoelectr.*, **20**, 387-402, 1968.

ROCHESTER, M. G. and D. E. SMYLIE, Geomagnetic core-mantle coupling and the Chandler wobble, *Geophys. J. R. astr. Soc.*, **10**, 289-315, 1965.

ROCHESTER, M. G., J. A. JACOBS, D. E. SMYLIE, and K. F. CHONG, Can precession power the geomagnetic dynamo?, Geophys. J. R. astr. Soc., **43**, 661-678, 1975.

RODEN, R. B., Electromagnetic core-mantle coupling, *Geophys. J. R. astr. Soc.*, **7**, 361-374, 1963.

RODI, W. L., A technique for improving the accuracy of finite element solutions for magnetotelluric data, *Geophys. J. R. astr. Soc.*, **44**, 483-506, 1976.

ROKITYANSKY, I. I., Geomagnetic variations behavior in Central Europe, *J. Geomag. Geoelectr.*, **32**, SI125-SI132, 1980.

ROONEY, D. and V. R. S. HUTTON, A magnetotelluric and magnetovariational study of the Gregory Rift Valley, Kenya, *Geophys. J. R. astr. Soc.*, **51**, 91-119, 1977.

ROSTOKER, G., Polar magnetic substorms, *Rev. Geophys. Space Phys.*, **10**, 157-211, 1972.

ROY, R. F., D. D. BLACKWELL, and E. R. DECKER, Continental heat flow, in *The Nature of the Solid Earth*, edited by E. C. Robertson, pp. 506-543, McGraw-Hill, New York, 1972.

RUNCORN, S. K., Palaeomagnetic comparisons between Europe and North America, *Phil. Trans. Roy. Soc. Lond. A*, **258**, 1-12, 1965.

RUNCORN, S. K., On the interpretation of lunar magnetism, *Phys. Earth Planet. Inter.*, **10**, 327-335, 1975.

RUNCORN, S. K., The ancient lunar core dynamo, *Science*, **199**, 771-773, 1978.

RUNCORN, S. K. and H. C. UREY, A new theory of lunar magnetism, *Science*, **180**, 636-638, 1973.

RUNCORN, S. K., D. W. COLLINSON, W. O'REILLY, M. H. BATTEY, A. STEPHENSON, J. M. JONES, A. J. MANSON, and P. W. READMAN, Magnetic properties of Apollo 11 lunar samples, *Geochim. Cosmochim. Acta*, **34**, Suppl. 1, 2435–2451, 1970.

RUNCORN, S. K., D. W. COLLINSON, W. O'REILLY, A. STEPHENSON, M. H. BATTEY, A. J. MANSON, and P. W. READMAN, Magnetic properties of Apollo 12 lunar samples, *Proc. Roy. Soc. Lond. A*, **325**, 157–174, 1971.

RUNCORN, S. K., L. M. LIBBY, and W. F. LIBBY, Primeval melting of the moon, *Nature*, **270**, 676–681, 1977.

RUSSEL, C. T., The magnetic moment of Venus: Venera-4 measurements reinterpreted, *Geophys. Res. Lett.*, **3**, 125–128, 1976.

RUSSEL, C. T., Update of the Bode's law of planetary magnetism, *Nature*, **273**, 147, 1978.

RUSSEL, C. T., Planetary magnetism, *Rev. Geophys. Space Phys.*, **17**, 295–301, 1979.

RUSSEL, C. T., P. J. COLEMAN, Jr. and G. SCHUBERT, Lunar magnetic field: permanent and induced dipole moments, *Science*, **186**, 825–826, 1974.

RUSSEL, C. T., R. C. ELPHIC, and J. A. SLAVIN, Initial Pioneer Venus magnetic field results: nightside observations, *Science*, **205**, 114–116, 1979.

RUSSEL, C. T., R. C. ELPHIC, and J. A. SLAVIN, Limits on the possible intrinsic magnetic field of Venus, *J. Geophys. Res.*, **85**, 8319–8332, 1980.

SASAI, Y., Spatial dependence of short-period geomagnetic fluctuations on Oshima Island (1), *Bull. Earthq. Res. Inst., Univ. Tokyo*, **45**, 137–157, 1967.

SASAI, Y., Spatial dependence of short-period geomagnetic fluctuations on Oshima Island (2), *Bull. Earthq. Res. Inst., Univ. Tokyo*, **46**, 907–926, 1968.

SASAI, Y., The piezomagnetic field associated with the Mogi model, *Bull. Earthq. Res. Inst., Univ. Tokyo*, **54**, 1–29, 1979.

SASAI, Y., Application of the elasticity theory of dislocations to tectonomagnetic modelling, *Bull. Earthq. Res. Inst., Univ. Tokyo*, **55**, 387–447, 1980.

SASAI, Y. and Y. ISHIKAWA, Tectonomagnetic event preceding a M 5.0 earthquake in the Izu Peninsula—aseismic slip of a buried fault?, *Bull. Earthq. Res. Inst., Univ. Tokyo*, **55**, 895–911, 1980.

SCHLAPP, D. M., Lunar daily geomagnetic variations at Tucson, *Geophys. J. R. astr. Soc.*, **55**, 183–191, 1978.

SCHMIDT, A., Der magnetische Zustand der Erde zur Epoche 1885.0, *Arch. Dewt. Seewarte*, **21** (2), 76 pp., 1898.

SCHMIDT, A., Tafeln der Normieten Kugelfunktionen, Engelhard-Reyher, Gotha, 52 pp., 1935.

SCHMUCKER, U., Erdmagnetische Tiefensondierung in Deutschland 1957–1959: Magnetogramme und erste Auswertung, Abhandl. Akad. Göttingen, *Math. Phys. Kl., Beitr. I. G. J.*, **5**, 1–51, 1959.

SCHMUCKER, U., Anomalies of geomagnetic variations in the southwestern United States, *J. Geomag. Geoelectr.*, **15**, 193–221, 1964.

SCHMUCKER, U., Conductivity anomalies, with special reference to the Andes, in *The Application of Modern Physics to the Earth and Planetary Interiors*, edited by S. K. Runcorn, pp. 125–138, Wiley-Interscience, London, 1969.

SCHMUCKER, U., Anomalies of geomagnetic variations in the southwestern United States, *Bull. Scripps Inst. Oceanogr.*, **13**, 165 pp., Univ. California Press, Berkeley and Los Angeles, 1970.

SCHMUCKER, U., O. HARTMANN, A. A. GIESECKE, Jr., M. CASAVERDE, and S. E. FORBUSH, Electrical conductivity anomalies in the earth's crust in Peru, *Carnegie Inst. Washington Yearbook*, **63**, 354–362, 1964.

SCHMUCKER, U., S. E. FORBUSH, O. HARTMANN, A. A. GIESECKE, Jr., M. CASAVERDE, J. CASTILLO, R. SALGUEIRO, and S. DEL POZO, Electrical conductivity anomaly under the Andes, *Carnegie Inst. Washington Yearbook*, **65**, 11-28, 1966.

SCHOBER, M., The electrical conductivity of some samples of natural olivine at high temperatures and pressures, *Z. Geophys.*, **37**, 283-292, 1971.

SCHOLZ, C. H., L. R. SYKES, and Y. P. AGGARWAL, Earthquake prediction: a physical basis, *Science*, **181**, 803-810, 1973.

SCHUSTER, A., The diurnal variation of terrestrial magnetism, *Phil. Trans. Roy. Soc. Lond. A*, **208**, 163-204, 1908.

SHAMSI, S. and F. D. STACEY, Dislocation models and seismomagnetic calculations for California 1906 and Alaska 1964 earthquakes, *Bull. Seismol. Soc. Am.*, **59**, 1435-1448, 1969.

SHANKLAND, T. J. and H. S. WAFF, Partial melting and electrical conductivity anomalies in the upper mantle, *J. Geophys. Res.*, **82**, 5409-5417, 1977.

SHANKLAND, T. J., R. J. O'CONNELL, and H. S. WAFF, Geophysical constraints on partial melt in the upper mantle, *Rev. Geophys. Space Phys.*, **19**, 394-406, 1981.

SHARP, L. R., P. J. COLEMAN, Jr., and B. R. LICHTENSTEIN, Orbital mapping of the lunar magnetic field, *Moon*, **7**, 322-341, 1973.

SHARPE, H. N. and D. W. STRANGWAY, The magnetic field of Mercury and models of thermal evolution, *Geophys. Res. Lett.*, **3**, 285-288, 1976.

SHIRAKI, M., Monitoring of the time change in transfer functions in the central Japan conductivity anomaly, *J. Geomag. Geoelectr.*, **32**, 637-648, 1980.

SHIRAKI, M., Seasonal dependence of Lunar daily geomagnetic variations in different regions of the world, *J. Geomag. Geoelectr.*, **33**, 467-501, 1981.

SHIRAKI, M. and K. YANAGIHARA, Transfer functions at Kakioka (part II): reevaluation of their secular changes, *Mem. Kakioka Mag. Obs.*, **17**, 19-25, 1977 (in Japanese with English abstract).

SHUEY, R. T., D. K. SCHELLINGER, E. H. JOHNSON, and L. B. ALLEY, Aeromagnetics and the transition between the Colorado Plateau and Basin Range Provinces, *Geology*, **1**, 107-110, 1973.

SHUEY, R. T., D. K. SCHELLINGER, A. C. TRIPP, and L. B. ALLEY, Curie depth determination from aeromagnetic spectra, *Geophys. J. R. astr. Soc.*, **50**, 75-101, 1977.

SILVESTER, P. and C. R. S. HASLAM, Magnetotelluric modelling by the finite element method, *Geophys. Prospect.*, **20**, 872-891, 1972.

SMITH, B. E. and M. J. S. JOHNSTON, A tectonomagnetic effect observed before a magnitude 5.2 earthquake near Hollister, California, *J. Geophys. Res.*, **81**, 3556-3560, 1976.

SMITH, E. J., L. J. DAVIS, and D. E. JOHNS, Magnetic field measurements near Mars, *Science*, **149**, 1241, 1965.

SMITH, P. J., The intensity of the ancient geomagnetic field: a review and analysis, *Geophys. J. R. astr. Soc.*, **12**, 321-362, 1967.

SMITH, P. J. and J. NEEDHAM, Magnetic declination in mediaeval China, *Nature*, **214**, 1213-1214, 1967.

SMITH, R. B., R. T. SHUEY, J. R. PELTON, and J. P. BAILEY, Yellowstone hot spot: contemporary tectonics and crustal properties from earthquake and aeromagnetic data, *J. Geophys. Res.*, **82**, 3665-3676, 1977.

SMOLUCHOWSKI, R., Magnetism of the moon, *Moon*, **7**, 127-131, 1973.

SONETT, C. P. and D. S. COLBURN, The principle of solar wind induced planetary dynamos, *Phys. Earth Planet. Inter.*, **1**, 326-346, 1968.

SONETT, C. P., D. S. COLBURN, and R. G. CURRIE, The intrinsic magnetic field of the moon,

J. Geophys. Res., **72**, 5503–5507, 1967.

Sonett, C. P., G. Schubelt, B. F. Smith, K. Schwartz, and D. S. Colburn, Lunar electrical conductivity from Apollo 12 magnetometer measurements: compositional and thermal inferences, *Proc. Lunar Sci. Conf. 2nd*, 2415, 1971.

Sonett, C. P., D. S. Colburn, B. F. Smith, G. Schubelt, and K. Schwartz, The induced magnetic field of the moon: conductivity profiles and inferred temperature, *Proc. Lunar Sci. Conf. 3rd*, 2309, 1972.

Soward, A. M., A kinematic theory of large magnetic Reynolds number dynamos, *Phil. Trans. Roy. Soc. Lond. A*, **272**, 431–462, 1972.

Srivastava, B. J. and H. Abbas, An interpretation of the induction arrows at Indian stations, *J. Geomag. Geoelectr.*, **32**, Suppl. I, SI187–SI196, 1980.

Srivastava, B. J., D. S. Bhaskara Rao, and S. N. Prasad, Geomagnetic variation anomalies in the Koyna and the Bhadrachalam seismic areas in Peninsula India, *J. Geomag. Geoelectr.*, **26**, 247–255, 1974.

Stacey, F. D., *Physics of the Earth*, 324 pp., John Wiley, New York, 1969.

Stacey, F. D., The coupling of the core to the precession of the earth, *Geophys. J. R. astr. Soc.*, **33**, 47–55, 1973.

Stacey, F. D. and S. K. Banerjee, *The Physical Principles of Rock Magnetism*, 195 pp., Elsevier, Amsterdam, 1974.

Stacey, F. D. and M. J. S. Johnston, Theory of the piezomagnetic effect in titanomagnetite-bearing rocks, *Pure Appl. Geophys.*, **97**, 146–155, 1972.

Steenbeck, M., F. Krause, and K. -H. Rädler, A calculation of the mean electromotive force in an electrically conducting fluid in turbulent motion, under the influence of Coriolis forces, *Z. Naturforsch*, **21a**, 369–376, 1966 (English translation, Roberts and Stix, 1971).

Steenbeck, M., I. M. Kirko, A. Gailitis, A. P. Klawina, F. Krause, I. J. Laumanis, and O. A. Lielausis, An experimental verification of the α-effect, *Monats. Dt. Akad. Wiss., Berlin*, **9**, 714–719, 1967 (English translation, Roberts and Stix, 1971).

Steinhauser, P. and S. A. Vincenz, Equatorial paleo poles and behavior of the dipole field during polarity transitions, *Earth Planet. Sci. Lett.*, **19**, 113–119, 1973.

Stephenson, A., Crustal remanence and magnetic moment of Mercury, *Earth Planet. Sci. Lett.*, **28**, 454–458, 1976.

Strangway, D. W., E. E. Larson, and G. W. Pearce, Magnetic studies of lunar samples—breccia and fines, *Geochim. Cosmochim. Acta*, **34**, Suppl. 1, 2435–2451, 1970.

Strangway, D. W., G. W. Pearce, W. A. Gose, and R. W. Timme, Remanent magnetization of lunar samples, *Earth Planet. Sci. Lett.*, **13**, 43–52, 1971.

Strangway, D. W., W. A. Gose, G. W. Pearce, and J. G. Carnes, Magnetism and the history of the moon, *Proceedings of 18th Annual Conference on Magnetism and Magnetic Materials*, Am. Inst. Physics Conf. Proc. 10, pp. 1178–1196, Am. Inst. Physics, New York, 1974.

Sumitomo, N., Geomagnetic variation anomaly in the vicinity of Tottori facing the Japan Sea, in the south-western Japan, *Contribution Geophys. Inst. Kyoto Univ.*, **12**, 117–128, 1972.

Suzuki, A., A new analysis of the geomagnetic *Sq* field, *J. Geomag. Geoelectr.*, **25**, 259–280, 1973.

Takeuchi, H. and M. Saito, Electromagnetic induction within the earth, *J. Geophys. Res.*, **68**, 6287–6291, 1963.

Takeuchi, H. and Y. Shimazu, On a self-exciting process in magneto-hydrodynamics, *J. Phys. Earth*, **1**, 1–9, 1952a.

TAKEUCHI, H. and Y. SHIMAZU, On a self-exciting process in magneto-hydrodynamics (II), *J. Phys. Earth*, **1**, 57–64, 1952b.

TAKEUCHI, H. and Y. SHIMAZU, On a self-exciting process in magneto-hydrodynamics, *J. Geophys. Res.*, **58**, 497–518, 1953.

TAKEUCHI, H. and Y. SHIMAZU, On a self-exciting process in magneto-hydrodynamics (III), *J. Phys. Earth*, **2**, 5–12, 1954.

TALWANI, M., Computation with the help of a digital computer of magnetic anomalies caused by bodies of arbitrary shape, *Geophysics*, **30**, 797–817, 1965.

TALWANI, M., C. C. WINDISCH, and M. G. LANGSETH, Reykjanes ridge crest: a detailed geophysical study, *J. Geophys. Res.*, **76**, 473–517, 1971.

TALWANI, P. and R. L. KOVACH, Geomagnetic observations and fault creep in California, *Tectonophysics*, **14**, 245–256, 1972.

TANAKA, H., Intensity variation of the geomagnetic field in Japan during the last 30,000 years determined from volcanic rocks and potteries, Ph.D. Thesis, Univ. Tokyo, 104 pp., 1980.

THELLIER, E. and O. THELLIER, Sur l'intensité du champ magnétique terrestre dans le passé historique et géologique, *Ann. Géophys.*, **15**, 285–376, 1959.

TOLLAND, H. G., Electrical properties of a synthetic lunar pyroxenite, and the internal temperature of the moon, *Phys. Earth Planet. Inter.*, **8**, 292–294, 1974.

TOMODA, Y., K. KOBAYASHI, J. SEGAWA, M. NOMURA, K. KIMURA, and T. SAKI, Linear magnetic anomalies in the Shikoku Basin, northeastern Philippine Sea, *J. Geomag. Geoelectr.*, **28**, 47–56, 1975.

TOUGH, J. G., Nearly symmetric dynamos, *Geophys. J. R. astr. Soc.*, **13**, 393-396, 1967.

TOUGH, J. G. and R. D. GIBSON, The Braginskii dynamo, in *The Application of Modern Physics to the Earth and Planetary Interiors*, edited by S. K. Runcorn, pp. 555–569, Wiley-Interscience, London, 1969.

TOZER, D. C., The interpretation of upper-mantle electrical conductivities, *Tectonophysics*, **56**, 147–163, 1979.

TOZER, D. C. and J. WILSON, The electrical conductivity of the moon's interior, *Proc. Roy. Soc. Lond. A*, **296**, 320–329, 1967.

TREUMAN, R., Electromagnetic induction problem in plates with two-dimensional conductivity distribution I. General theory, *Geomag. Aeron.*, **10**, 376–389, 1970a.

TREUMAN, R., Electromagnetic induction problem in plates with two-dimensional conductivity distribution II. Approximate method and solutions, *Geomag. Aeron.*, **10**, 464–472, 1970b.

TSAY, L. J., A spatial analysis of upward continuation of potential field data, *Geophys. Prospect.*, **26**, 822–840, 1978.

UNO, S., Non-steady state of Bullard-Gellman-Lilley dynamo model, *J. Geomag. Geoelectr.*, **24**, 203–222, 1972.

UNTIEDT, J., Conductivity anomalies in central and southern Europe, *J. Geomag. Geoelectr.*, **22**, 131–149, 1970.

UYEDA, S., Thermoremanent magnetism as a medium of palaeomagnetism, with special reference to reverse thermoremanent magnetism, *Jap. J. Geophys.*, **2**, 1–123, 1958.

UYEDA, S. and T. RIKITAKE, Electrical conductivity anomaly and terrestrial heat flow, *J. Geomag. Geoelectr.*, **22**, 75–90, 1970.

VACQUIER, V., *Geomagnetism in Marine Geology*, 185 pp., Elsevier, Amsterdam, 1972.

VACQUIER, V. and S. UYEDA, Palaeomagnetism of nine seamounts in the western Pacific and three volcanoes in Japan, *Bull. Earthq. Res. Inst., Univ. Tokyo*, **45**, 815–848, 1967.

VAN DER VOO, R., Paleomagnetism related to continental drift and plate tectonics, *Rev.*

Geophys. Space Phys., **17**, 227–235, 1979.

VANYAN, L. L., The electrical conductivity of the moon, *Geophys. Surveys*, **4**, 173–185, 1980.

VANYAN, L. L., Deep geoelectrical models: geological and electromagnetic principles, *Phys. Earth Planet. Inter.*, **25**, 273–279, 1981.

VASSEUR, G., K. BABOUR, M. MENVIELLE, and J. C. ROSSIGNOL, The geomagnetic variation anomaly in the northern Pyrenees: study of the temporal variation, *Geophys. J. R. astr. Soc.*, **49**, 593–607, 1977.

VASSEUR, G. and P. WEIDELT, Bimodal electromagnetic induction in non-uniform thin sheets with an application to the northern Phrenean induction anomaly, *Geophys. J. R. astr. Soc.*, **51**, 669–690, 1977.

VESTINE, E. H., On the analysis of surface magnetic fields by integrals, 1, *Terr. Mag. Atmos. Electr.*, **46**, 27–41, 1941.

VESTINE, E. H., On variations of the geomagnetic field, fluid motions, and the rate of the earth's rotation, *J. Geophys. Res.*, **58**, 127–145, 1953.

VESTINE, E. H., The world magnetic survey and the earth's interior, *J. Geomag. Geoelectr.*, **17**, 165–171, 1965.

VESTINE, E. H., Main geomagnetic field, in *Physics of Geomagnetic Phenomena 1*, edited by S. Matsushita and W. H. Campbell, pp. 181–234, Academic Press, New York, 1967.

VESTINE, E. H. and A. B. KAHLE, On the small amplitude of magnetic secular change in the Pacific area, *J. Geophys. Res.*, **71**, 527–530, 1966.

VESTINE, E. H. and A. B. KAHLE, The westward drift and geomagnetic secular change, *Geophys. J. R. astr. Soc.*, **15**, 29–37, 1968.

VESTINE, E. H., L. LAPORTE, I. LANGE, and W. E. SCOTT, The geomagnetic field: its description and analysis, *Carnegie Inst. Washington Publ.*, **580**, 1–390, 1947.

VINE, F. J. and D. H. MATTHEWS, Magnetic anomalies over oceanic ridges, *Nature*, **199**, 947–949, 1963.

WAFF, H. S., Theoretical consideration of electrical conductivity in a partially molten mantle and implications for geothermometry, *J. Geophys. Res.*, **79**, 4003–4010, 1974.

WAFF, H. S., Relations of electrical conductivity to physical conditions within the asthenosphere, *Geophys. Surveys*, **4**, 31–41, 1980a.

WAFF, H. S., Effects of the gravitational field on liquid distribution in partial melts within the upper mantle, *J. Geophys. Res.*, **85**, 1815–1825, 1980b.

WAFF, H. S. and J. R. BULAU, Equilibrium fluid distribution in an ultramafic partial melt under hydrostatic stress conditions, *J. Geophys. Res.*, **84**, 6109–6114, 1979.

WAFF, H. S. and D. F. WEILL, Electrical conductivity of magmatic liquids: effects of temperature, oxygen fugacity and composition, *Earth Planet. Sci. Lett.*, **28**, 254–260, 1975.

WAIT, J. R., On the relation between telluric and the earth's magnetic field, *Geophysics*, **19**, 281–289, 1954.

WANG, C. Y., R. E. GOODMAN, P. N. SUDARAM, and H. F. MORRISON, Electrical resistivity of granite in frictional sliding: application to earthquake prediction, *Geophys. Res. Lett.*, **2**, 525–528, 1975.

WARE, R. H., High-accuracy magnetic field difference measurements and improved noise reduction techniques for use in tectonomagnetic studies, *J. Geophys. Res.*, **84**, 6291–6295, 1979.

WARREN, R. E., J. G. SCLATER, V. VACQUIER, and R. F. ROY, A comparison of terrestrial heat flow and transient geomagnetic fluctuations in the southwestern United States, *Geophysics*, **34**, 463–478, 1969.

WASILEWSKI, P. J., H. H. THOMAS, and M. A. MAYHEW, The Moho as a magnetic boundary,

Geophys. Res. Lett., **6**, 541–544, 1979.

WATANABE, H., Measurements of electrical conductivity of basalt at temperatures up to 1500°C and pressures to about 20 kilobars, *Spec. Contrib., Geophys. Inst., Kyoto Univ.*, **10**, 159–170, 1970.

WATANABE, H., Bounds on the fluid velocity and the magnetic field in the earth's core imposed by hydromagnetic consideration of an α-ω dynamo, *J. Geomag. Geoelectr.*, **29**, 191–209, 1977.

WATANABE, H., Non-steady state of the hydromagnetic $\alpha\omega$-dynamo and its application to the geomagnetic reversals, *J. Geomag. Geoelectr.*, **33**, 531–543, 1981.

WATANABE, H. and T. YUKUTAKE, Electromagnetic core-mantle coupling associated with changes in the geomagnetic dipole field, *J. Geomag. Geoelectr.*, **27**, 153–173, 1975.

WATANABE, N., Secular variation in the direction of geomagnetism as the standard scale for geomagnetochronology in Japan, *Nature*, **182**, 383–384, 1958.

WATANABE, N., The direction of remanent magnetism of baked earth and its application to chronology and archaeology in Japan, *J. Fac. Sci., Univ. Tokyo, Sect. V*, **2**, 1–188, 1959.

WATKINS, N. D., Non-dipole behavior during an Upper Miocene geomagnetic polarity transition in Oregon, *Geophys. J. R. astr. Soc.*, **17**, 121–149, 1969.

WATTS, A. B., J. K. WEISSEL, and R. L. LARSON, Sea-floor spreading in marginal basins of the western Pacific, *Tectonophysics*, **37**, 167–181, 1977.

WEAVER, J. T., The electromagnetic field within a discontinuous conductor with reference to geomagnetic micropulsations near a coastline, *Can. J. Earth Sci.*, **41**, 484-495, 1963.

WEAVER, J. T. and C. R. BREWITT-TAYLOR, Improved boundary conditions for the numerical solution of *E*-polarization problems in geomagnetic induction, *Geophys. J. R. astr. Soc.*, **54**, 309–317, 1978.

WEAVER, J. T. and D. J. THOMSON, Induction in a non-uniform conducting half-space by an external line current, *Geophys. J. R. astr. Soc.*, **28**, 163–185, 1972.

WEIDELT, P., The inverse problem of geomagnetic induction, *Z. Geophys.*, **38**, 257–289, 1972.

WEIDELT, P., Electromagnetic induction in three-dimensional structures, *J. Geophys.*, **41**, 85–109, 1975a.

WEIDELT, P., Inversion of two-dimensional conductivity structures, *Phys. Earth Planet. Inter.*, **10**, 282–291, 1975b.

WEISSEL, J. K. and A. B. WATTS, Tectonic complexities in the South Fiji marginal basin, *Earth Planet. Sci. Lett.*, **28**, 121–126, 1975.

WHITE, A., A geomagnetic variation anomaly across the northern Gulf of California, *Geophys. J. R. astr. Soc.*, **33**, 1–25, 1973a.

WHITE, A., Anomalies in geomagnetic variations across the central Gulf of California, *Geophys. J. R. astr. Soc.*, **33**, 27–46, 1973b.

WHITE, A. and O. W. POLATAJKO, The coast effect in geomagnetic variations in South Australia, *J. Geomag. Geoelectr.*, **30**, 109–120, 1978.

WHITHAM, K., The relationships between the secular change and the non-dipole fields, *Can. J. Phys.*, **36**, 1372–1396, 1958.

WHITHAM, K., An anomaly in geomagnetic variations at Mould Bay in the Arctic archipelago of Canada, *Geophys. J. R. astr. Soc.*, **8**, 26–43, 1963.

WHITHAM, K., Geomagnetic variation anomalies in Canada, *J. Geomag. Geoelectr.*, **17**, 481–498, 1965.

WIESE, H., Geomagnetische Tiefentellurik. I. Die elektrische Leitfähigkeit der Erdkruste und des oberen Erdmantels, *Geofis. Pura Appl.*, **51**, 59–78, 1962a.

WIESE, H., Geomagnetische Tiefentellurik 2: Die Streichrichtung der Untergrundstrukturen des elektrischen Widerstandes, Erschlossen aus geomagnetischen Variationen, *Geofis.*

Pura Appl., **52**, 83–103, 1962b.

WILLIAMSON, K., C. HEWLETT, and H. Y. TAMMEMAGI, Computer modelling of electrical conductivity structures, *Geophys. J. R. astr. Soc.*, **37**, 533–536, 1974.

WILSON, J. T., A new class of faults and their bearing on continental drift, *Nature*, **207**, 343–347, 1965.

WINCH, D. E., Evaluation of geomagnetic dynamo integrals, *J. Geomag. Geoelectr.*, **26**, 87–94, 1974.

WOODS, D. V. and F. E. M. LILLEY, Geomagnetic induction in central Australia, *J. Geomag. Geoelectr.*, **31**, 449–458, 1979.

WOODS, D. V. and F. E. M. LILLEY, Anomalous geomagnetic variations and the concentration of telluric currents in south-west Queensland, Australia, *Geophys. J. R. astr. Soc.*, **62**, 675–689, 1980.

XU Wen-Yao, QI Kui and WANG Shi-Ming, On the short period geomagnetic variation anomaly of the eastern Kansu Province, *Acta Geophysica Sinica*, **21**, 218–224, 1978 (in Chinese with English abstract).

YAGI, T. and S. AKIMOTO, Electrical conductivity jump produced by the α-β-transformation in Mn_2GeO_4, *Tech. Rep., Inst. Solid State Phys., Univ. Tokyo, Ser. A*, **620**, 1–20, 1973.

YAMAKAWA, N., On the strain produced in a semi-infinite elastic solid by an interior source of stress, *Zisin*, (ii) **8**, 84–98, 1955.

YAMAZAKI, Y. and T. RIKITAKE, Local anomalous changes in the geomagnetic field at Matsushiro, *Bull. Earthq. Res. Inst., Univ. Tokyo*, **48**, 637–643, 1970.

YANAGIHARA, K., Secular variation of the electrical conductivity anomaly in the central part of Japan, *Mem. Kakioka Mag. Obs.*, **15**, 1–11, 1972.

YANAGIHARA, K. and T. NAGANO, Time change of transfer function in the Central Japan Anomaly, *J. Geomag. Geoelectr.*, **28**, 157–163, 1976.

YASKAWA, K., T. NAKAJIMA, N. KAWAI, M. TORII, N. NATSUHARA, and S. HORIE, Palaeomagnetism of a core from Lake Biwa (I), *J. Geomag. Geoelectr.*, **25**, 447–474, 1973.

YOKOYAMA, I., Anomalous changes in geomagnetic field on Oosima Volcano related with its activities in the decade of 1950, *J. Phys. Earth*, **17**, 69–76, 1969.

YOSHII, T., Regionality of group velocities of Rayleigh waves in the Pacific and thickening of the plate, *Earth Planet. Sci. Lett.*, **25**, 305–312, 1975.

YOSHIMURA, H., Nonlinear astrophysical dynamos: autonomous and sporadic field reversals of steady dynamos and the polarity transition phenomena of the geodynamo, *Astrophys. J.*, **235**, 625–649, 1980.

YUKUTAKE, T., The influence of the magnetic field on spectra of seismic core waves, *Bull. Earthq. Res. Inst., Univ. Tokyo*, **35**, 641–658, 1957.

YUKUTAKE, T., Attenuation of geomagnetic secular variation through the conducting mantle of the earth, *Bull. Earthq. Res. Inst., Univ. Tokyo*, **37**, 13–32, 1959.

YUKUTAKE, T., Archaeomagnetic study on volcanic rocks in Oshima Island, Japan, *Bull. Earthq. Res. Inst., Univ. Tokyo*, **39**, 467–476, 1961.

YUKUTAKE, T., The westward drift of the magnetic field of the earth, *Bull. Earthq. Res. Inst., Univ. Tokyo*, **40**, 1–65, 1962.

YUKUTAKE, T., The solar cycle contribution to the secular change in the geomagnetic field, *J. Geomag. Geoelectr.*, **17**, 287–309, 1965.

YUKUTAKE, T., Electromagnetic induction in a conductor bounded by an inclined surface, *Publ. Dominion Obs. Ottawa*, **35**, 317–353, 1967.

YUKUTAKE, T., The drift velocity of the geomagnetic secular variation, *J. Geomag. Geoelectr.*, **20**, 403–414, 1968a.

YUKUTAKE, T., Two methods of estimating the drift rate of the earth's magnetic field, *J. Geomag. Geoelectr.*, **20**, 427–428, 1968b.

YUKUTAKE, T., Geomagnetic secular variation, *Comments Earth Sci.: Geophys.*, **1**, 55–63, 1970.

YUKUTAKE, T., Spherical harmonic analysis of the earth's magnetic field for the 17th and 18th centuries, *J. Geomag. Geoelectr.*, **23**, 11–31, 1971.

YUKUTAKE, T., The effect of change in the geomagnetic dipole moment on the rate of the earth's rotation, *J. Geomag. Geoelectr.*, **24**, 19–47, 1972.

YUKUTAKE T., Fluctuations in the earth's rate of rotation related to changes in the geomagnetic dipole field, *J. Geomag. Geoelectr.*, **25**, 195–212, 1973.

YUKUTAKE, T., A stratified core motion inferred from geomagnetic secular variations, *Phys. Earth Planet. Inter.*, **24**, 253–258, 1981.

YUKUTAKE, T. and H. TACHINAKA, Geomagnetic variation associated with stress change within a semi-infinite elastic earth caused by a cylindrical force source, *Bull. Earthq. Res. Inst., Univ. Tokyo*, **45**, 785–798, 1967.

YUKUTAKE, T. and H. TACHINAKA, The non-dipole part of the earth's magnetic field, *Bull. Earthq. Res. Inst., Univ. Tokyo*, **46**, 1027–1074, 1968a.

YUKUTAKE, T. and H. TACHINAKA, The westward drift of the geomagnetic secular variation, *Bull. Earthq. Res. Inst., Univ. Tokyo*, **46**, 1075–1102, 1968b.

YUKUTAKE, T. and H. TACHINAKA, Separation of the earth's magnetic field into the drifting and the standing parts, *Bull. Earthq. Res. Inst., Univ. Tokyo*, **47**, 65–97, 1969.

YUKUTAKE, T., K. NAKAMURA, and K. HORAI, Magnetization of ash-fall tuffs of Oshima volcano, Izu II: Application to archaeomagnetism and volcanology, *J. Geomag. Geoelectr.*, **16**, 183–193, 1964.

ZHAO Yü-Lin and QIAN Fu-Ye, Electrical resistivity anomaly observed in and around the epicentral area prior to the Tangshan earthquake of 1976, *Acta Geophysica Sinica, 21*, 181–190, 1978 (in Chinese with English abstract).

ZIETZ, I., P. C. BATEMAN, J. E. CASE, M. D. CRITTENDEN, Jr., A. GRISCOM, E. R. KING, R. J. ROBERTS, and G. R. LORENTZEN, Aeromagnetic investigation of crustal structure for a strip across the western United States, *Geol. Soc. Am. Bull.*, **80**, 1703–1714, 1969.

SUBJECT INDEX